Lecture Notes in Computer Science   2414

Edited by G. Goos, J. Hartmanis, and J. van Leeuwen

W0227324

Springer
Berlin
Heidelberg
New York
Barcelona
Hong Kong
London
Milan
Paris
Tokyo

Friedemann Mattern
Mahmoud Naghshineh (Eds.)

Pervasive Computing

First International Conference, Pervasive 2002
Zurich, Switzerland, August 26-28, 2002
Proceedings

 Springer

Series Editors

Gerhard Goos, Karlsruhe University, Germany
Juris Hartmanis, Cornell University, NY, USA
Jan van Leeuwen, Utrecht University, The Netherlands

Volume Editors

Friedemann Mattern
Department of Computer Science, ETH Zürich IFW
Haldeneggsteig 4, 8092 Zürich/Switzerland
E-mail: mattern@inf.ethz.ch

Mahmoud Naghshineh
Director, Emerging Markets and IP Creation
IBM Technology Group
E-mail: mahmoud@us.ibm.com

Cataloging-in-Publication Data applied for

Die Deutsche Bibliothek - CIP-Einheitsaufnahme

Pervasive computing : first international conference, pervasive 2002,
Zurich, Switzerland, August 26 - 28, 2002 ; proceedings / Friedemann Mattern ;
Mahmoud Naghshineh (ed.). - Berlin ; Heidelberg ; New York ; Barcelona ;
Hong Kong ; London ; Milan ; Paris ; Tokyo : Springer, 2002
    (Lecture notes in computer science ; Vol. 2414)
    ISBN 3-540-44060-7

CR Subject Classification (1998): C.2.4, C.3, C.5.3, D.4, H.3-5, K.4, K.6.5, J.7

ISSN 0302-9743
ISBN 3-540-44060-7 Springer-Verlag Berlin Heidelberg New York

Springer-Verlag Berlin Heidelberg New York
a member of BertelsmannSpringer Science+Business Media GmbH

http://www.springer.de

© Springer-Verlag Berlin Heidelberg 2002

Typesetting: Camera-ready by author, data conversion by PTP-Berlin, Stefan Sossna e. K.
Printed on acid-free paper   SPIN: 10873748   06/3142   5 4 3 2 1 0

Preface

This volume contains the proceedings of Pervasive 2002, the first in a series of international conferences on Pervasive Computing. The conference took place at ETH Zurich from August 26 to 28, 2002. Its objective was to present, discuss, and explore the latest technical developments in the emerging field of pervasive computing, as well as potential future directions.

Pervasive Computing is a cross-disciplinary area that extends the application of computing to diverse usage models. It covers a broad set of research topics such as low power, integrated technologies, embedded systems, mobile devices, wireless and mobile networking, middleware, applications, user interfaces, security, and privacy. The great amount of interest we are witnessing in Pervasive Computing is driven by relentless progress in basic information technologies such as microprocessors, memory chips, integrated sensors, storage devices, and wireless communication systems that continue to enable ever smaller, lighter, and faster systems. Such systems are also becoming affordable due to their high integration and mass production, paving the way for their adoption.

Over the past five years, Pervasive Computing has emerged as an area with tremendous potential where new computing models and associated applications and usage scenarios are driving many new research topics as well as business models. We are entering an era beyond large hosts, PCs, and laptop computers where we find computers in every aspect of our lives. Computers are becoming so ubiquitous that they are integrated inside all sorts of appliances or can be worn unobtrusively as part of clothing and jewelry. Pervasive Computing is the catalyst for the Internet being extended to billions of new hosts. This explosive game-changing phenomenon can be compared to the spread of electric motors over the past century, but promises to revolutionize life much more profoundly than, say, elevators ever did. While this transition is exciting, it bears profound and tremendous new challenges including technical, safety, social, legal, political, and economic issues. We hope to have covered at least some of these topics with a diverse collection of papers here.

Pervasive Computing has become an active and popular research area, which is witnessed by the 162 submissions we received from authors all over the world. The program committee chose 21 papers for inclusion in the conference. It was a difficult choice, based on several hundred reviews produced by the program committee and many outside referees, where each paper was typically reviewed by three reviewers.

In addition to the papers contained in these proceedings, the conference program included a special short-paper session presenting late-breaking results, a demo and exhibition track, and a panel discussion on security and privacy. We also had two invited presentations: Randy Katz, from the University of California, Berkeley, delivered the invited talk on the first day titled "Pervasive Computing: It's all about Network Services". Ralf Guido Herrtwich from DaimlerChrysler Research gave the second invited talk on "Ubiquitous Computing in the Automotive Domain".

Moreover, the conference offered six half-day tutorials:

- Data Management for Pervasive Computing (Anupam Joshi, University of Maryland Baltimore County)
- Personal Privacy in Pervasive Computing (Marc Langheinrich, ETH Zurich)
- Developing i-mode Services and Mobile Applications with Java Technologies (Olivier Liechti, Karim Mazouni, Kim Kauffmann, Sun Microsystems/ Mobile Webzone)
- Context Aware Communication (Bill Schilit, Intel Research Lab Seattle)
- Introduction to Wearable Computing (Thad Starner, Georgia Tech)
- Pervasive Portals and WebServices (Martin Welsch, IBM Boeblingen Development Lab)

In closing, we would like to express our sincere appreciation to all authors who submitted papers. We deeply thank all members of the program committee and the external reviewers for their time and effort as well as their valuable input. Finally we would like to thank Springer-Verlag for their excellent cooperation, our sponsoring institutions, and the organizing committee, in particular Jürgen Bohn, Vlad Coroama, Phil Janson, and Marc Langheinrich.

June 2002   Friedemann Mattern
                                                            Mahmoud Naghshineh

Organization

Pervasive 2002, the first in a series of international conferences on Pervasive Computing, took place in Zurich, Switzerland from August 26 to 28, 2002. It was jointly organized by IBM Research and ETH Zurich, the Swiss Federal Institute of Technology.

Executive Committee

Conference Chair: Mahmoud Naghshineh (IBM Technology Group, USA)

Program Chair: Friedemann Mattern (ETH Zurich, Switzerland)

Tutorial Co-chairs: Bernt Schiele (ETH Zurich, Switzerland) and Marisa Viveros (IBM Research, USA)

Exhibition/Demo Chair: Oliver Haase (Bell Labs Research, USA)

Organization Chair: Phil Janson (IBM Research, Switzerland)

Financial Co-chairs: Phil Janson (IBM Research, Switzerland) and Marc Langheinrich (ETH Zurich, Switzerland)

Technical Editor: Jürgen Bohn (ETH Zurich, Switzerland)

Publicity Co-chairs: Vlad Coroama (ETH Zurich, Switzerland) and Marc Langheinrich (ETH Zurich, Switzerland)

Registration and Local
Arrangements Chair: Vlad Coroama (ETH Zurich, Switzerland)

Program Committee

Emile Aarts (Philips Research Laboratories, The Netherlands)
Arndt Bode (TU Munich, Germany)
Gaetano Borriello (University of Washington and
    Intel Seattle Research Lab, USA)
Dave DeRoure (University of Southampton, UK)
Oliver Haase (Bell Labs Research, USA)
Stefan Hild (IBM Research, Switzerland)
Dirk Husemann (IBM Research, Switzerland)
Pertti Huuskonen (Nokia, Finland)
Alan Jones (AT&T Laboratories Cambridge, UK)
Kazuhiko Kato (University of Tsukuba, Japan)
Tim Kindberg (HP Laboratories, USA)
Kazushi Kuse (IBM Research, Japan)
Gerald Maguire (KTH Stockholm, Sweden)
Joachim Posegga (SAP Corporate Research, Germany)
Apratim Purakayastha (IBM Research, USA)
Jun Rekimoto (Sony, Japan)

Kurt Rothermel (University of Stuttgart, Germany)
Larry Rudolph (Massachusetts Institute of Technology, USA)
Bernt Schiele (ETH Zurich, Switzerland)
Bill Schilit (Fuji-Xerox Palo Alto Laboratory, USA)
Roy Want (Intel Research, USA)

Sponsoring Institutions

Bell Labs, USA
ETH Zurich, Switzerland
IBM Zurich Research Laboratory, Switzerland
Nokia, Finland
Philips Research, The Netherlands

Table of Contents

Invited Talks

System Design

Applications

Identification and Authentication

Device Independence and Content Distribution

The SAHARA Model for Service Composition across Multiple Providers

Bhaskaran Raman, Sharad Agarwal, Yan Chen, Matthew Caesar, Weidong Cui,
Per Johansson*, Kevin Lai, Tal Lavian[†], Sridhar Machiraju, Z. Morley Mao,
George Porter, Timothy Roscoe[‡], Mukund Seshadri, Jimmy Shih, Keith Sklower,
Lakshminarayanan Subramanian, Takashi Suzuki, Shelley Zhuang,
Anthony D. Joseph, Randy H. Katz, and Ion Stoica

University of California at Berkeley

Abstract. Services are capabilities that enable applications and are of crucial importance to pervasive computing in next-generation networks. *Service Composition* is the construction of complex services from primitive ones; thus enabling rapid and flexible creation of new services. The presence of multiple independent service providers poses new and significant challenges. Managing trust across providers and verifying the performance of the components in composition become essential issues. Adapting the composed service to network and user dynamics by choosing service providers and instances is yet another challenge. In SAHARA[1], we are developing a comprehensive architecture for the creation, placement, and management of services for composition across independent providers. In this paper, we present a layered reference model for composition based on a classification of different kinds of composition. We then discuss the different overarching mechanisms necessary for the successful deployment of such an architecture through a variety of case-studies involving composition.

1 Introduction

Pervasive computing demands the all-encompassing exploitation of services inside the network. By services, we mean both the components of distributed applications and the glue that interconnects them as they function across the network. Services range from providing basic network reachability to creating overlay networks with enhanced qualities like predictable latencies and sustained bandwidths. Services also include instances of application building blocks, requiring processing and storage, judiciously placed in the network to control connection latencies and to achieve scale through load sharing. Such services may be simple format translators, interworking functions, or major subsystems for content distribution or Internet search, which are often regarded as applications in their own right. Composition via interconnection of services allows more sophisticated services and applications to be constructed hierarchically from more primitive ones. Since economics makes it unlikely that any single service provider will be able to provide all of the connectivity, applications building blocks, processing, and

---

* Ericsson Berkeley Wireless Center; [†] Nortel Networks; [‡] Intel Berkeley Research Lab

[1] Project supported by Sprint, Ericsson, NTTDoCoMo, HRL, and Calif. Micro Grant #01-042.

F. Mattern and M. Naghshineh (Eds.): Pervasive 2002, LNCS 2414, pp. 1–14, 2002.

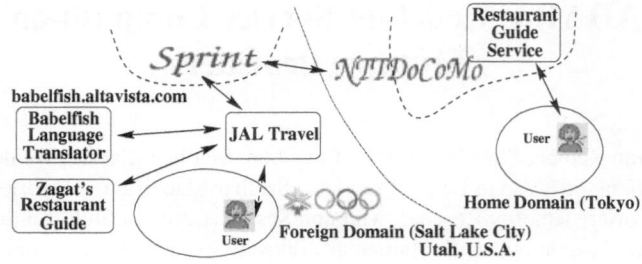

Fig. 1. Service Composition: A Multi-Provider Scenario

storage resources to effectively deploy a globe-spanning application, the composition of services across independent providers is essential. This paper proposes a comprehensive reference model for composed services in support of pervasive computing.

To illustrate our concept of service composition across service providers, consider the following scenario (Figure 1). Ms. Tanaka travels from Tokyo to Salt Lake City to attend the Winter Olympics. Her cellular provider, NTTDoCoMo, maintains roaming agreements with several foreign network operators, such as Sprint, so she can make and receive calls in the U.S.. But her new information appliance, built by Ericsson, is much more capable than just a phone – it is a gateway to extensive information, entertainment, and messaging services. In Japan, she subscribes to NTTDoCoMo's restaurant recommendation service, and would like to use the same application in Utah. In the U.S., Sprint has an arrangement with Zagat's Guide to present such information to its subscribers, but the text is in English and formatted for presentation on a different kind of display than Ms. Tanaka's. A third party, JAL Travel, assembles a special new service for Japanese tourists at the Olympics from component services: Zagat's Restaurant Guide, Japanese translation using Babelfish, and reformatting for Japanese-style information appliance displays. Ms. Tanaka subscribes to this service for two weeks, and the usage charges appear on her NTTDoCoMo bill back in Japan.

This scenario exemplifies several key points about services composition. In next generation networks, users will demand enhanced services like restaurant recommendations, but they expect them to look and feel the same whether they are at home or in a foreign network. A new service, *composed* from localized information sources, appropriate language translators, and content reformatters, makes the underlying differences transparent to the user. Such a service can be created quickly by simply connecting the components across the network. But many entities participate in the realization of this service, and they must be managed, authenticated and compensated in some way.

In the context of object-based systems, programming by composition across the network is hardly new. Yet there are critical new challenges. The first is compositions across independent service providers: Zagat owns the database and the web site, NTTDoCoMo maintains the relationship with the client and collects her monthly bill, and Sprint provides the local wireless access as well as the gateway access to local services. Creating the necessary mechanisms to support cooperative composition of services across indepen-

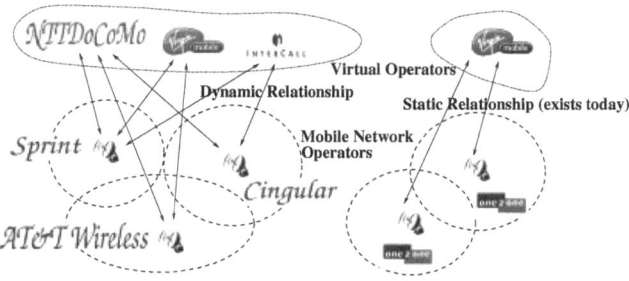

Fig. 2. Multi-provider Scenario in Wireless Connectivity Service

dent service providers, each with its own authentication, authorization, and accounting mechanisms, is an essential challenge.

JAL Travel created a new service from pieces, some of which were provided by other third parties, such as Babelfish, and all of which need to run on machines interconnected across the network, spanning Internet Data Center providers and Internet Service Providers. Herein lies the second challenge: the ability for third parties to discover components and to broker new services from constituent pieces, some of which may not even be aware of the composition in which they are participating. As the qualities of a composed service are no better than its weakest component, an essential need is for brokers to be able to verify the performance and behavior of the assembled components, whether or not these underlying participants are aware of their role in compositions. If a component does not meet its performance or behavioral specification, it must be "composed out", and a new instance from a different provider "composed in".

A third challenge for service providers is the need for an extensive set of new service composition management tools. From a provisioning viewpoint, sufficient instances of the components need to be placed at locations within the network to ensure scalable performance and high availability even in the face of site failures or network outages. Such placement also needs to ensure appropriate network and processing latencies to achieve adequate responsiveness for the supported applications. Such tools include a policy management mechanism for service providers to inform service composers about how their instances in their network should be used for providing fault tolerant and load balanced behavior. A pervasive monitoring and measurement infrastructure is needed to detect changing access patterns and shifting workloads, to drive redirection to unloaded service instances or to change the number and placement of deployed instances. Network topology-awareness is important, for availability as well as performance. Placement and connectivity issues are complicated since some service instances are anchored to fixed locations, while lighter weight services can be placed close to the user community.

Further challenges arise when we consider user dynamics. A large number of foreign roamers like Ms. Tanaka converge in Salt Lake City, yielding flash crowds and over-utilized spectrum. We need new ways of efficiently allocating resources in the context of new service provider business models. The 3rd Generation Partnership Project (3GPP) defines the concept of a Mobile Virtual Network Operator (MVNO) [1], an entity with subscribers but no network. The MVNO provides wireless connectivity service composed from the physical network resources of underlying wireless operators. In our example, NTTDoCoMo acts as a virtual operator using spectrum for its subscribers from the already established network operators such as Sprint, Cingular, and AT&T Wireless. We view such a multi-provider relationship as just another case of *service composition*. Today, the relationships between the virtual operator and the Mobile Network Operator (MNO) are static and negotiated long in advance (e.g., between *Virgin Mobile* and *One2One* in UK [1], see Fig. 2). But this is inefficient when user dynamics are considered; and we expect much more dynamic formation and dissolution of relationships in the future. Thus, as a fourth challenge, we address issues in efficient resource allocation across providers, considering the dynamics of user communities. In our MVNO scenario, this translates to dynamically selecting ("roaming") among co-located MNOs. Figure 2 illustrates such dynamic allocations in the MVNO context. We envision that dynamic relationships will last for short time-scales of minutes to hours, thus allowing for load balancing, and efficient resource usage.

Our overall goal is to define *a comprehensive reference model that is able to describe the assembly from components of end-to-end services with desirable, predictable, enforceable properties, yet spanning potentially uncooperating service providers*. We are developing these concepts in the context of the SAHARA[2] project, which is also the name of our prototype architecture. The next section summarizes the discussion above with the technical issues in composition. Sec. 3 presents a classification of the different kinds of composition and Sec. 4 presents a layered reference model for composition. We describe the different mechanisms we employ to address the technical issues in composition in Sec. 5. We discuss related work in Sec. 6 and conclude in Sec. 7.

2   Technical Issues

The scenario in our prior section can be understood along three dimensions in the choices in service composition: (a) what set of services to use for composition, (b) which service providers' resources to use, and (c) which instances of each service to use for a particular client session. In addition, we also have the issue of who makes these decisions. Considering these three dimensions, the technical issues that must be addressed in a reference model for service composition are:

- *Trust management and behavior verification:* When multiple providers interact, it is important to establish mutual trust. This is not only for the purpose of user authentication and billing, but also to verify the behavior of the components in composition. Does a component meet its promises in terms of functionality, protocols, performance, availability, or other properties?
- *Adapting to network dynamics:* With network dynamics, workloads can shift, congestion can arise, and reachability to services can be lost. This implies the need for performance monitoring, modeling and prediction, and a performance-sensitive choice of providers and instances. We term this the *service selection* problem.

---

[2] SAHARA: Service Architecture for Heterogeneous Access, Resources, and Applications; http://sahara.cs.berkeley.edu

- *Adapting to user dynamics:* User dynamics can cause different providers to face varying demand for physical resources from their users. Allocation of resources (spectrum, network bandwidth, CPU, etc.) across providers, based on current demand is important to achieve fair/utility-driven resource allocation.
- *Resource Provisioning and Management:* For a given community of users and a set of performance, availability, and administrative constraints, how many instances of a service are needed? How can this be optimized given a knowledge of the provisioning done by other service providers, and network topology information? We call this the *service placement* problem. Also, how can service providers enforce their local policies of the use of their service instances in a composition performed by a third party? We term this the *policy management* issue.
- *Interoperability across multiple service providers:* When composing services across providers, we have to deal with heterogeneity in protocols and data formats, as well as authentication and authorization mechanisms.

3   Service Composition Models

We now classify service composition into two models based on the type of interaction between composed component service providers and analyze their pros and cons: (1) **Cooperative Model:** Service providers interact in a distributed fashion, with distributed responsibility, to provide an end-to-end composed service; (2) **Brokered Model:** A single provider, the broker, uses the functionalities provided by underlying service providers and encapsulates these to compose the end-to-end service.

In either case, the end-user subscribes to only one provider. However, the difference lies in the way the responsibility for the composed service is apportioned. The two possibilities represent different business models for composition. In the cooperative model, the properties of the composed service such as functionality, performance, and availability, are guaranteed by the design of the distributed interaction, and through service-level agreements between the interacting entities. Each service provider is only responsible for providing guarantees for the portion of the composed service within its domain. In the brokered model, the broker assumes responsibility for the properties of the composed service. We can imagine a broker entering into contracts with service providers, and using these to construct end-to-end composed services. The broker verifies the functionality of the individual pieces in the service path. This is because individual component providers may not trust each other – they may limit the information about the state of their service they expose to the other providers in the composition, or may actively seek to cheat on the quality of service they provide.

An example of cooperative composition is cellular roaming as a service, composed from the resources of multiple mobile network operators, as in Sec. 1. The distributed interaction between operators (NTTDoCoMo and Sprint) enables roaming. Another example is end-to-end connectivity service in the Internet. An inter-domain routing protocol allows cooperation between domains in a distributed fashion. In these examples, there are long-term, static, negotiated contracts between the participant providers.

An example of brokered composition is the restaurant guide service assembled by JAL Travel in our earlier scenario. It assumes responsibility for the functionality and

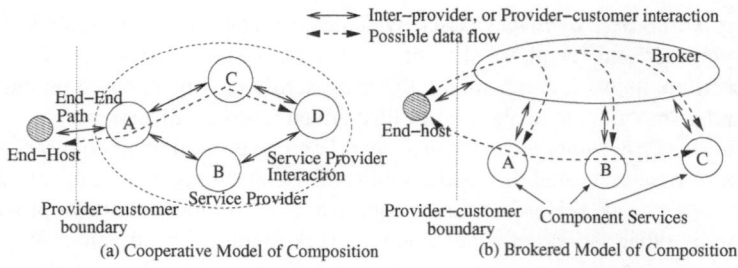

Fig. 3. Service Composition Models

performance of the composed service. Another example is the Yahoo portal service that composes third party services such as the Google search engine, stock ticker and news.

Fig. 3 illustrates these alternate composition models. The models of composition say nothing about the data flow, only the nature of interaction, or the business model, between providers. As an example, in the brokered model, we could have data flow through the broker, who assembles it (shown by the dotted lines on the top). Instead, the broker could set up the data exchange and not be in the data path (shown by the dotted line at the bottom of the figure).

The same composed service could be implemented in either model. Consider the provision of a connectivity service with QoS guarantees between two points on the Internet. In a cooperative model, ISPs enter into service level agreements that specify mutual QoS guarantees. These may be stitched together in a distributed fashion to offer end-to-end guarantees (www.merit.edu/working.groups/i2-qbone-bb). In a brokered model, a provider like InterNAP (www.internap.com) purchases pipes with specified guarantees from individual ISPs and uses them to provide QoS to its customers.

Each of these models is suited to a particular environment. In the cooperative model, providers work together and can share performance information to ensure end-to-end properties of the composed service. However, they must rely on each other, which leads to issues of trust. Since the responsibility for the composed service is distributed, each provider must continuously verify that the others with whom it has agreements meet their service specifications. These specifications are in terms of functionality, protocol, performance, and availability. Such comprehensive verification is absent in the cooperative composition of Internet connectivity service across network domains today.

In the brokered model, the broker composes an end-to-end service by selecting individual services residing in different domains. This simplifies service deployment since the members of the composition need not agree among themselves, only with the broker. This also enables the composition of services across competing service providers. The broker assumes responsibility for constructing the entire end-to-end service path. Brokering is a powerful tool to construct services from providers who are not necessarily aware that they are participating in a larger end-to-end service (e.g., in Figure 1, Zagat's restaurant guide service does not know it is being composed with Babelfish's language translation service). However, because the broker has limited visibility into the underlying provider resources, sub-optimal utilization of provider resources may result. The

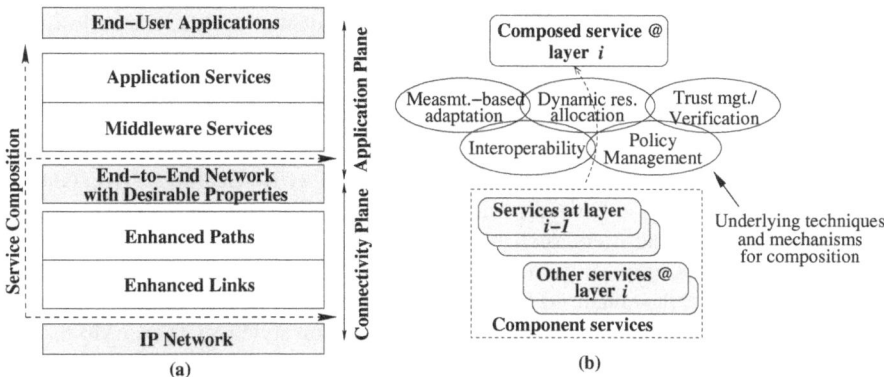

Fig. 4. A layered reference model for Service Composition

nature of the composition and the relationships of the underlying service providers will determine which model is most appropriate.

4 Service Composition: A Layered Reference Model

The previous section classified service composition by how responsibility is shared across providers. Here we present a layered reference model for composition across different layers of concern.

Composed services build on top of connectivity provided by the IP layer, the bottom slice in Fig. 4(a). We assume that IP provides reachability, and build several "desirable" properties in the *connectivity plane*. These include features such as performance guarantees (e.g., latency, bandwidth, loss-rate bounds), availability guarantees (e.g., available 99.99% of the time), as well as functionality or protocol guarantees (e.g., guaranteeing that a particular advertised Internet route is valid). We achieve them through composition at the connectivity plane, across multiple providers. This results in an "end-to-end network with desirable properties", as shown in the middle slice of the figure.

The connectivity plane is further divided into two layers: *Enhanced links*, and *Enhanced paths*. The enhanced links abstraction is built between two interacting entities: between two service providers, or between a provider and the end-user. This abstraction achieves desirable performance oriented properties, and verification of routing protocols between peering entities. Verification checks if the routes advertised by a peer are valid. An example of this composition is of the MVNO choosing between multiple MNOs. Here the enhancement is the improved performance through reduced call blocking rate, due to dynamic load sharing across multiple MNOs.[3]

Enhanced paths build on enhanced links, and provide desirable properties in an end-to-end path between points on the Internet. The path spans multiple service domains, and can be chosen adaptively via resource allocation across providers to meet performance or

---

[3] Although it is not obvious here how this example builds on IP technology, it will be clear after a discussion of the mechanisms for resource allocation, in Sec. 5.

availability constraints. Our reference model is independent of whether the performance guarantees are strict, or simply "enhancements" to the best-effort Internet. Alternative enhancements might be appropriate for different end applications.

The top half of the figure represents the application plane. These layers support end-user applications and are in turn built on top of the end-to-end network. The *middleware services* are enablers, such as the Babelfish language translation service, content-distribution networks or video/audio transcoders. The *application services* layer consists of services useful to end-users, such as the Zagat's restaurant guide, search engines or a voice-mail service. Composition at the application plane results in enhanced functionality. In our example, the enhanced functionality is that of the restaurant guide appearing in the user's native language (Japanese) and presentation style (NTTDoCoMo user interface) in a foreign network.

Service composition can be applied within and across these layers, as illustrated in Fig. 4(a). Fig. 4(b) shows this more explicitly. A composed service at a higher layer is composed of multiple services in the same layer, or of services at the layers below. In Fig. 1, composition takes place across the application (restaurant guide), middleware (language translation), and connectivity plane (roaming service enabled by Sprint) layers. In Fig. 2, composition is at the enhanced link layer, using the component connectivity services offered by co-located MNOs.

We note that this layerization is only a reference model, and compositions need not strictly adhere to it. Some application services can be composed directly of enhanced links, without using the enhanced paths abstraction or middleware services. This will occur typically in performance sensitive applications, where the composer needs full visibility via a flat rather than a hierarchical composition. We next discuss the techniques and mechanisms used in enabling composition as shown in Fig. 4(b).

5   Mechanisms for Service Composition

The issues we listed in Sec. 2 appear in different flavors in the alternate models of composition. SAHARA is our architectural prototype to explore the mechanisms required to address these issues. To understand the various mechanisms, we are working on several case studies of composition. These cover the dimensions of classifications in last two sections. We now describe the various mechanisms that we are designing to address the challenges presented earlier and how they are used in the individual case studies. We have performed in-depth evaluations of several of these mechanisms. Due to space limitations, we summarize the evaluations of only a subset of these mechanisms.

5.1   Measurement-Based Adaptation

For service composition, it is desirable to dynamically choose service providers, and service instances based on current network and server loads. Measurements can be carried out by a third party measurement service – a common element of the service infrastructure, or by the composer itself. This applies to both the cooperative and brokered models of composition. We now describe two services that we are developing involving measurement-based adaptation.

We are developing a general end-to-end Internet host distance estimation service. Such a service is especially useful in a brokered composition since the broker does not have insight into the network characteristics of individual providers. Given the potential large numbers of service providers and instances, to scale the measurement service, we cluster the end hosts to be monitored based on the similarity of their perceived distance to the measurement points. The cluster center is then used as a single measurement target for future monitoring. Simulations with real Internet measurement data show that our scheme has good prediction accuracy and stability with a small communication and computation cost [2].

An application service that we have developed is the Universal Inbox [3], a metaphor for any-to-any communication across heterogeneous devices and networks. Data transformation services such as audio/video transcoders and text ↔ speech engines are extensively used to adapt content between communicating devices. For example, in our earlier scenario, Ms. Tanaka's email service could be composed with a text-to-speech conversion service so that she can listen to emails over her cellular-phone. We use the brokered model of composition here. To adaptively choose service providers and instances, we have designed a middleware measurement layer that exchanges network and server load using a link-state algorithm [4]. This exchange takes place across service execution platforms, enabling a dynamic choice of service instances, possibly in the middle of the user sessions, to hide network and server failures from end-users.

A critical challenge that we address in the context of such an application is *availability*. When composed services span multiple providers, data could traverse the wide-area Internet. We detect and recover from Internet path failures *quickly* by using the middleware measurement layer to choose alternate service instances for the client session. Our measurements in [4] show that Internet path failures that happen due to congestion or other factors can be detected reliably within about *2 seconds*. Further, subsequent recovery by using alternate service instances can be completed in a *few hundred milliseconds*. Thus, network path failures lasting several tens of seconds to minutes [5] can be completely masked from the end client.

In addition to these two services, we have also designed and implemented a measurement methodology [6] to improve DNS-based server selection, which is a common technique used by Content Distribution Networks (CDNs) today. Our technique enables the collection of client to local DNS server mappings to allow more accurate server selection based on a client's local DNS server. Understanding the distribution of HTTP requests corresponding to local DNS servers also enables better load prediction given a DNS request and thus improved server selection mechanisms.

5.2   Utility-Based Resource Allocation Mechanisms

In a multi-provider environment, different providers may experience different demands for resources due to user dynamics. Demand or utility-based resource allocation can be applied within a service provider to manage its instances. In the brokered model, it can be used to allocate resources across providers. In SAHARA, we are exploring two resource allocation mechanisms.

Auctions are one way of constructing a marketplace where a resource, such as bandwidth or physical spectrum, can be dynamically allocated. Auctions allocate resources

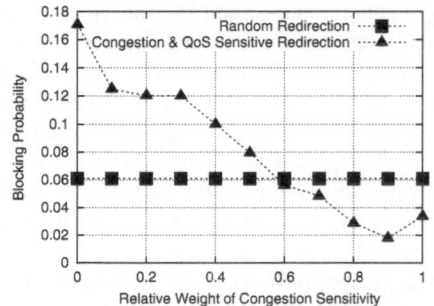

Fig. 5. Congestion-sensitivity in QoS-based redirection

to consumers based on their bids, which represent the value of the good to them. Furthermore, the resource can be subdivided into units, and multiple bidders can be allocated the resource until the resource pool is exhausted. Auctions can occur in rounds, where the allocation determined by each round can be for some future time period. Auction-based allocations in a multi-provider environment can provide the mechanism for demand-based resource allocation. For instance, in the MVNO scenario in Section 1, spectrum resources could be auctioned off every few minutes to competing virtual operators based on their current user-load in the area of coverage.

Congestion pricing is an allocation mechanism that assigns scarce resources to consumers using the abstraction of price as a means to moderate demand. During high demand, such a market ensures that the price increases. Only those consumers with the greatest need and having sufficient currency will obtain the needed resource. During low demand, the price drops, and access to the resource with be cheap and plentiful. This approach should yield an assignment of resources (supply) to the need (demand) that adapts to instantaneous demands [7].

As an application that uses the congestion pricing mechanism, we have looked at the selection of Voice-over-IP (VoIP) gateways across multiple providers [8]. These gateways are deployed by independent entities, and exchange dynamic pricing information as well as the peering relationships between the provider entities based on the IETF TRIP (Telephony Routing over IP) protocol. The price is decided based on the load, or congestion, at each gateway; and the user gets to choose between several gateways based on the price and the required quality of service. This achieves pricing-based load-sharing among the gateways. In [9], we looked at the trade-off between using QoS-based redirection (nearest VoIP gateway) and congestion-based redirection. This is shown in Figure 5 where we see that incorporating congestion-sensitivity in QoS-based redirection can improve call blocking rate by as much as a factor of three in comparison to random redirection. We also observe that increasing congestion-sensitivity does not significantly degrade the QoS of the VoIP calls.

5.3   Trust Management and Verification of Service/Usage

An important issue for composition is the establishment and monitoring of trust relationships between inherently untrusting entities. This is important in cooperative composition

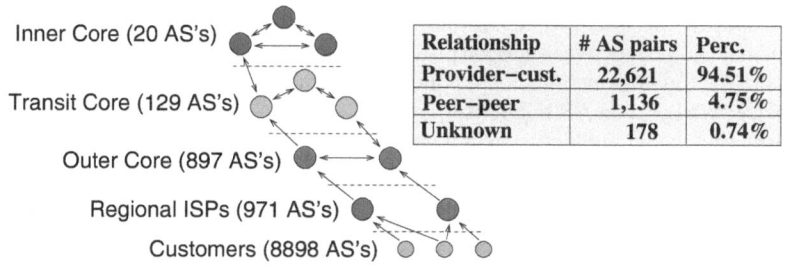

Relationship	# AS pairs	Perc.
Provider–cust.	22,621	94.51%
Peer–peer	1,136	4.75%
Unknown	178	0.74%

Fig. 6. Internet AS Hierarchy and Inferred Relationships for 23,935 AS Pairs

where providers have pairwise service agreements between them. This is also important in brokered composition where the agreements are with the broker. Typically a AAA (*Authentication, Authorization and Accounting*) server governs service instances and users within one administrative domain. However, in our scenario, we need to compose services across domains governed by multiple, different AAA servers.

We are investigating an authorization control scheme with credential transformations to enable cross-domain service invocation [10]. Federated administrative domains form credential transformation rules based on established peering agreements. These are used by a AAA server to make authorization decisions for a service request from an affiliated domain.

Another important issue in service composition is to verify whether the provided service adheres to the desirable properties advertised by its provider. Such properties can be specified in a bilateral Service Level Agreement (SLA) between provider and requester. We use parameter verification and usage monitoring as mechanisms to ensure that the properties specified in the SLA are being honored. For instance, in a case-study of connectivity composition across domains, we have border routers monitoring control traffic from different providers to detect malicious route advertisements.

5.4 Policy Management

An advantage of cooperative composition is that each provider has visibility into its network of services, while a broker does not. The disadvantage of this distributed form of composition is the lack of central control over the composition that the broker enjoys. This disadvantage can be minimized if some form of distributed policy management is in place. Specifically, in cooperative composition, the service composition policies of one service provider can be made visible to and applied at distant service providers further along the composed path. Such policies may include which service instances are for primary use and which are solely for use in various failure modes, and policies that govern load balancing between instances.

A case study involving this principle is the distributed application of policies in inter-domain routing on the Internet. The BGP protocol was not designed to allow local policies to be imposed at distant points in the network. The Internet infrastructure has been plagued of late with pathological routes that attempt to impose such inter-domain routing policies by tricking distant service providers' route selections [11].

To address this issue, we are developing an architecture [12] that allows distributed policy management between service providers. By negotiating policy changes at various points in the topology using a map, policy agents can improve load balancing and fault tolerance. We build an AS relationship map in [13] by using BGP measurements from multiple vantage points. The map indicates the nature of inter-provider relationship that exists between neighbor ASes: peer-to-peer or customer-provider. Figure 6 shows a summary of the results from this study.

5.5   Interoperability through Transformation

Interoperability is important for service composition across different service providers. Data and protocols formats may need to be transformed for interoperability across heterogeneity. This is an important issue for composition in the connectivity plane and in the application plane where services have more complex interfaces. Defining the service interface and propagating it are two key challenges here. We have investigated one such interoperability service at the connectivity plane. In our *broadcast federation* work [14], a global multicast service is composed from the multicast implementations in different provider domains. We use protocol transformation gateways between administrative domains that have non-interoperable implementations of multicast.

5.6   Service Deployment

In any composition model, a service provider has to decide how many instances to place and where to place them. In SAHARA, we are exploring two mechanisms to enable such decisions. (1) First, we infer inter-provider relationships between Internet ASes by passively monitoring inter-domain routes from multiple locations [13]. Such a relationship map allows us to understand traffic flows in the underlying network topology by giving more information about reachability between points in the network, than a simple connectivity graph. This richer reachability information helps in service instance placement – the provider can ensure that there are sufficient instances with good reachability from various points on the Internet. (2) Next, we have designed a dynamic replica placement protocol that enables an application level multicast tree for Web content distribution. This can meet client QoS constraints and server capacity constraints, while retaining *efficient* and *balanced* resource consumption [15]. To provide a scalable solution, we aggregate the access patterns of clients that are topologically close in the network. We use incremental clustering and distribution based on client access patterns to adaptively add new documents and purge old ones from the content clusters [16].

6   Related Work

Related work falls into two categories: (a) Architectures for seamless integration of devices and services, and (b) Internet-based web-services initiatives.

Architectures for Seamless Device/Service Integration: The UMTS model [17] admits of a sophisticated accounting, billing, and settlements architecture to support third-party brokering between subscriber needs for service and multiple service providers.

However, there is no explicit consideration of where service provision and service mediation should exist in this architecture, other than in the core network that ties together various access networks. The Virtual Home Environment (VHE) concept of IMT2000 [18] permits users to roam away from home, seeing the same service interface (service mobility). ICEBERG [19] looks at extensible personal and service mobility. Nevertheless, these efforts do not consider composition of services across multiple providers, efficiency through network awareness, or resource allocation issues.

TINA [20] is a CORBA-based [21] service architecture. The TINA reference architecture contributes the conceptual separation of the business model, the informational model, and the computational model. Its three layer model of applications, distributed processing, and network environment has influenced our layerization of composition in SAHARA. Key differences are that SAHARA adds elements of composition across heterogeneous providers, with a greater awareness and management of the underlying network topology. SAHARA also considers resource management via placement, allocation, redirection to services and resources.

Internet-Based Web-Services Initiatives: There are several industrial initiatives to enable web-services which "integrate PCs, other devices, databases, and networks into one virtual computing fabric that users could work with via browsers" [22]. These include HP's e-speak (www.e-speak.net) and web-services platform, Microsoft's .NET, and Sun ONE. These are based on a language for description of web services (WSDL), a common wire format for these descriptions (SOAP), and a registry to support service location (UDDI). Microsoft's .NET also defines a language independent software platform for easy and secure interoperation of applications [22]. However, these do not define a wide-area service architecture; they are complementary to SAHARA's goals of service placement, resource allocation, and network awareness aspects.

7 Conclusions

In this paper, we have presented a vision of distributed systems composed from services placed in the wide-area Internet, and spanning service providers at different levels. SAHARA is our evolving architectural prototype for the creation, placement, and management of services in next generation networks. Our goal is to enable end-to-end service composition with desirable, predictable and enforceable properties spanning multiple potentially distrusting service providers. We investigate two forms of service composition under different business models with varying degrees of cooperation and trust among providers. We classify component services and composed services into a layered hierarchy. The overarching themes in the various techniques and mechanisms that we use for composition include (a) measurement-based adaptation through dynamic choice among service providers and service instances, (b) utility-based resource allocation for demand-driven load sharing across provider resources, and (c) a trust-but-verify approach to management of trust and behavior verification when multiple providers interact to provide a composed service. We continue to develop these mechanisms through prototype distributed applications spanning wide-area networks, and constructed through service composition at various layers under different models.

References

1. Curley, F.: Mobile Virtual Network Operators (Part IV). In: Eurescom Summit, 3G Technologies and Applications. (2001)
2. Chen, Y., Lim, K., Overton, C., Katz, R.H.: On the Stability of Network Distance Estimation. In: ACM SIGMETRICS PAPA Workshop. (2002)
3. Raman, B., Katz, R.H., Joseph, A.D.: Universal Inbox: Providing Extensible Personal Mobility and Service Mobility in an Integrated Communication Network. In: WMCSA. (2000)
4. Raman, B., Katz, R.H.: Emulation-based Evaluation of an Architecture for Wide-Area Service Composition. In: SPECTS. (2002)
5. Labovitz, C., Ahuja, A., Abose, A., Jahanian, F.: An Experimental Study of Delayed Internet Routing Convergence. In: ACM SIGCOMM. (2000)
6. Mao, Z., Cranor, C., Douglis, F., Rabinovich, M., Spatscheck, O., Wang, J.: A Precise and Efficient Evaluation of the Proximity between Web Clients and their Local DNS Servers. In: USENIX. (2002)
7. Shih, J., Katz, R.H., Joseph, A.D.: Pricing Experiments for a Computer-Telephony-Service Usage Allocation. In: IEEE GLOBECOM. (2001)
8. Caesar, M., Balaraman, S., Ghosal, D.: A Comparative Study of Pricing Strategies for IP Telephony. In: IEEE GLOBECOM. (2000)
9. Caesar, M., Ghosal, D., Katz, R.H.: Resource Management in IP Telephony Networks. In: IWQOS. (2002)
10. Suzuki, T., Katz, R.H.: An Authorization Control Framework to Enable Service Composition Across Domains. In: ACM WWW (poster). (2002)
11. Huston, G.: Analyzing the Internet's BGP Routing Table. Cisco Internet Protocol Journal (2001)
12. Agarwal, S., Chuah, C., Katz, R.H.: An Overlay Policy Protocol to Augment BGP (2002) Work in progress.
13. Subramanian, L., Agarwal, S., Rexford, J., Katz, R.H.: Characterizing the Internet Hierarchy From Multiple Vantage Points. In: IEEE INFOCOM. (2002)
14. Chawathe, Y., Seshadri, M.: Broadcast Federation: An Application-layer Broadcast Internetwork. In: NOSSDAV. (2002)
15. Chen, Y., Katz, R.H., Kubiatowicz, J.: Dynamic Replica Placement for Scalable Content Delivery. In: IPTPS. (2002)
16. Chen, Y., Qiu, L., Chen, W., Nguyen, L., Katz, R.H.: On the Clustering of Web Content for Efficient Replication (2002) Submitted for publication.
17. Mohr, W., Konhauser, W.: Access Network Evolution Beyond Third Generation Mobile Communications. IEEE Communications Magazine (2000)
18. Pandya, R., Grillo, D., Lycksell, E., Mieybegue, P., Okinaka, H.: IMT-2000 Standards: Network Aspects. IEEE Personal Communications Magazine (1997)
19. Wang, H.J., Raman, B., Chuah, C.N., Biswas, R., Gummadi, R., Hohlt, B., Hong, X., Kiciman, E., Mao, Z., Shih, J., Subramanian, L., Zhao, B.Y., Joseph, A.D., Katz, R.H.: ICEBERG: An Internet-core Network Architecture for Integrated Communications. IEEE Personal Communications Magazine (2000)
20. TINA-C: Telecommunications Information Networking Architecture Consortium. http://www.tinac.com/ (web-site)
21. Mowbray, T.J., Zahavi, R.: The Essential CORBA: System Integration Using Distributed Objects. Wiley Computer Pub. (1997)
22. Vaughan-Nichols, S.J.: Web Services: Beyond the Hype. IEEE Computer (2002)

Ubiquitous Computing in the Automotive Domain (Abstract)

Ralf G. Herrtwich

DaimlerChrysler AG
Alt-Moabit 96a
D-10559 Berlin
Germany

`ralf.herrtwich@daimlerchrysler.com`

Abstract. Examples for ubiquitous computing applications usually come from the household domain. Typical lists include microwave ovens with integrated web-pads, refrigerators or washing machines with remote Internet connections for maintenance access, and even instrumented coffee mugs or clothes. While many of these examples have substantial entertainment value, the likelihood of their realization and pervasive deployment in the not too distant future is questionable. There is, however, another application domain for ubiquitous computing which holds substantial promise, but is often overlooked: the automotive sector.

Cars are fairly attractive protagonists for ubiquitous computing: They are large enough to have communication devices integrated in them, in fact, a substantial portion of them has integrated phones today. They come with their own power source which can also feed their communications equipment. Their price is some orders of magnitude higher than that of the device to be included, so the relative price increase to make them communicate is small. And, perhaps most importantly, some services such as mayday, remote tracking, or telediagnosis make vehicle connectivity desirable for car buyers and car manufacturers alike.

In this talk, we discuss how ubiquitous computing in the automotive domain can become a reality. We investigate the principal services resulting from network-connected cars, focussing on vehicle-originated rather than passenger-related communication as we believe that ubiquitous computing is more about communicating machines than communicating humans. Within the vehicle-centric services identified, we distinguish between client/server and peer-to-peer applications, resulting in different communication requirements and system setups. We outline some network solutions to meet these requirements, including technologies for car-to-infrastructure and car-to-car communication in different regions of the world. We conclude by discussing the overall effect which these developments may have on the automotive industry.

F. Mattern and M. Naghshineh (Eds.): Pervasive 2002, LNCS 2414, p. 15, 2002.
© Springer-Verlag Berlin Heidelberg 2002

Building Applications for Ubiquitous Computing Environments*

Christopher K. Hess, Manuel Román, and Roy H. Campbell

Department of Computer Science
University of Illinois at Urbana-Champaign
Urbana, IL 61801

Abstract. Ubiquitous computing embodies a fundamental change from traditional desktop computing. The computational environment is augmented with heterogeneous devices, choice of input and output devices, mobile users, and contextual information. The design of systems and applications needs to accommodate this new operating environment. In this paper, we present our vision of future computing environments we term *User Virtual Spaces*, the challenges facing developers, and how they motivate the need for new application design. We present our approach for developing applications that are portable across ubiquitous computing environments and describe how we use contextual information to store and organize application data and user preferences. We present an application we have implemented that illustrates the advantages of our techniques in this new computing environment.

1 Introduction

Ubiquitous computing challenges the conventional notion of a user logged into a personal computing device, whether it is a desktop, a laptop, or a digital assistant. When the physical environment of a user contains hundreds of networked computer devices each of which may be used to support one or more user applications, the notion of personal computing becomes inadequate. Further, when a group of users share such a physical environment, new forms of sharing, cooperation and collaboration are possible and mobile users may constantly change the computers with which they interact. We believe this requires a new abstraction for computing which we term a *User Virtual Space*. The User Virtual Space is composed of data, tasks and devices associated to users, it is permanently active and independent of specific devices, moves with the users, and proactively maps data and tasks into the users' ubiquitous computing environment according to current context.

User Virtual Spaces pose several challenges regarding data availability, application mobility and adaptability, context management, and resource coordination. As a result, applications must provide functionality to be partitioned

---

* This research is supported by a grant from the National Science Foundation, NSF
0086094, NSF 98-70736, and NSF 99-72884 CISE.

F. Mattern and M. Naghshineh (Eds.): Pervasive 2002, LNCS 2414, pp. 16–29, 2002.

among different devices, move from device to device, and adapt as demanded by users and their environment. We believe that this environment requires support from an external software middleware infrastructure that provides a standard set of mechanisms and interfaces to coordinate the contained resources. We use the term *active space* to refer to any ubiquitous computing environment orchestrated by such a middleware infrastructure.

In this paper, we discuss our proposed solution for constructing user-centric, space-aware, multi-device applications, which meet the requirements imposed by User Virtual Spaces and describe an example application that we have implemented in our prototype active space. The remainder of the paper continues as follows: Section 2 discusses the motivation for our application framework. Section 3 details the framework and Section 4 discusses how data is managed in our environment. Section 5 gives an overview of our development platform and Section 6 describes an application developed with our framework, which includes how the application is created, instantiated, and used. Sections 7 and 8 discuss related work and conclusions, respectively.

2   Motivation

Aspects unique to User Virtual Spaces motivate the need for a different approach to application construction. Below we discuss some of the inherent difficulties in the environment that affect applications. In later sections, we present our model for building applications that accommodates these challenges.

Context. A distinguishing feature of ubiquitous computing is context-driven application adaptation. Context can trigger changes in applications to adapt to the surroundings of a user in order to facilitate the use of the computational environment [10,2]. User Virtual Spaces maintain a collection of data for each user that may be remotely distributed and constitutes user's application data, configurations, and preferences. For any given task, only a portion of this personal data may be required; context can be used to organize data such that the information pertinent to the current context is easily accessible. Limiting the amount of data is even more important when a space is populated by a number of users that have come together to perform a specific task, where each user may contribute some amount of information to the shared space, and the amount of aggregate information can become large. By pruning irrelevant material not necessary for the current task, locating information may be simplified. In addition, continually running applications may not have the luxury of human intervention to manually search for data. If relevant information, which may change over time, is known to exist in a particular location, the application is relieved from performing costly searches over the entire collection of data.

Binding. Ubiquitous environments are often task-driven, which often precludes applications from being permanently associated to a single device, or to a collection of well-defined devices. Applications must be constructed so they can be partitioned and mapped into the most appropriate resources according to physical context, personal configurations, and other attributes, thereby transferring the application binding from particular devices to users and spaces. Consider a music jukebox application, in which the application automatically finds and uses the most appropriate resources to continue playing and controlling the music as context changes (e.g., as the user moves from office to car) The emphasis is on the application itself, instead of the devices used to execute and control the application.

Mobility. Mobility affects the application architecture by requiring the ability to migrate parts of their functional components at runtime. User mobility also has implications on data availability; users should not be burdened with manually transferring files or data, be it configurations, preferences, or application data. Personal data should be available to them regardless of their physical location. Data becomes implicitly linked to a user and can follow them around, becoming available whenever they enter a new active space.

Adaptability. Application developers should not be concerned with the complexities of data format conversions; they should gain access to data in a particular format by simply opening the data source as the specific desired type and the system should be responsible for automatically adapting content to the desired format.

User preferences or influences from the environment may also affect the internal structure of an application. Applications must expose a means to configure this structure in order modify the way in which a user interacts with an application or the way in which the application data is presented. For example, control of an application running in an active space may be passed between group members who have different devices, new controllers may be attached to enable parallel manipulation of the application, or a user may switch to a new type of controller, such as from pen-based to voice-activated.

3   Application Framework

Ubiquitous environments require a user-centric multi-device application model where applications are partitioned across a group of coordinated devices, receive input events from different devices, and present their state using different types of devices. Applications must be designed such that their composition may be changed based on the context of the current situation. Therefore, the structure of an application must be described by a generic set of components that can be customized based on the available resources or preferences. This gives users more flexibility in deciding how to interact with applications by allowing them to choose among a number of input, output, and processing devices.

To address the challenges inherent in developing applications for ubiquitous computing environments, we have developed an application framework that assists in the creation of generic loosely coupled distributed component-based applications. The framework reuses the application partitioning concepts described by the traditional Model-View-Controller and introduces new functionality to export and manipulate the bindings of the application components. Exposing information about component bindings allows reasoning about the composition of the application components, and enables modification of this composition according to different properties, such as context and user preferences.

Our application framework defines three basic components that constitute the building blocks for all applications: the Model, Presentation (a generalization of the View), and Controller, where the Model implements the application logic, the Presentation exports the model's state, the Controller maps input events (e.g., input sensors and context changes) into messages sent to the model. To coordinate these components, the application framework introduces a fourth element, called the Coordinator, which is responsible for storing the application component bindings, as well as exporting mechanisms to access and alter these bindings, such as (dis)connecting a controller or presentation, and listing current presentations. The Model, Presentation, and Controller are strictly related to application functionality, while the Coordinator implements the meta-level functionality of the application. The application framework (MPCC) defines methods for performing common tasks, such as application mobility.

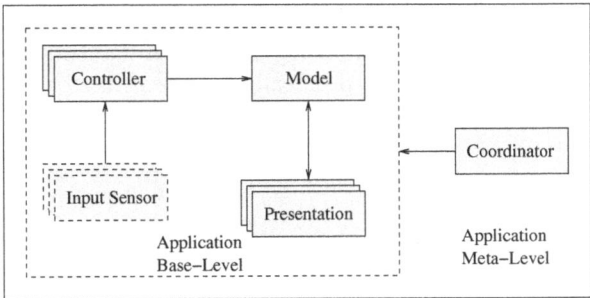

Fig. 1. The Model-Presentation-Controller-Coordinator application framework decouples application components and exposes the internal structure of the application

Applications are constructed independent of a particular active space by using generic application descriptions that can be customized for the resources available in a specific space. For example, a calendar application running in an active office may use a plasma display to present the appointments for the week, a handheld to display the appointments for the day, and may use a controller running in the desktop PC to enter data. However, the same calendar running in a vehicle may use the sound system to broadcast information about the next appointment, and use a controller based on speech recognition to query the calendar or to enter and delete data. The framework defines two types of application description files: the application generic description (AGD), and the application customized description (ACD). The AGD is independent of the space in which the application runs and contains information regarding the required components that compose the application, such as the name and type of components and the number of instances allowed. The AGD acts as a template from which concrete application configurations can be generated. The ACD is an application description that customizes an AGD to the resources of a specific active space. The ACD consists of information about what components to use, how many instances to create, and where to instantiate the components. The application framework offers tools that allow users to create and store ACDs. The resulting ACD is a script that coordinates the application instantiation process.

4   Application Data

Generated ACDs are typically associated with the context for which they are created. Therefore, we store application configurations in the context of the active space, particular to the owner of the configurations. We have developed a data management system that is context aware and plays the role of a file system in a traditional environment. It allows application data to be organized based on context to limit the scope of a user's full collection of data to what is important for the current task and provides dynamic data types to accommodate device characteristics and user preferences. The system uses context to alleviate many of the tasks that are traditionally performed manually or require additional programming effort. More specifically, context is used to 1) automatically make personal storage available to applications, conditioned on user presence, 2) organize data to simplify the location of data important for applications and users, and 3) retrieve data in a format appropriate to user preferences or device characteristics.

The file system is composed of distributed file servers and mount servers. Each file server manages data on the machine on which it is executing. An active space maintains a single mount server, which stores the current storage layout (i.e., namespace) of the space. The mount server contains mappings to data relevant to the active space, as well as the personal data of users. The user mappings are dynamic, changing as users move between spaces; when a user enters an active space, their personal data is dynamically added, making it locally available.

There are occasions when data is generated in a different context from which it will be used. In such cases, the context in which the data is important must be explicitly attached to the data to make it available in the specified context. Our system uses standard file system primitives (i.e., *rename*, *remove*, *copy*, *mkdir*) to attach and detach context to files and directories.

The system constructs a virtual directory hierarchy based on what types of context are available and what context values are associated with particular files or directories. What data items are available depends on the current context of the active space, which may change over time as users move, situations change,

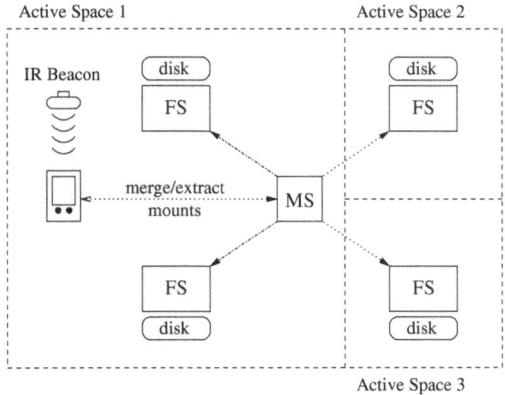

Fig. 2. The file system architecture consists of distributed file servers (FS) and mount servers (MS). A mobile handheld may be used to carry the mount points of a user, which can be merged into an active space to make remote personal storage available to the space

or new tasks are initiated. The layout of the directory hierarchy is implemented using the mounting mechanism, where mount points are owned by users and contain context indicators. Mount points may be automatically retrieved from a home server or can be carried with a user via a mobile handheld device and injected into the current environment to make personal storage available to applications and other users, as shown in Figure 2. In our current implementation, we employ the latter approach. A special directory called "current" aggregates data that is important to the current context. This supports the concept of the User Virtual Space by automatically configuring an active space to the needs of the user and the tasks being performed. Even though a user's data may be dispersed among several remote machines, that data is presented as a single source with only pertinent information visible.

We have implemented a shell and graphical directory browser, shown in Figure 3, that can be used to view the virtual directory structure, can be used to associate context with files and directories, and can launch applications.

5   Infrastructure

We have developed a middleware infrastructure, called *Gaia* [9], that exports and coordinates the resources contained in a physical space and provides a generic computational environment. *Gaia* converts physical spaces and the ubiquitous computing devices they contain into a programmable computing system. *Gaia* is similar to traditional operating systems by managing the tasks common to all applications built for physical spaces. Each space is self contained, but may interact with other spaces. *Gaia* provides core services, including events, entity

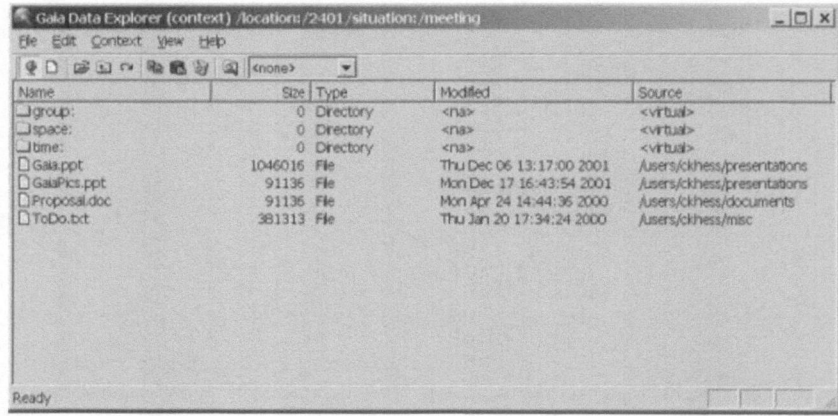

Fig. 3. Screenshot of the context file system browser. The virtual hierarchy is used to attach context to files and directories

presence (devices, users, and services), context notification, discovery, and naming. By specifying well-defined interfaces to services, applications may be built in a generic way that are able to run in arbitrary active spaces. The core services are started through a bootstrap protocol that starts the *Gaia* infrastructure.

6   Presentation Manager

In this section, we present the Gaia Presentation Manager (GPM), an application based on the application framework that provides functionality for creating and presenting synchronized slide shows. These slide shows exploit multiple input and output devices contained in a ubiquitous computing environment, can present multiple (possibly different) slides simultaneously in multiple form factors, can attach and detach controllers, and can migrate slides views to different displays.

The functionality to edit synchronized slide shows allows users to define the number of steps for the presentation, select multiple slides from existing Power Point files, and specify synchronization rules. These synchronization rules define which slides to present in every step, and what display to use for every slide. The functionality to present the synchronized slide shows allows users to navigate through the different steps of the presentation. As an MPCC application, the GPM also exports functionality to move and duplicate controllers and presentations, adapt to different active spaces, and partition the application among multiple devices. Figure 4 shows the application running in our prototype active space. We have implemented and currently use the GPM to present slide shows in our experimental active space, which implements the full functionality described in this section.

Fig. 4. Example of an active meeting room in which the presentation manager is deployed

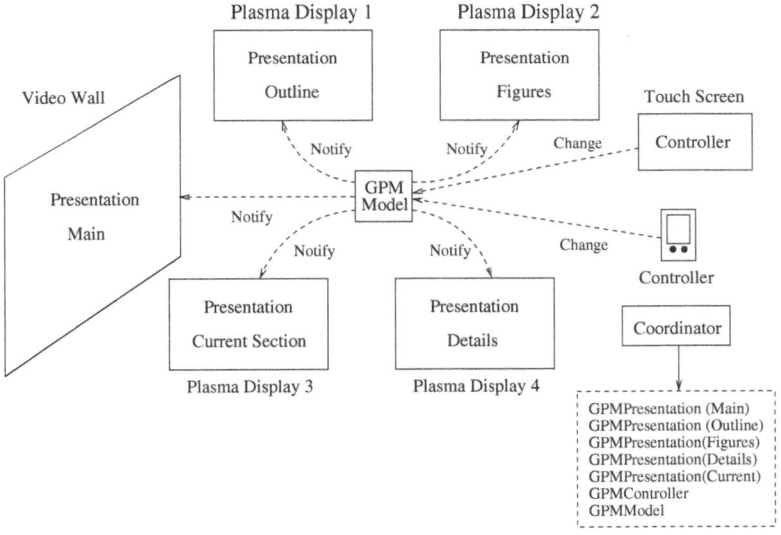

Fig. 5. Schematic of the GPM application architecture

6.1 Application Design and Implementation

The GPM is composed of five components: Model, Presentation, Controller, Editor, and Coordinator, as shown in Figure 5. The GPM implements the first four components and reuses the default Coordinator provided by the application framework; the GPM does not require any special behavior from the Coordinator.

The *GPM model* uses an acyclic directed graph to model the synchronized slide show presentation. Nodes in the graph store information about the slides to present and the displays to use, while arcs define the transition order. The GPM model defines an abstraction called a *virtual display* that decouples the slide show views from specific environmental resources. While editing, users assign the slides to the virtual displays, which will be mapped to real displays contained in the ubiquitous computing environment. When the state of the model is changed by a controller, the model sends a notify event to all presentations. Presentations affected by the model's state change contact the model to retrieve the new changes and update their presented information. The GPM Model leverages the event functionality implemented by the *Gaia* middleware infrastructure to notify the GPM presentations.

The *GPM presentation* provides functionality to display the contents of a slide. When instantiating a GPM presentation, the application assigns it an id that corresponds to one of the virtual display ids used by the GPM model. During the presentation of a synchronized slide show, the GPM model sends update events including a virtual display id and information specific to the update. GPM presentations compare the received virtual display id with their assigned id to decide whether or not to they need to update their assigned slide.

The *GPM controller* is used during the presentation of the synchronized slide show and provides functionality to start and stop the presentation, and move to the next and previous step.

The *GPM editor* is a design-time tool used to edit the synchronized slide show presentation. It provides functionality to create and delete steps, create and delete virtual displays, select existing Power Point presentations, assign slides to virtual displays, and save the synchronized slide show presentation.

Table 1 presents the AGD defined for the GPM. The model, for example, is implemented by a component named CORBA/GPMModel, requires the name of a previously created synchronized slide show as an input parameter, has a cardinality of one (a GPM application has exactly one model), and requires an ExecutionNode device (device with functionality to execute components) running on Windows 2000.

6.2   Using the Application

This section describes the different mechanisms involved in using the GPM application in a real environment: a meeting room equipped with a variety of resources including four 61" plasma displays, 6 touch screens, two Sound Web audio systems, badge detectors, and IR beacons. Mobile devices communicate with the applications and services in the active space via 802.11 wireless. We describe the usage of this application for a group presentation in the active meeting room, consisting of the following actions: creating an ACD for the active meeting room, entering the meeting room, registering a handheld device,

Table 1. Examples of a presentation manager AGD (left) and ACD (right)

```
Model {                                      Application = {
  ClassName    CORBA/GPMModel                  Model = { {
  Params       -f <fileName>                   ClassName ="CORBA/GPMModel",
  Cardinality  1 1                             Hosts = {{ "amr1.as.edu" },}
  Requirements device=ExecutionNode          } },
               and OS=Windows2000              Presentation = {
}                                                ClassName ="CORBA/GPMPresentation",
Presentation {                                   Hosts = { { "plasma1.as.edu","-i Outline" },
  ClassName    CORBA/PPTPresentation                     { "plasma2.as.edu","-i Current" },
  Params        -i<VirtualDisplayID>                     { "projector.as.edu","-i Main" },
  Cardinality  1 *                                       { "plasma3.as.edu","-i Figures" },
  Requirements device=Display                            { "plasma4.as.edu","-i Details" }
               and OS=Windows2000                      }
}                                                } },
Controller {                                   Controller = { {
  ClassName    Exec/VCRController                Classname ="Exec/GPMNavigationController",
  Cardinality  1 *                               Hosts = {{ "touchscreen1.as.edu"},}
  Requirements device=Touchscreen             } },
               and OS=Windows2000              Coordinator = { {
}                                                ClassName ="CORBA/Coordinator",
Coordinator {                                    Hosts = {{ "amr2.as.edu","" },}
  ClassName    CORBA/Coordinator              } },
  Cardinality  1 1                           }
  Requirements device=ExecutionNode
               and OS=Windows2000
}
```

mounting personal data storage, starting the application, and interacting with
the application.

Creating an ACD for the Active Meeting Room. The application framework provides a tool to generate ACDs from an AGD based on the resources
available in an active space, as shown in Figure 6. This tool parses the GPM
AGD and presents a list of devices contained in the active meeting room compatible with the requirements specified by each application component. When the
specialization is finished, the specialization tool generates an ACD customized
for the active meeting room where the presentation will be held.

ACDs are space dependent and user specific. Different users may have different configuration preferences for the same application and the same active space.
The specialization mechanism stores the resulting ACD in the users' personal
storage and attaches information specific to the context for which the AGD was
customized. The customized application description presented in Table 1 is the
result of customizing the AGD to the active meeting room.

Entering the Active Meeting Room. The active meeting room is equipped
with active badge detectors, iButtons, and infrared beacons to detect people entering the room. The speaker walks in the room with a handheld device (equipped

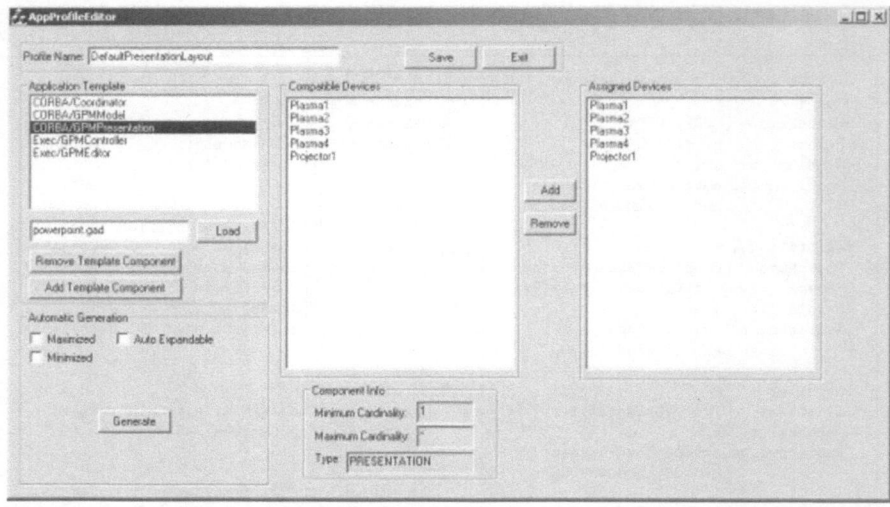

Fig. 6. The specialization tool is used to customize an AGD to the resources available in an active space

with an infrared port) and points it at the infrared beacon. The handheld device receives a reference to the active meeting room that allows the speaker to access available services via the handheld's wireless interface. The handheld uses this reference to register itself with the the active space, thereby becoming a resource of the active meeting room. The speaker then uses the handheld to merge the personal data mount points into the mount server of the room. Based on the current context, the synchronized presentation and the ACD files are visible in the "current" directory.

Starting the Application. Using the graphical directory browser, the speaker finds the file for the presentation in the "current" context directory. The context file system browser allows mapping file extensions to applications. Although this is a common practice in almost every traditional file browser (e.g., Windows Explorer, and KDE File Manager), ubiquitous computing environments require additional functionality for the mapping mechanism. Specifying a single default application for every file extension is not sufficient because an active space allows multiple layouts for the same application. Furthermore, different users may have different preferred configurations. The file browser solves this problem by selecting the ACDs in the speaker's "current" directory and allowing the speaker to choose one of these ACDs. This mechanism presents the speaker with ACDs associated to the file extension and relevant to the current context. Once the application is started, the four plasma displays and the projector have an instance of a GPM presentation and the GPM controller is displayed on the touch screen.

The GPM controller is a GUI with four push buttons and is used to start, stop, and navigate through the slides.

Interacting with the Application. The speaker starts the presentation using the controller, which triggers the four plasma displays and the projector to display the first slide of the presentation, according to the synchronization information stored in the model. The flexibility of the application framework allows the speaker to move or duplicate both controllers and presentations to any display in the room, including handhelds owned by attendees. For example, if the speaker wishes to start a video that requires an occupied display, the slide rendered on the display can be moved to an auxiliary display. Since both the video and presentation applications have the same control interface, the controller may be reused by simply attaching it to the video application.

7   Related Work

Our work is related to work involving ubiquitous computing, application construction, and data management. The Easy Living project at Microsoft [6] focuses on home and work environments and includes an infrastructure that allows user interfaces to move, according to user location. We differ in that we change the way in which applications are built by partitioning them across devices. Our work is similar to the Interactive Workspaces [4] from Stanford in that we believe there is a need for a supporting infrastructure or operating system in workspaces. However, we consider the concept of the virtual space in which applications and data are associated to the user and can move with them. The Aura architecture [12] allows tasks to be bound to users, similar to our virtual space concept. We are also considering application partitioning and dynamic composition. The Virtual Home Environment (VHE) model [3] proposes an architecture where mobile users may access their environment (e.g., services) from different locations and devices. The model considers device and network heterogeneity with the goal of presenting a consistent look and feel to services. Our model is more related to an operating system, by treating a space as a programmable entity to assist in the development of interactive ubiquitous applications.

PIMA [1] and I-Crafter [8] propose models for building platform independent applications. Developers define an abstract application that is automatically customized at run-time to particular devices. PIMA and I-Crafter generate applications for a single device, while we consider applications partitioned across devices. The Pebbles [7] project is investigating partitioning user interfaces among a collection of devices. While Pebbles is mostly concerned with issues related to GUIs, our application model focuses on the application structure (logic, control, presentation, and meta-level management), application life-cycle, and application adaptability and configurability.

Early work in integrating context with file access was investigated as part of the ParcTab project [11]. The tab allowed access to files that were meaningful to a particular location. As users moved between office spaces, the file browser would change to display relevant data. While they only considered location in their file system, this seminal work was important in establishing the relevance of context in data access and application adaptation. Research in tangible interfaces has proposed tying digital information to physical objects, which can trigger some action (e.g., file transfer) when they are discovered by a new environment [5]. We expand on this idea by treating the *user* as the physical object that triggers the addition of information into a space. The FUSE research group has developed applications for teamwork support based on active documents [13]. Our work differs in that we allow arbitrary context to be attached to data and we consider user configurations and preferences as part of the user's context data.

8   Conclusions and Future Work

This paper has presented our framework for building applications in future computing environments in which a user's environment is implicitly linked to them and is available as they move between physical spaces. Such an environment diverges from the traditional desktop environment due to factors such as context, mobility, and device heterogeneity. Our framework deals with both the structure of the application and the management of the data, which includes application data, configurations, and user preferences. Application logic is separated from presentation and control and the framework introduces a meta-level component that is used to expose the internal structure of the application so that it may be manipulated. Our framework is built on top of a ubiquitous computing middleware infrastructure we have developed that has allowed us to experiment with several applications. Applications built using the framework are described in generic terms and may be customized for the resources that are available in a particular space. As a user moves between spaces, the application can (un)bind new hardware resources and interfaces when they become (un)available. We believe that future applications will have increased flexibility in terms of choice of devices, partitioning, and migration to accommodate the requirements of User Virtual Spaces. This degree of suppleness is amenable to applications being constructed with a component-based approach, where portions of the application may be swapped, removed, or added.

Our future work will involve increasing the ease with which our applications may be used and configured. We have built several applications that have fit well into the framework. However, we will be constructing more applications to validate our approach to building a wide range of ubiquitous applications. Gaining more experience with building and instantiating ubiquitous applications will help us to refine the framework and the management of application data. Our current implementation does not include access control or authentication. We are currently developing a security architecture to control access to devices, applications, and data.

Acknowledgments. We would like to thank Herbert Ho for his work on the implementation of the GPM. We would also like to the thank the anonymous reviewers for providing valuable comments.

References

1. Guruduth Banavar, James Beck, Eugene Gluzberg, Jonathan Munson, Jeremy B. Sussman, and Deborra Zukowski. Challenges: an application model for pervasive computing. In *Mobile Computing and Networking*, pages 266–274, 2000.
2. Anind K. Dey, Gregory D. Abowd, and Daniel Salber. A Context-based Infrastructure for Smart Environments. In *Proceedings of the 1st International Workshop on Managing Interactions in Smart Environments (MANSE '99)*, pages pp. 114–128, 1999.
3. EURESCOM. Realizing the Virtual Home Environment (VHE) concept in ALL-IP UMTS networks. http://www.eurescom.de.
4. Armando Fox, Brad Johanson, Pat Hanrahan, and Terry Winograd. Integrating Information Appliances into an Interactive Workspace. *IEEE Computer Graphics and Applications*, 20(3), May/June 2000.
5. Hiroshi Ishii and Brygg Ullmer. Tangible Bits: Towards Seamless Interfaces between People, Bits and Atoms. In *Proceedings of the ACM Conference on Human Factors in Computing Systems (CHI'97)*, pages 234–241, Atlanta, GA, March 22-27 1997.
6. Microsoft Corp. Easyliving. http://www.research.microsoft.com/easyliving.
7. B. A. Myers. Using Hand-Held Devices and PCs Together. In *Communications of the ACM*, volume 44, pages 34–41, 2001.
8. S. R. Ponekanti, B. Lee, A. Fox, P. Hanrahan, , and T. Winograd. ICrafter: A Service Framework for Ubiquitous Computing Environments. In *Ubiquitous Computing, Third International Conference (Ubicomp 2001)*, Atlanta, GA, 2001. Springer.
9. Manuel Roman, Christopher K. Hess, Renato Cerqueira, Klara Narhstedt, and Roy H. Campbell. Gaia: A Middleware Infrastructure to Enable Active Spaces. Technical Report UIUCDCS-R-2002-2265 UILU-ENG-2002-1709, University of Illinois at Urbana-Champaign, February 2002.
10. Daniel Salber, Anind K. Dey, and Gregory D. Abowd. The Context Toolkit: Aiding the Development of Context-Enabled Applications. In *Proceeding of CHI'99*, Pittsburgh, PA, May 15-20 1999. ACM Press.
11. Bill N. Schilit, Norman Adams, and Roy Want. Context-Aware Computing Applications. In *IEEE Workshop on Mobile Computing Systems and Applications*, Santa Cruz, CA, 1994.
12. Joao Pedro Sousa and David Garlan. Aura: an Architectural Framework for User Mobility in Ubiquitous Computing Environments. In *Working IEEE/IFIP Conference on Software Architecture*, Montreal, August 25-31 2002.
13. Patrik Werle, Fredrik Kilander, Martin Jonsson, Perter Lonnqvist, and Carl Gustaf Jansson. A Ubiquitous Service Environment with Active Documents for Teamwork Support. In *Ubiquitous Computing, Third International Conference (Ubicomp 2001)*, pages 139–155, Atlanta, GA, September 30-October 2 2001. Springer.

Systems Support for Ubiquitous Computing: A Case Study of Two Implementations of Labscape

Larry Arnstein[1,2], Robert Grimm[1], Chia-Yang Hung[1], Jong Hee Kang[1],
Anthony LaMarca[3], Gary Look[4], Stefan B. Sigurdsson[3], Jing Su[3], and
Gaetano Borriello[1,3]

[1]Department of Computer Science & Engineering, University of Washington,
[2]Cell Systems Initiative, Department of Bioengineering, University of Washington,
[3]Intel Research Laboratory @ Seattle
[4]Artificial Intelligence Laboratory, Massachusetts Institute of Technology,

Abstract. Labscape, a ubiquitous computing environment for cell biologists, was implemented twice: once using only standard tools for distributed systems (TCP sockets and shared file systems) and once using *one.world*, a runtime system designed specifically to support ubiquitous applications. We analyze Labscape in terms of the system properties that are required to provide a fluid user experience. Though the two implementations are functionally and architecturally similar, we found a significant difference in the degree to which they each exhibited the required properties. The fact that *one.world* was not designed specifically with Labscape in mind yet was found to support the application's requirements well suggests that ubiquitous applications have many aspects in common, and can benefit from a system support layer for coping with dynamic environments. We present, in detail, the concepts embodied in *one.world* that we have found to be most important for Labscape, and how some of these concepts might be extended.

1 Introduction

Labscape is a ubiquitous laboratory assistant that helps cell biologists in several ways: it presents needed information in the context of the experiment; it records experiment data and observations as the work is performed; and it provides ubiquitous access to the experiment record to support communication and collaboration. This project was not undertaken explicitly to evaluate systems technologies for ubiquitous computing, yet we have learned a great deal about the role that systems technology should play in realizing high performance, robust smart environments.

Labscape was initially developed using standard tools for distributed systems (TCP sockets and shared file systems). Although we achieved some degree of success in our first development effort, the resulting system failed to deliver adequate performance and reliability to support continued development. As a result, Labscape was re-implemented on *one.world* [1], a comprehensive run-time system that imposes a programming model specifically designed to support ubiquitous applications. The purpose of this paper is not to evaluate *one.world* against other systems for distributed programming; rather it is to offer some conclusions, based on our application

F. Mattern and M. Naghshineh (Eds.): Pervasive 2002, LNCS 2414, pp. 30-44, 2002.

experience, about the proper division of labor between the applications and systems communities in the development of ubiquitous computing environments.

In Section 2, we describe the user requirements of Labscape that place it squarely in the domain of ubiquitous computing, and which distinguish it from typical distributed applications. In Section 3, we review the system support alternatives that we had considered before deciding on *one.world*. Our purpose is to argue that our choice of *one.world* is reasonable, and that our experience demonstrates that many of the concepts it embodies are likely to be useful for a wide range of ubiquitous computing applications. In Section 4, we describe the architecture and the two implementations of Labscape, followed by a detailed comparison of the two implementations. In Section 5, we focus on the implementation strategies we employed in both cases, and how they were driven by user requirements. We conclude in section 6 with a summary of lessons learned.

2 The Application Requirements

An abstract flow-graph representation of a biology experiment is at the center of the Labscape user interface. The flow graph is a convenient abstraction for keeping track of planned and completed laboratory work. As laboratory work proceeds, the flow-graph is visually transformed into a record of the actual experiment, which is annotated with parameters used, observations made, and data files produced. Most importantly, the resulting electronic record can easily be shared or searched, thus providing new opportunities for collaboration and communication. To achieve the high degree of adoption that is required for the resulting record to have significant value, it is important that the interface itself be designed to deliver benefits to the biologist in all phases of laboratory work. However, good interface design is not sufficient – our aim is for Labscape to be perceived as a basic laboratory utility like purified water, laboratory gasses, ventilation, and electricity, which are all ubiquitously available and easily accessed throughout the laboratory.

Some characteristics of laboratory work that strongly influenced our design are: 1) information access is highly interleaved with the physical activities required of the experiment; and 2) the biologists' hands are typically full or busy when they move about the environment, and 3) information access is not currently well integrated into the laboratory workflow [2,3]. Our goal is to build a system that improves the overall fluidity of the work process with respect to information access and recording.

To increase the accessibility of information throughout the laboratory without adding additional items for the biologist to carry, we have equipped the environment with several pen/touch tablet computers (Fujistu Stylistic 3400), and radio frequency ID (RFID) and barcode scanners at each of the work areas. Access to the Labscape application is controlled by a system that keeps track of user proximity to the work areas. Proximity is detected either through direct touch interaction with the tablet computers, or through body-worn short-range infrared badges (similar to active badges [4]) that communicate with wireless sensors [5] arrayed along the lab bench. When our location system detects that the user has entered the region of influence of a device, such as a display, scanner, or digital camera, the user is associated with that region unless another user is currently present. If the device has a display, then the user's application interface appears, showing the evolving record of the current experiment.

Establishing the physical environment is, of course, the relatively easy part. Developing the software that orchestrates these devices into a *fluid* user experience is another matter. By fluidity, we mean that the tools and devices introduced into the lab should actually enhance the biologist's ability to focus on the biology. For Labscape, there are three major elements of a fluid user experience.

1. *Ubiquitous accessibility.* The biologists should be able to access their personal Labscape interface with low latency and high responsiveness at any work area they visit.
2. *Flexible control of I/O resources.* Work areas in the laboratory are task-specific rather than user-specific. And, not all work areas will necessarily be served by a separate CPU. This means that control of devices must be granted to users based on location and context rather than on how those devices are physically connected to computers. This includes scanners and cameras as well as microphones, keyboards, and specialized laboratory instruments.
3. *Robustness.* Our goal is that the biologists should not be concerned with file I/O, the risk of losing work due to system failures, or the need to be aware of the state and topology of the network. This means that we must provide transaction-level persistence and support for disconnected operation.

Taken individually, these requirements are not unique. It is the combination of all three that is characteristic of ubiquitous computing, and which presents implementation challenges that go beyond traditional distributed systems. Take, for example, the dynamic nature of our network: machines may come and go from our system frequently since the pen tablet computers can easily be removed from their dock. This level of dynamism is not unusual; the set of clients connected to the Internet is highly dynamic. The fact that Labscape has computation spread across a large number of machines is also not unique. Thousands of machines are commonly employed to solve computationally intense problems like weather simulation. Finally, the fact that Labscape has a variety of different types of components that interact in non-trivial ways is also not new. Most enterprise environments have a variety of component types including gateways and servers for web pages, files, and mail.

But, each of the above examples takes advantage of design trade-offs that simplify implementation in ways that are acceptable to their users and their needs. The Internet works, despite its dynamics, in part because it has a simple interaction model that exposes delays and errors to the user. Parallel computations take advantage of regularity to simplify programming and deployment of large numbers of homogeneous interacting components. Finally, complex heterogeneous enterprise and supply chain systems are feasible in practice because the networks are stable – changes do occur, but they are carefully managed by skilled staff. In contrast, Labscape must allow for dynamic reconfiguration of the system in the presence of complex interactions between heterogeneous software and hardware components, without an army of support personnel, and yet, it must appear stable and reliable to its users.

Though we had a clear understanding of the user requirements from the outset, we did not have a clear understanding of what sort of systems support would be most useful in our case. For this reason, we first chose to develop the application entirely from scratch using networked file I/O for storage, and low-level networking APIs such as sockets [6] for communication. Though we were able to complete crucial user studies in a real user environment with this implementation, it was necessary for us to seek higher-level systems support for the next implementation so that we could focus our effort on application level concerns rather than detailed system level interactions. In Section 3, we outline the basis for our selection of *one.world*.

3 System Support for Ubiquitous Computing

Software engineers developing ubiquitous computing applications have a number of choices of platforms on which to build. This section compares and contrasts the programming models and capabilities offered by these platforms relative to *one.world*, our chosen platform.

Some of the most common types of software platforms for developing distributed applications are remote procedure call (RPC) systems. RPC systems have been around for decades [7] and have proven to be flexible and long-lived. Sun/RPC is an example of a classic RPC system [8]; more recent examples include XML/RCP [9] and SOAP [10]. Distributed object systems such as CORBA [11], Modula-3's network objects [12] and Java RMI [13] are the object-oriented counterparts of RPC systems. The fundamental benefit of these systems is that they allow developers to easily build simple distributed systems using familiar programming models, namely, procedure calls and object/method invocation. Ironically, while this transparency provides for a familiar programming model, it is what makes standard RPC systems a poor choice for building ubiquitous applications. Whenever the thread of control is executing remotely, the execution context of the caller is at the mercy of network connections that may be unreliable. Because the programming model intentionally hides the difference between local and remote execution, it is difficult to reason about the conditions under which a given component can safely execute. Furthermore, the inherently synchronous nature of RPC interactions limits the responsiveness of applications, as they need to wait for remote services to complete processing each procedure call. As a result, it is difficult to build robust and responsive ubiquitous applications using RPC systems. [1]

The problems with RPC are actually just symptoms of a more general issue: *ubiquitous computing environments are dynamic*. Thus, we have to find a way to write software components that respond appropriately to change. *one.world* addresses this problem in a general way: it insulates software components from changes that can be handled by the system, and it notifies them of changes that should be addressed at the application level. This, in turn, gives the application the opportunity to treat the user in a similar way. The application can insulate users from changes when possible, and interact with them when input is needed.

---

[1] Jini [14] provides much of what we want: a discovery service that enables dynamic composition and leases to control allocation of resources, and a discovery service to expose the dynamic set of services. But, because it is built on Java RMI, it suffers from the same limitations, described above, that apply to all RPC systems.

34   L. Arnstein et al.

one.world is a Java-based run-time system that executes on a standard JVM. Components are objects that are prohibited from accessing system resources directly. In the *one.world* programming model, RMI is prohibited, as are application-level ownership and control of threads. Direct access to local resources such as the file system and the native operating system is discouraged but can be enabled by the developer. By restricting software components to consist of nothing but a collection of asynchronous event handlers, all interactions can be mediated by the *one.world* system. In this way, the system has the opportunity to handle changes or failures, and it can notify the component when necessary. Some examples of system notification events are activation and deactivation of components, event delivery failure, and relocation of a component from one node to another. By exposing change to the application, *one.world* allows the developer to determine how best to respond in terms of benefits to the user.

one.world provides alternatives to mitigate what is prohibited or restricted: a location independent tuple-store instead of file system access, event queues and timer events instead of threads, remote event passing instead of RMI, and a variety of other system-like utilities. By adhering to these alternatives, *one.world* guarantees that a component can execute on any node given enough physical resources. Because all interactions between components are through asynchronous events, *one.world* provides a rich event delivery infrastructure including early- and late-binding discovery, multicast, and lookups on the events themselves.

In addition, *one.world* allows components to be dynamically organized into hierarchical execution *environments*. Each environment contains a tuple-store for persistent data, and can contain running components and subordinate environments. The tuple-store can be used by components for data storage and for checkpointing execution state. Environments, not components, are the unit of migration in *one.world*, thus ensuring the availability of all state that is essential for continued execution after migration.

Our choice of *one.world* was due, in part, to the fact that it was developed and supported by our colleagues at UW/CSE. However, it is also clear that no other available system provided such a comprehensive set of features designed specifically to support ubiquitous computing applications. Our primary concerns regarding the use of *one.world* were: 1) whether or not a real application, like Labscape, could be supported within the restricted programming model it requires; and 2) as a new system, *one.world* could well have had performance and reliability problems of its own. While the first issue has proved to be somewhat problematic, we have found the system to be largely stable and responsive.

Though the ensuing comparison paints a positive picture of our experience with *one.world*, we are comparing it only to our negative experience of having no systems support beyond basic sockets and distributed file systems. The point of this discussion is to argue, in general, for the role of systems support in developing ubiquitous applications, rather than to promote *one.world* as the best possible solution.

4 Comparison of Two Implementations

The high-level component decomposition of the Labscape system remains largely unchanged between our two implementations, as does the primary method of composition—asynchronous events. In the first implementation, events are sent through socket connections between components. In the second, we rely on the late-binding discovery event delivery mechanism provided by *one.world*. Figure 1 shows the major components of Labscape. The following discussion describes each component and mentions important differences between the two implementations.

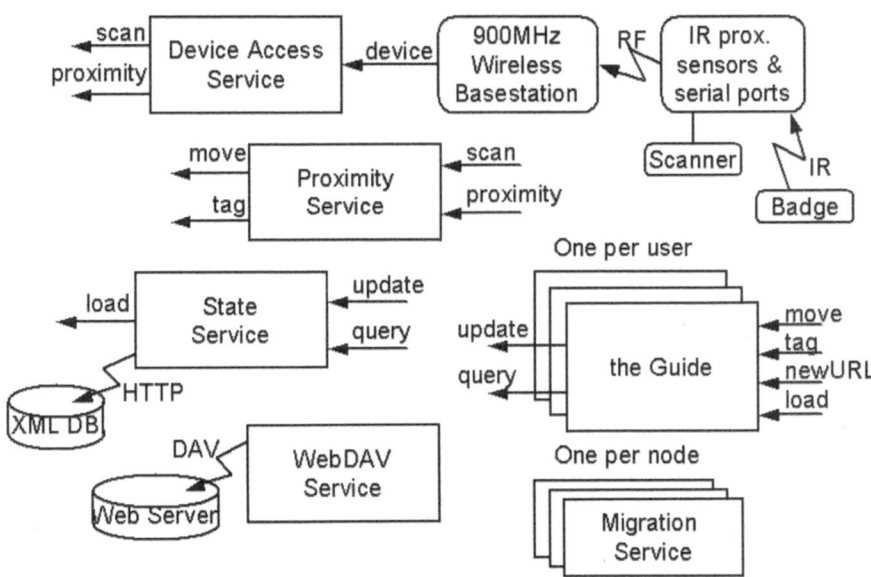

Fig. 1. The major components of Labscape

The Guide is the main GUI component of the system; it presents the biologist with the flow-graph representation of current and past experiments. Incoming events notify the Guide of user migrations (*move*), new data files (*newURL*), tags that have been scanned (*tag*), and responses from queries to the state service (*load*). As the user updates the experimental record, or asks for more information, the Guide makes corresponding requests to the state service The persistence oriented events, *update*, *query*, and *load*, are only part of the second implementation since the first relied directly on native file I/O.

The Device Access Service listens on the serial ports for incoming events from networked sensors such *proximity* events from IR sensors, and *scan* events from RFID and barcode scanners. The incoming serial data is decoded, cast into appropriate *one.world* events, and then forwarded to the Proximity service via late-binding discovery.

The Proximity Service maintains associations between users and locations based on *proximity* events generated by the IR detectors or user interactions with displays. When the Proximity service concludes that the user has moved to a new location, it emits a *move* event. The Proximity service also associates incoming *scan* events with users based on the sensor ID through which it was received. Incoming *scan* events are transformed into *tag* events and forwarded to the appropriate Guide through discovery.

The WebDAV Service watches specific directories in the file system for files created by users, either directly or through the use of laboratory equipment. When a file appears in the user's directory, it is copied to a web server through a DAV interface, and it emits a *newURL* for that user.

The State Service, which exists only in our *one.world* implementation, provides a gateway to the long-term persistence engine, which is currently an XML database for representing laboratory procedures and data. The State service consumes *update* and *query* events and emits *load* events.

The Migration Service, which exists only in our original implementation, listens for incoming object streams from migrating Guide components, and maintains a control connection with the proximity service. The Migration service must always be running on each CPU in the system. In our second implementation, the *one.world* runtime environment eliminated the need for this service.

In our first development effort, we knew that we could not address all three of the fluidity requirements described in Section 2. Instead, we focused our efforts on ubiquitous accessibility and flexible control of I/O resources, as these were more important than robustness for initial evaluation of the user interface. The infrastructure that we built consists of ~3000 non-commenting source statements (NCSS) that required three full-time developers about three months to produce. This NCSS count does not include the application-specific components (mostly user interface and flow-graph objects), which consist of another ~7700 NCSS. Despite our efforts, this implementation fell short in terms of both ubiquitous accessibility and flexibility, and even more than expected in terms of robustness. (Refer to Section 5 for the details as to how and why.)

The second implementation of Labscape, using *one.world*, required two skilled developers approximately two months to complete. One of the developers was also on the development team for the sockets implementation. This effort included porting the main application specific component (the Guide) and re-developing all of the other components to conform to the *one.world* programming model. As before, the code base can be split into two parts: the application-specific components consisting of ~9700 NCSS and the infrastructure components, requiring ~4700 NCSS. Though there is more code in the *one.world* infrastructure components than in the sockets-based infrastructure components, there are three important differences that are not reflected in these numbers:

- The *one.world* version has more functionality in both the GUI and in the infrastructure, as well as increased stability and performance. The extra functionality in the current *one.world* version is primarily for transaction-level persistence.

- The *one.world* components are expressed primarily as event handlers that map directly to concepts in the application domain. Thus, it is easy to understand and modify the code without creating unintended interactions.
- The infrastructure components in the *one.world* implementation are generic and independent enough to be used by other applications running in the same environment. This cannot be said of the original effort.

These are important properties of the *one.world* implementation that demonstrate its benefits for application developers: higher-level abstractions and greater code reusability result in more robust applications. Though the two architectures are very similar, the results of our development efforts were quite different with respect to our fluidity goals. The reasons for these differences are discussed in detail below.

5 Evaluation

In this section, we evaluate, in detail, the two separate implementations with respect to the three user requirements defined in Section 2: ubiquitous accessibility, flexible control of I/O resources, and robustness. Furthermore, we present a user scenario that highlights these capabilities along with additional system-level properties that should eventually be provided in middleware for ubiquitous computing.

5.1 Ubiquitous Accessibility

We have considered three alternative strategies for delivering the application interface to the touch-tablet computers scattered throughout the environment: virtual terminal, on-demand state migration, and proactive state replication. These alternatives have distinctly different characteristics with respect to latency, responsiveness, resource utilization, and robustness. Virtual terminal technologies such as X-windows, Microsoft Terminal Server and VNC offer relatively low initial latency and low local resource requirements in exchange for high network bandwidth, poor or variable responsiveness, and no support for disconnected operation. State migration suffers from higher initial latency and increase use of local resources, but it offers better response times and can run disconnected after migration. Proactive replication achieves very low latency in exchange for memory capacity and network bandwidth to maintain and synchronize each copy. Choosing the right approach is a legitimate application level concern.

Hoping to keep it simple, we first implemented a VNC strategy for interface migration. Though initial latency under VNC was adequate, our users considered the responsiveness of the UI to be unsatisfactory, even over a reliable 100Mb local area network. This was primarily due to our extensive use of pop-up windows and menus in addition to smooth animation in the movement of screen objects. Our experience is consistent with the comparison of various thin client architectures reported by Nieh [15]. To improve responsiveness, we implemented a state migration scheme over TCP sockets that works as follows:

- The Proximity service sends a *move* event to the Guide when the user's presence is detected at a new work area. The *move* event includes the IP address and port number of the Migration service running at the destination work area.
- Upon receipt of a *move* event, the Guide releases local resources (files, socket connections, database connections, etc); serializes itself through a new socket connection to the destination Migration service; and then terminates locally.
- The newly instantiate Guide re-establishes needed socket connections, environment variables, and file descriptors and then resumes execution.

This scheme worked after a significant development effort, but we suffered from both performance and maintenance problems. To attack the performance issues, we carefully identified the minimal amount of application state that must be preserved to provide the user with a strong sense of continuity. In the end, we left all of the GUI components behind in favor of rebuilding them on demand at the new location. This effort reduced the size of the transported data set by an order of magnitude. Unfortunately, our migration scheme consisted of over 500 NCSS lines of code distributed across several components, which required review and modification as new features where added to the system. As an example, the addition of a simple utility that logged user interactions to a file created a number of migration failures that had to be painstakingly resolved. There were two major lessons from this experience:

- It is natural in traditional application development to be somewhat cavalier about creation and use of state. This includes both open and closed files in the file system, network connections, local environment variables, database connections, and the execution context of every thread in a multi-threaded system. As a result, state migration is difficult to add to an existing system, and one must continuously reconsider the effects of migration when adding new features.
- Realization of a state migration mechanism is beyond the scope of what application developers should focus on. Application developers should decide *what* bits to move and *when*, but they should not be concerned with *how* to safely and quickly move the bits. Systems support should offer several methods with differing performance and reliability guarantees, and it should notify the application about changing conditions that might affect the choice of which mechanism to apply.

The restricted programming model of *one.world* directly addresses the state management issue by encapsulating all essential state within environments, which move as a unit. As application developers, our only responsibility was to declare soft state as transient and to fill in template methods for activation and deactivation of the component. The *one.world* platform allows us to completely eliminate our own custom migration service. But most importantly, it provided a migration scheme that was insensitive to the implementation of other aspects of our system, as long as we stayed within the *one.world* programming model.

Table 1 shows the migration data sizes and times for three flow-graphs of different sizes on the *one.world* implementation. The reported times are averages of hundreds of migrations. A 64-sample flow-graph contains approximately 640 nodes and 1200 edges, and represents a large experiment. The migration times are measured from the receipt of the move event to the time that the complete GUI becomes available at the destination. Because the data sizes are small, most of the time is spent in deserialization and reconstruction of the GUI components, which were not serialized, rather than in data transmission.

Table 1. Data set sizes and migration times for the *one.world* implementation

Data Structure Size	Size of Migrating Data Stream	Average Time to Complete Migration
Empty Guide Migration	19.5 Kbytes	2448ms
32 Sample Guide Migration	174.8 Kbytes	4476ms
64 Sample Guide Migration	323.0 Kbytes	7089ms

There is still a possibility that these migration latencies could become unacceptable. If current user studies bear this out, then we will have to implement a more proactive strategy. Currently, this would require us to implement a replication layer on top of *one.world*'s existing tuple-store. Fortunately, replication is a focus of a new set of services being developed within *one.world*. Though perhaps we could get direct support for replication from another storage subsystem now, it is likely it would not conform to the *one.world* programming model and would involve lots of effort to port. This is one of the risks of working in a restrictive system.

5.2 Flexible Control of I/O Resources

The application requirement is to allow users to access I/O devices, such as keyboards, cameras, laboratory instruments, and barcode scanners, based on context and location regardless of how the device is physically connected to the system. Flexibility in our first implementation was realized by tearing down and rebuilding socket connections between Guide components and proxies for bench-top I/O devices. This constant reconfiguration was accomplished by maintaining permanent control connections between the Proximity service and each user's Guide component. Predictably, this approach led to serious scalability and robustness problems.

The scalability problem stems from our need to maintain large numbers of active socket connections. Every open socket requires operating systems resources. While there might be system-level solutions to this resource utilization problem that would still allow large numbers of open sockets, such solutions should be beyond the concern of application developers. The robustness problem stems from our reliance on long-term socket connections without extensive failure detection and recovery programming. Once again, this problem can be solved with careful implementation of low-level protocols that should not be required of application developers. At its zenith, our sockets-based system had to be completely rebooted after any single component failure. The mean time between failures (MTBF) was about 30 minutes with just four work areas under light user load.

The second implementation avoids the scalability and robustness problems of the first through the use of *one.world*'s communication infrastructure. By using asynchronous event delivery through late-binding discovery, resource utilization scales with activity rather than with the number of components. And, because we do not maintain any permanent connections, failures in the system tend to be transient rather than permanent. Though we still have to deal with the potential for lost events, this can be done selectively, only in the cases where some action is necessary. For example, no response is necessary for lost *proximity* events because they represent transient information that will soon be refreshed, whereas the user should be notified in the case that the system could not deliver a barcode *scan* event.

The Proximity service plays essentially the same role in both our sockets and *one.world* implementations. But, instead of maintaining hundreds of open socket connections, the new Proximity service simply exports one event handler of the right type. Our current implementation has to process an average load of 170Kbytes/sec of incoming proximity event data to support approximately 100 users each generating one *proximity* event per second. Because this is an entirely manageable load, the Proximity service can now be viewed as a utility that can be shared by many applications running in the same environment, instead of as a private component of the Labscape system.

There was one useful property of sockets that we lost when we moved to *one.world*: in-order delivery of events. In some cases order is not important, such as for the IR *proximity* events. But order is essential to the user in barcode scanning. Lack of direct support for ordered event semantics placed an unexpected burden on the application development team. This issue is discussed in more detail in the next section.

5.3 Robustness

In Section 2, we listed two aspects of robustness that are important to our users: transaction-level persistence, and disconnected operation. Of these, we have only addressed persistence in our current *one.world* implementation as we expect this to have the largest initial impact on the user's sense of fluidity and security.

For persistence, we implemented an asynchronous interface to an XML database to eliminate the need for user awareness of file I/O, and to hide database transaction latency from the users. The State service is a gateway that translates *update* and *query* events from the Guide components into the (HTTP) connection-based interface to the XML database system. The State service emits *load* events containing information returned by the XML repository. The asynchrony of the interaction is the primary reason that we were able to achieve our combined goal of continuous persistence and high responsiveness. But, because *one.world* prohibits the use of application-level threads, the programmer must explicitly maintain enough context for proper processing of the eventual response or timeout event. This requirement is an inconvenience for the developers and can result in code that is difficult to trace and understand.

Two other problems that we had to address at the application level were the possibility of out-of-order event delivery and lost events that could be devastating to the integrity of the persistent experiment record. The solution to the first problem was to maintain a reordering buffer in our State service for each active Guide. The solution to the second problem was to make use of *operation* primitives provided by *one.world*. Operations are just like regular events, except that they are guaranteed to receive exactly one (asynchronous) response. If the response is an acknowledgement, then the event has safely arrived at the destination. However, if the response is a timeout, then the event may or may not have arrived. Thus, a resend could result in a duplicate event arriving at the destination. We solved this problem by having our reordering buffer reject such duplicates.

The need for this pattern has proved to be the rule rather than the exception in our system. The *proximity* event is the only one in our system that is naturally insensitive to order and duplication. Though the *one.world* operation primitive proved to be useful, more direct support from the system for ordering and idempotency would have kept our application developers from implementing their own, likely imperfect, protocols. Several patterns like these are providing the *one.world* developers with potential enhancements to their system.

The incremental network activity generated by a single user, including migration, persistence events, and proximity activity is approximately 50Kbytes/sec per user. Thus, a single 100BaseT subnet should be able to support approximately 60 users before overloading the network (which we consider to occur at utilizations greater than 33%). Since one biologists requires approximately 150 square feet of space, our basic system should be able to support approximately 9,000 square feet of laboratory space.

5.4 A User Scenario

It is common for real laboratory work to cross physical and organizational boundaries for access to special equipment, facilities, or biochemistry that cannot easily be reproduced. The following realistic scenario highlights how we imagine that ubiquitous access, flexibility, and robustness can work together to provide a fluid user experience even in the wide area. And, it raises a potential maintenance and interoperability problem associated with our system. The scenario depends on support for re-synchronization after disconnected operation, which is currently under development.

Before leaving her home environment, the biologist would migrate her plan for the day to her PDA (just the Guide's environment). Given sufficient resources, the Guide could run stand-alone on the PDA, allowing the biologist to view and modify her plan en-route.

Upon arrival, her PDA joins the local network and collaborates with the local *one.world* nodes to synchronize the two discovery services. This is an automatic procedure that is already supported by *one.world*. As long as it is allowed by the host facility, her Guide can now migrate to any of the host laboratory's display resources and access devices through that environment's proximity system. *If the Labscape event ontology is respected*, then everything works as it did in the home environment, except for perhaps one thing: the local State service should reject the user's database *update* requests because she does not have permission, and any attempt for those events to reach the home state service might be blocked by a firewall. Yet, through checkpointing and update backups in the local tuple-store, the Guide can continue to operate and cache the work locally as it migrates around the host's facility.

Upon departure, the Guide is pulled back to the user's PDA, and the pending updates are committed as soon as the home State service becomes accessible again. It is no problem if the PDA runs out of batteries on the way home, as periodic checkpointing will ensure that the work is saved in the device's nonvolatile storage. This scenario exemplifies how migration, flexibility, and robustness should work together to deliver a fluid user experience. But it also points out how our application is susceptible to incompatibilities between the ontologies of different versions of the same application, or between completely different applications that share services in the same environment.

In the end, we did not completely succeed in obeying all of restrictions of *one.world*'s programming model. We still rely on access to the file system to detect when data files are created by the user through interactions with legacy laboratory equipment, and we still need to make native system calls for external utilities such as the third-party graph layout tool that we rely on in our GUI. To the extent that we violate environment encapsulation, *one.world* cannot provide its portability and change-notification guarantees. It is entirely reasonable that, in such cases, the responsibility falls back to the application developer to ensure that state is properly managed at activation and deactivation time. Most of these special cases could have been eliminated if *one.world* offered a standard wire protocol for exchanging asynchronous events with external components.

6 Status and Conclusions

The current *one.world*-based implementation of Labscape has sufficient performance and robustness to meet the needs of our users for authentic evaluation of the user experience. We are currently engaged in a formal user study that will assess the impact that Labscape has on the overall fluidity of the laboratory workflow. But most importantly, we are confident that we can continue to maintain and extend this application into the future while we learn more about how *one.world*'s guarantees can provide unique capabilities. In the future, we expect to have several applications running over the same set of basic services. It may well be that some of the services we have created for Labscape will be so universal that they will be incorporated into the core system just as discovery already has been. The following three major themes have emerged in our experience with two implementations of Labscape.

1. Minor failures should not become major failures. In our original implementation, a lost socket connection resulted in the complete loss of the system. But, a lost socket connection is a minor failure that should be recoverable. Insulating users from minor failures requires sophisticated algorithms operating at a fairly low level. These algorithms should be packaged and made available to developers in the form of system support. In fact, *one.world* uses sockets to move events, but this is complete hidden from the application components, which only see asynchronous message delivery. If the system can't solve the problem, then it should notify the application components. As an example, a warning to the user might be the most appropriate response to an application whose serialized footprint has grown too large for safe checkpointing on the current device.
2. Diversity and orthogonality are essential. The key difference between the two development efforts was that, in the *one.world* case, we were able to address each of our concerns independently. Ideally, systems that support ubiquitous computing should provide several alternative mechanisms that can be mixed and matched with predictable results. As an example, once our migration system was working, it

never failed due to some other feature that we added to the system, such as when we added transaction-level persistence. This is counter to our experience with the first system, in which the addition of new features frequently had adverse interactions with existing capabilities.

3. Standardization on the runtime environment must be coupled with support for heterogeneity. *one.world* presents a restrictive programming model in exchange for certain guarantees on portability and adaptability that we have, in fact, found to be important for our application. But, not all components in the system should be forced to make this same tradeoff or to run in the same environment. In the case of Labscape, the webDAV and Device Access services both need to access native system resources and should run in the native environment. But, we were prevented from implementing such a hybrid system because *one.world* did not offer an external event interface.

Labscape running on *one.world* is an experiment into how applications can be made more adaptive to changes in their execution environment. The biggest surprise to our development team was the relative ease with which we could adhere to the seemingly restrictive programming model and how much it helped us build a fluid system. The fact that *one.world* was not designed specifically with Labscape in mind supports the idea that ubiquitous applications have many requirements in common. Our experience is that such applications are indeed facilitated by asynchronous event models, careful delineation of essential volatile and non-volatile state, and run-time support for appropriate handling of changes and failures that arise in dynamic environments.

Acknowledgements. We would like to acknowledge the NSF under Grant REC-0112997, the Portolano Expedition funded by the DARPA Ubiquitous Computing Program, Intel Research Seattle, and the Intel Corporation Research Council for financial support; Sunny Consolvo of Intel Research Seattle for the user study data; and Qing Hong Zhou, Michael Look, and Bob Franza of CSI for collaboration on the design and evaluation of the system.

References

1. Grimm, R., Davis, J., Lemar, E., Macbeth, A., Swanson, S., Gribble, S., Anderson, T., Bershad, B., Borriello, G., Wetherall, D., Programming for Pervasive Computing Environments, University of Washington Technical Report UW-CSE-01-06-01, June, 2001.
2. Consolvo, S., Arnstein, L., Franza, B.R., *User Study Techniques in the Design and Evaluation of a Ubicomp Environment.* Intel Research Seattle Technical Report IRS-TR-02-012.
3. Arnstein, L. F., and Borriello, G., *Labscape: The Design of a Smart Environment,* Intel Research Seattle Technical Report IRS-TR-02-008.
4. Want,R., Hopper, A., Falcao, V., Gibbons, J., *The Active Badge Location System,* Technical Report 92.1, 1992, ORL, 24a Trumpington Street, Cambridge, CB2, 1QA.

5.   Hill, J., Szewczyk, R., Woo, A., Hollar, W., Culler, D., Pister, K., "System architecture directions for network sensors", ASPLOS 2000.
6.   Stevens, R. W., UNIX Network Programming, Prentice Hall, vol. 1, 2nd edition, 1998, ISBN 0-13-490012-X
7.   Birrell, A. D., Nelson, B. J., "Implementing Remote Procedure Calls", *ACM Transactions on Computer Systems*, vol. 2, no. 1, pp. 49-59, Feb, 1984.
8.   Srinivasan, R., *Remote Procedure Call Protocol Specification Version 2*, Internet Engineering Task Force, RFC No. 1831, Aug 1995, http://www.ietf.org/rfc/rfc1831.txt?number=1831
9.   Winer, D., XML-RPC Specification, http://www.xmlrpc.com/spec, UserLand, Burlingame California, October 1999.
10.   Box, D., Ehnebuske, D., Kakivaya, G., Layman, A., Mendelsohn, N., Frystyk, N., Thatte, S., Winer, D., Simple Object Access Protocol (SOAP) 1.1, World Wide Web Consortium Note, Cambridge Massachusetts, May, 2000.
11.   Siegel, J., CORBA Fundamentals and Programming, John Wiley & Sons, April, 1996, ISBN 0-471-12158-7.
12.   Birrell, A., Nelson, G., Owicki, S., Wobber, E., Network Objects, Technical Report, Digital Equipment Systems Research Center, #115, Palo Alto, California, Feb, 1994
13.   Sun Microsystems Corporation, Java Remote Method Invocation Specification, Rev 1.7, Palo Alto, California, December 1999.
14.   Arnold, K., O'Sullivan, B., Scheifler, R.W., Waldo, J., Wollrath, A., *The Jini Specification*, Addison Wesley, 1999, ISBN 0-201-61634-3.
15.   Nieh, J., Yang, S.J., Novik, N., *A Comparison of Thin-Client Computing Architectures*, Network Computing Laboratory, Columbia University, Technical Report CUCS-022-00, November 2000.

On the Gap between Vision and Feasibility

Christopher Lueg

Faculty of Information Technology
University of Technology, Sydney
PO Box 123, Broadway NSW 2007, Australia
lueg@it.uts.edu.au

Abstract. Information appliances, user interfaces, and context-aware devices are necessarily based on approximations of potential users and usage situations. However, it is not an unusual experience for developers that in some areas, appropriate approximations are extremely difficult to realize. Often, these difficulties are not apparent from the beginning. Nevertheless, difficulties are rarely addressed in the pervasive computing literature as they appear to be peripheral compared to the technical challenges. In this paper, we argue that the field would largely benefit from addressing these issues explicitly. First, focussed discussions would help identify areas that have already shown to be difficult or even intractable in related disciplines, such as AI or CSCW. Second, it would help developers become aware of the difficulties and would allow them to deliberately circumvent such areas. We use example scenarios from the pervasive computing literature to illustrate these points. Difficulties to describe and to analyze impacts of pervasive computing applications indicate a need for an analysis framework providing a specific terminology.

1 Introduction

In his invited contribution to the IEEE Personal Communications special issue on pervasive computing, Satyanarayanan discusses that pervasive computing can be seen as a major evolutionary step based on ground breaking work in fields, such as distributed systems and mobile computing [1]. Issues investigated in distributed systems were, for example, remote communication, fault tolerance, high availability, remote information access and last but not least security. In mobile computing, enabled by the appearance of full-function laptop computers and wireless LAN in the early Nineties, additional constraints were introduced, such as unpredictable variations in network quality, lowered trust and robustness of mobile elements, limitations on local resources imposed by size and weight constraints, and energy consumption problems.

Clearly, the background of pervasive computing is mainly technical. However, there is also little doubt that the field penetrates areas in which not only technical considerations but also social and cognitive aspects are relevant. The long term expectation is that technologies weave themselves into the fabric of everyday life until they are indistinguishable from it [2]. Everyday life is shaped by people

F. Mattern and M. Naghshineh (Eds.): Pervasive 2002, LNCS 2414, pp. 45–57, 2002.

and what they do, how they do it, and how they perceive what they are doing, after all.

Roy Want and his colleagues were early witnesses of non-technical issues in pervasive computing when fielding their active badge location system in a research lab [3]. They report that the most common usage of the system was by the receptionist who routinely used it when forwarding phone calls to the location of a recipient's current location. Want et al. report that staff wearing badges found it useful to have phone calls accurately directed to their current location. However, staff also wanted to be able to exhibit some control over when calls were forwarded to them. Want et al. conclude that "amongst professional people responsible for their own work time, it is a very useful and welcome office system."

A decade later we know that despite offering certain benefit, something like the active badge location system did not become standard office equipment; although useful, the technology did not get weaved into the fabric of office life or everyday life. Considering the technical progress over the past ten years, it is reasonable to assume that it is not technical problems that have prevented such systems from becoming standard office equipment.

We take this observation as motivation for looking more closely at the area where the "technical world" and the "human world" overlap. In particular, we are interested in those areas where technology is used to approximate human activities and usage situations. By approximation we mean that developers of technology make assumptions about which aspects of human activities and their physical and social environment are important in future usage situations. Understanding the nature of approximations is particularly important if we consider Weiser's vision that embedded and invisible technology calms our lives by removing the annoyances [2]. After all, removing annoyances not only involves developing smart technology but also requires a good understanding of the nature of these annoyances.

A look at the original active badge location system in terms of approximations suggests that determining the specific usage of the system was largely left to the participants in the trial. This means that the system's developers did not implement too many approximations of how they thought the system should be used. This is partly due to the focus on the technical constraints present in the design space, such as size, power consumption and data transmission issues. Want et al. actually mention that the active badge location system was a technology-driven research project in which the enabling technology has been low-cost IR emitters and some other bits and pieces. However, they state that the most important result of their work was not 'Can we build a location system?' but 'Do we want to be part of a location system?' [3].

An extended version of the active badge location system [4] already features more work on approximation. The extended system allowed users to write personal control scripts that would control phone forwarding based on aspects of the environment, such as location or time. These scripts implement approximations as they are formal descriptions of what the user considers appropriate behavior in future situations. In a way, these scripts can be used to implement what would be called context awareness in these days. Context awareness means that computational artifacts are to some extent able to sense the situation in which they are used in order to adapt their functionality. In the case of the extended active badge location system, this notion of context awareness could mean that a phone call is not forwarded if the user's current location is in a meeting room.

An analysis of more recently developed gadgets, such as the Internet fridge or the media cup [5], reveals interesting aspects in terms of the approximations they implement. In the case of the media cup, the approximation of the user's "coffee behavior" is that the user basically wants the same kind of coffee again and again. The fridge is slightly more complex in this respect. The fridge implements a similar approximation ("The user wants more of the food items that were removed from the fridge."). In addition, the fridge may automatically order these items from the next grocery store. This means that the fridge is able to make decisions with an impact beyond the actual usage situation of removing items from the fridge. Indeed, such a notion of pro-activeness is seen as an integral part of the pervasive computing vision (e.g., [1]). For a number of years, such ideas were almost exclusively pursued in artificial intelligence research in the context of intelligent agents. In these days, pro-activeness can be found in most example scenarios in the pervasive computing literature (see below).

So far, the focus of the pervasive computing field clearly is on exploring the technical challenges. Satyanarayanan, for example, argues that "[...] the real research [in pervasive computing] is in the seamless integration of component technology into a system [...] The difficult problems lie in architecture, component synthesis, and system-level engineering." [1]. Similarly, MIT's project Oxygen features technology-oriented research. The project web site states that for Oxygen to succeed, it must address many challenges which are typically technical challenges (from URL http://oxygen.lcs.mit.edu/):

– Hardware must become adaptable, scalable and stream-efficient to provide computational resources that are both energy efficient and powerful for a variety of computational tasks.
– Software and protocols must become adaptable to provide flexibility and spontaneity, for example, by supporting smooth vertical hand-offs among communication technologies.
– Services and software objects must be named by intent, for example, "the nearest printer," as opposed to by address.
– Software must become eternal and embedded, yet replaceable on the fly.
– Software must free itself from hardware constraints imposed by bounded resources and address instead system constraints imposed by user demands, available energy and power, and available communication bandwidth.

The focus on technology is also clearly reflected in what the US National Institute of Standards and Technology (NIST) sees as the synthesis of pervasive computing (from URL http://www.nist.gov/pc2001/):

"Pervasive computing is a term for the strongly emerging trend toward: numerous, casually accessible, often invisible computing devices,
frequently mobile or embedded in the environment, connected to an increasingly ubiquitous network infrastructure composed of a wired core
and wireless edges."

While it certainly is important that the technical challenges are addressed,
we see additional aspects of pervasive computing that need to be investigated.
In particular, we argue that some of the problems that turn attractive example
scenarios into science fiction are characteristic for situations in which technology is used to approximate human activities and usage situations. Investigating
this issue complements discussions of other non-technical issues, such as privacy
concerns [6,7] or usability issues [8].

So far, approximation issues are often addressed only implicitly in the technology-oriented pervasive computing literature. We argue that the field could
largely benefit from addressing these issues explicitly. First, focussed discussions
would help identify "difficult" areas and benefit from work on such areas in
related disciplines, such as artificial intelligence (AI) and computer supported
cooperative work (CSCW). Second, it would help developers become aware of
the difficulties and would allow them to deliberately circumvent such areas.

In what follows, we look in more detail at some of the issues that make
the weaving of technology into the fabric of everyday life so difficult. First, we
outline what we mean by approximation. Then, we discuss example scenarios
from the pervasive computing literature and try to pin down why many of them
seem like science fiction. Next, we focus on context awareness as one of the
core areas and discuss why it is so hard to operationalize context awareness
in devices. Difficulties to describe and to analyze the specific characteristics of
pervasive computing applications indicate the need for a framework providing
specific terminology. Finally, we provide an outlook on future research.

2   What Do We Mean by Approximation?

We use the term approximation to denote that developers of technology (and
artifacts in general) have make assumptions about which aspects of human activities and their physical and social environment will be important in future
usage situations. Our usage of the term is inspired by work on the gap between
social requirements and technical feasibility in computer-supported cooperative
work (CSCW) [9]. The term is adopted from fluid dynamics as a metaphor to
denote tractable solutions that solve specific problems with known trade-offs.

We are particularly interested in clarifying the specific limitations of technology when used to approximate human activities and usage situations. The
already mentioned media cup and Internet fridge can be used to illustrate the
use of approximations in pervasive computing. The media cup approximates the
user's "coffee behavior" and the fridge approximates of the user's grocery good
buying behavior (roughly, the user does the same again. This may not perfectly

reflect reality but it doesn't really matter). Disregarding the fridge's capability of automatically ordering items from the next grocery store, these gadgets have little impact on the user as the user is not required to adapt his or her behavior to the behavior approximations presumably implemented in these gadgets (at least not to an extent that matters). Either the user likes what these gadgets are doing or the user overrides their actions (e.g., select a different kind of coffee or dump the shopping list generated by the fridge).

Related areas relying on approximations are information retrieval and information filtering (approximations of human information seeking behaviors are used in interactive information retrieval to make relevance estimations), intelligent meeting support systems (approximations of how people behave in meetings and how meetings are conducted) and intelligent desktop agents (approximations of interaction styles). Quality of approximations in information retrieval and intelligent agents are of particular relevance to work on information appliances and content distribution in pervasive computing.

The outstanding importance of the quality of approximations suggests to investigate in detail the relationship between computational artifacts, embedded approximations of human activities and usage situations, and what people actually do. In particular, we found the following dimensions to be helpful:

1. To what extent do users have to adapt their behavior to the approximation implemented in the computational artifact? Does the artifact still perform if users do *not* behave in the expected way?

   Electronic tags used to automatically charge cars crossing the Sydney Harbour Bridge use hardly any approximation: how the tag is moved over the bridge doesn't really matter as long as the tag can be sensed. The media cup only expects to be used like a regular coffee cup and users wouldn't expect that the media cup works if not used like a coffee cup. Aura (a pervasive computing system described in [1] and discussed below), however, expects much more and it is unclear what Aura does if the user's behavior does not comply to the user's schedule.

2. To what extent is the expected user behavior evident from the artifact's behavior?

   Again, the media cup only expects to be used like a regular coffee cup. In the case of Aura it is not apparent what behavior would help Aura perform. There is some work done in the areas of affordances [10] and persuasive computing [11] that is relevant to this dimension.

3. Does interacting with the artifact have implications beyond the actual interaction situation? Are these implications comprehensible?

   The Internet fridge ordering food items from the next grocery store and Aura sending documents to remote presentation locations are examples for artifacts that trigger such implications. The latter question has been addressed [12] but it is unclear if the proposed solution scales [13].

Analyzing computational artifacts along these dimensions is important but difficult. One of the reasons is that an appropriate framework along with a specific vocabulary are not readily available (see below).

3   Why Do Some Scenarios Seem Like Science Fiction?

A discussion of pervasive computing scenarios in the literature may help illustrate some the difficulties that arise when technology is used to approximate human activities. The motivation is Satyanarayanan's intriguing question "Why do these scenarios seem like science fiction rather than reality today?" [1].

In [1] two pervasive computing scenarios are described that embody many key ideas in pervasive computing. In the context of this paper, the second scenario is particularly interesting:

> "Fred is in his office, frantically preparing for a meeting at which he will give a presentation and software demonstration. The meeting room is a 10-minute walk across campus. It is time to leave, but Fred is not quite ready. He grabs his PalmXXII wireless handheld computer and walks out of the door. Aura [a pervasive computing system] transfers the state of his work from his desktop to his handheld, and allows him to make his final edits using voice commands during his walk. Aura infers where Fred is going from his calendar and the campus location tracking service. It downloads the presentation and the demonstration software to the projection computer, and warms up the projector. Fred finishes his edits just before he enters the meeting room. As he walks in, Aura transfers his final changes to the projection computer. As the presentation proceeds, Fred is about to display a slide with highly sensitive budget information. Aura senses that this might be a mistake: the room's face detection and recognition capability indicates that there are some unfamiliar faces present. It therefore warns Fred. Realizing that Aura is right, Fred skips the slide. He moves on to other topics and ends on a high note, leaving the audience impressed by his polished presentation." (from [1], page 12)

Satyanarayanan uses the example scenario to illustrate that the seamless integration of component technology is the most important research issue in pervasive computing [1]. This would mean that the difficult problems lie in architecture, component synthesis, and system-level engineering. Apart from these technical challenges, however, the example points to some of the problems resulting from approximations of human activities by technical systems, such as Aura.

Aura embodies an approximation of human activities in the sense that users are assumed to be reliably represented by electronically available information, such as location or schedule. Humans, however, have shown to act spontaneously and opportunistically. Plans (or, in this context, schedules) are used by humans to guide action but plans do not determine action (see [14] for a detailed discussion). This implies that even in the presence of detailed plans, such resources need to be interpreted and verified by systems like Aura. Such interpretations and verifications, however, require deep understanding of human behavior which means that the task is everything but trivial.

Other crucial aspects include the availability of detailed schedules. Experiences with shared calendar management systems reported in the CSCW literature suggest that users do not always keep their schedules updated. Reasons are similar to what was found in the context of the active badge locator system: in certain situations, employees do not want their colleagues to know every detail of their activities [3]. Systems like Aura, however, depend on preferably unambiguous information about the user and his or her activities.

Last but not least, the example points to some responsibility issues. Downloading the presentation and the demonstration software to the projection computer involves taking responsibility for the content in the sense that the transmission process and the target system have to be secure. The issue also concerns the face recognition system. What if the system did not recognize the unknown persons – who would be responsible for the potential leaking of highly sensitive information? How does the system know that the presentation contains highly sensitive information? Did Fred mark the documents as such? Is Aura required to be capable of sophisticated text understanding and reasoning about sensitivity of information? Who would be responsible if the document was marked wrongly?

In the other example scenario provided in [1], some of the aforementioned issues are avoided in a very elegant way:

"Jane is at Gate 23 in the Pittsburgh airport, waiting for her connecting flight. She has edited many large documents, and would like to use her wireless connection to e-mail them. Unfortunately, bandwidth is miserable because many passengers at Gates 22 and 23 are surfing the Web. Aura [a pervasive computing system] observes that at the current bandwidth Jane won't be able to finish sending her documents before her flight departs. Consulting the airport's network weather service and flight schedule service, Aura discovers that wireless bandwidth is excellent at Gate 15, and that there are no departing or arriving flights at nearby gates for half an hour. A dialog box pops up on Jane's screen suggesting that she go to Gate 15, which is only three minutes away. It also asks her to prioritize her e-mail, so that the most critical messages are transmitted first. Jane accepts Auras advice and walks to Gate 15. She watches CNN on the TV there until Aura informs her that it is close to being done with her messages, and that she can start walking back. The last message is transmitted during her walk, and she is back at Gate 23 in time for her boarding call." (from [1], page 12)

Clearly, this example also depends on the availability of detailed information about the environment and Jane's schedule. However, in contrast to the other example, the difficult issue of text understanding is circumvented as Aura leaves it to the user to prioritize her e-mail. Also, Aura's pro-activeness is limited to making suggestions which the user may accept or not. In the example discussed previously, Aura made the decision to forward documents without further inquiries.

So why do these scenarios actually seem like science fiction rather than reality today? One of the reasons seems to be that even supposedly simple scenarios easily cross the line between "regular" technical systems and systems that require human-like capabilities. The problem with the latter is that the "artificial intelligence problem" is still largely unsolved. As Michael Dertouzos, Director of the Laboratory for Computer Science at MIT pointed out in July 2000:

> "The AI problem, as it's called – of making machines close enough to how human beings behave intelligently – ... has not been solved. Moreover, there's nothing on the horizon that says, I see some light. Words like 'artificial intelligence', 'intelligent agents', 'servants' – all these hyped words we hear in the press – are statements of the mess and the problem we're in." (quote from [15]).

This statement – made by the head of the lab that is running MIT's Oxygen project – might be considered pessimistic but at least it suggests to carefully look at what can reasonably be expected from today's and tomorrow's technology.

4   Pervasive Computing and Context Awareness

Many argue (e.g., [16]) that achieving invisibility in pervasive computing will require tremendous progress in user interfaces, context awareness and other technologies.

The idea behind context awareness is that computational artifacts are enabled to sense the context in which they are being used so that they can adapt their functionality accordingly. A context-aware mobile phone, for example, would use context aspects, such as the user's identity, the user's location, and the user's current schedule, to determine the level of intrusiveness that would be appropriate when trying to notify the user of incoming calls. Notifications could range from ringing (quite intrusive) to buzzing or vibrating (less intrusive). The mobile even might suppress notifications of less important calls (not intrusive at all) [17]. First steps towards implementing such a mobile can be found in [18].

Yet another motivation for making use of context is providing users with exactly the information that is useful in a particular work situation. The idea of an awareness information environment, for example, is to "provide users with information that is related to their current context and therefore of most value for the coordination of the group activities" [19]. An agent-based example addressing context-dependent content distribution and delivery is described in [20]. "Just in time" information retrieval agents are software agents that monitor the user's writing activities and continuously scan databases for information that might be relevant to the current writing context. Indeed, software agents illustrate that context-dependent delivery of information is closely related to context awareness. Software agents can be viewed as early (software) approaches to context-aware artifacts. Examples discussed in [21] include an agent for electronic mail handling and an agent for electronic news filtering. Clearly, such agents require a degree of context awareness in order to be able to deliver what they promise.

The problem with implementing context awareness in artifacts is that features of the world are not context because of their inherent properties. Rather, they become context *through their use* in (human) interpretation [22]. Context is shaped by the specific activities being performed at a moment; these activities also influence what participants treat as *relevant* context [23].

According to [24] people use the various features of their physical environment as resources for the social construction of a place, i.e., it is through their ongoing, concerted effort that the place – opposed to space – comes into being. An artifact will be incapable of registering the most basic aspects of this socially constructed environment. Accordingly, a context-aware artifact may fail annoyingly as soon as a context-aware system's (wrong) choices become significant.

For context-aware artifacts it may be difficult or impossible to determine an appropriate set of canonical contextual states. Also, it may be difficult to determine what information is necessary to infer a contextual state [25]. Goodwin and Duranti even concluded that "it does not seem possible at the present time to give a single, precise, technical definition of context, and eventually we might have to accept that such a definition may not be possible" [23].

Elsewhere [17,13] we have outlined that context awareness in any non-trivial sense involves the frame problem (e.g., [26]) which is one of the hard problems in classical representation-based artificial intelligence. Roughly, the frame problem is about what aspects of the world have to be included in a sufficiently detailed world model and how such a world model can be kept up-to-date when the world changes. Indeed, the frame problem has shown to be intractable in realistic settings (e.g., [15]). The real world is constantly changing, intrinsically unpredictable, and infinitely rich [27].

The frame problem is said to be a more technical problem but it can also be understood as an ontological problem as aspects of the world included in a world model determine the understanding of the world based on the model: facts not included in the model and not derivable from the model cannot be explained based on the model. This means that the frame problem in AI is directly related to artifacts trying to understand any notion of context.

Lessons to be learned from investigations of the frame problem suggest that there is little hope that computational artifacts will finally become context aware in a non-trivial sense. However, the hardness of the frame problem does not suggest to abandon research on context-aware artifacts but to keep in mind that such artifacts may well fail when trying to recognize a situation. The potential for failures should be taken into account when designing such artifacts (e.g., [12, 17]) and contributes to answering the question why many pervasive computing scenarios still seem like science fiction rather than reality.

We have started to analyze how definitions of context found in the technical literature on context awareness relate to what has actually been implemented in context-aware artifacts [17,13]. Areas that need to be investigated in more detail are:

1. How are definitions of context (typically, the researcher's understanding of context as expressed in research papers) and the models of context implemented in context-aware artifacts related to each other? What simplifications have been used in order to gain technical feasibility? Are these simplifications implicit or explicit? Are implications addressed?

   Motivation: a lot of research in context-aware artifacts appears to be based on the assumption that in some application domains context is not continuously changing and that it is therefore feasible to represent context in rather static data structures. Knowing about these domains and what characterizes them would largely help research in context awareness.

2. In which situations do models of context work? Why?

   Motivation: it is hard to find analyses or even descriptions of context-aware artifacts that have been used over extended periods of time. A detailed discussion like in [4] of other fielded context-aware artifacts would allow others to benefit from the experiences and help them obtain a better understanding of the issues involved.

3. In which situations did models of context not work? Why?

   Motivation: problems that were detected when trying to operationalize context are rarely published although these insights would be of particular value.

   Knowing more details would support research in context awareness and would also help assess the potential of context awareness in pervasive computing.

5   Summary and Conclusions

In this paper, we have tried to shed a bit of light on Satyanarayanan's intriguing question "Why do [pervasive computing scenarios] seem like science fiction rather than reality today?" We looked in particular at the role of technical approximations and used Satyanarayanan's example scenarios to highlight areas where limitations of technical approximations prevent visions from becoming reality. The investigation indicates that devices may require "human-level intelligence" even in situations that appear to be rather simple. This experience has been made in other disciplines as well. In information retrieval, for example, estimating relevance has shown to be a hard problem (e.g., [28,29,30]). Experiences with information retrieval systems are in turn relevant to pervasive computing areas, such as content distribution and delivery. Other areas are CSCW and workflow management where experiences suggest that it is extremely difficult to represent work as such representations are inherently incomplete and ambiguous.

The main contribution of this paper is that we have highlighted the need to address the difficult areas in pervasive computing explicitly. First, focussed discussions would help identify difficult areas and would allow to benefit from work on similar areas in related disciplines, such as AI or CSCW. We have provided an example of such inter-disciplinary work by relating the context awareness issue in pervasive computing to previous research on the frame problem in AI. The discussion suggests that there is little hope that context awareness can be realized to the extent that would be required for realizing some of the example scenarios in pervasive computing. Given these difficulties, addressing these issues explicitly is even more important as it could help developers become aware of the difficulties and to deliberately circumvent such areas.

The second main contribution is that we have outlined some of the difficulties to describe and to analyze the specific characteristics of pervasive computing applications and devices. These difficulties indicate a need for a framework providing specific terminology. Such a terminology could foster inter-disciplinary cooperations that are required to further analyze the gap between application scenarios and technical feasibility in pervasive computing. An example for relevant work in CSCW is the proposal to re-consider Simon's idea of Sciences of the Artificial [31] as a foundation for analyzing the gap between social requirements and technical feasibility in CSCW [9]. An example for a comprehensive taxonomy for location in mobile computing is to be found in [32].

6   Future Research

We are working on analyzing further scenarios reported in the literature. As an interim outcome we expect a list of criteria that can be used to investigate scenarios for (perhaps not apparent) difficulties. A longer term goal is the already mentioned framework providing specific terminology to describe and to analyze the specific characteristics of pervasive computing applications and devices. This framework would provide means to express and share experiences in pervasive computing and related disciplines such as AI and CSCW.

In order to complement this more theoretical work, we just submitted a large research grant proposal to investigate the issues jointly with specialists in ethnographic work place studies and personalization technology.

Acknowledgments. The author is grateful to Barbara Gorayska, David McDonald, Toni Robertson and Wayne Brookes for enlightening discussions.

References

1. Satyanarayanan, M.: Pervasive computing: Vision and challenges. IEEE Personal Communications (2001) 10–17
2. Weiser, M.: The computer for the 21st century. Scientific American (1991)

3. Want, R., Hopper, A., Falcao, V., Gibbons, J.: The active badge location system. ACM Transactions on Information Systems **10** (1992) 91–102
4. Want, R., Hopper, A.: Active badges and personal interactive computing objects. IEEE Transactions on Consumer Electronics **38** (1992) 10–20
5. Beigl, M., Gellersen, H.W., Schmidt, A.: Mediacups: Experience with design and use of computer-augmented everyday objects. Computer Networks **35** (2001) 401–409 Special Issue on Pervasive Computing.
6. Mattern, F.: Ubiquitous computing: Vision und technische Grundlagen. Informatik/Informatique (2001) 4–7
7. Ackerman, M., Darrell, T., Weitzner, D.J.: Privacy in context. Human Computer Interaction **16** (2001)
8. Bellamy, R., Swart, C., Kellogg, W., Richards, J., Brezin, J.: Designing an e-grocery application for a Palm computer: Usuability and interface issues. IEEE Personal Communications (2001) 60–64
9. Ackerman, M.: The intellectual challenge of CSCW: the gap between social requirements and technical feasibility. Human Computer Interaction **15** (2000) 179–203
10. Gibson, J.: The Ecological Approach to Visual Perception. Houghton Mifflin, Boston (1979)
11. Fogg, B.J.: Persuasive computers: perspectives and research directions. In: Proceedings of the Annual ACM SIGCHI Conference on Human Factors in Computing Systems, New York, NY, USA, ACM Press (1998) 225–232
12. Bellotti, V., Edwards, K.: Intelligibility and accountability: Human considerations in context aware systems. Human Computer Interaction **16** (2001)
13. Lueg, C.: Operationalizing context in context-aware artifacts: Benefits and pitfalls. Informing Science **5** (2002) ISSN 1521-4672.
14. Suchman, L.A.: Plans and situated actions - The Problem of Human-Machine Communication. Cambridge University Press (1987)
15. Dreyfus, H.L.: On the Internet. Thinking in Action. Routledge, London, UK (2001)
16. Gupta, S., Lee, W.C., Purakayastha, A., Srimani, P.: An overview of pervasive computing. IEEE Personal Communications (2001) 8–9
17. Lueg, C.: On context-aware artifacts and socially responsible design. In Smith, W., Thomas, R., Apperley, M., eds.: Proceedings of the Annual Conference of the Computer Human Interaction Special Interest Group of the Ergonomics Society of Australia. (2001) 84–89 ISBN 0-7298-0504-2.
18. Schmidt, A., Van Laerhoven, K.: How to build smart appliances. IEEE Personal Communications (2001) 66–71
19. Gross, T., Prinz, W.: Gruppenwahrnehmung im Kontext. In: Tagungsband der Deutschen Computer-Supported Cooperative Work Konferenz, Teubner (2000) 115–126
20. Rhodes, B., Maes, P.: Just-in-time information retrieval agents. IBM Systems Journal **39** (2000) 685–704
21. Maes, P.: Agents that reduce work and information overload. Communications of the ACM **37** (1994) 31–40
22. Winograd, T.: Architectures for context. Human Computer Interaction **16** (2001)
23. Goodwin, C., Duranti, A.: Rethinking context: an introduction. In Duranti, A., Goodwin, C., eds.: Rethinking context: language as an interactive phenomenon. Cambridge University Press (1992)
24. Agre, P.: Changing places: contexts of awareness in computing. Human Computer Interaction **16** (2001)
25. Greenberg, S.: Context as dynamic construct. Human Computer Interaction **16** (2001)

26. Pylyshyn, Z.W., ed.: The Robot's Dilemma: The Frame Problem in Artificial Intelligence. Ablex Publishing Corporation, Norwood, NJ (1987)
27. Pfeifer, R., Rademakers, P.: Situated adaptive design: Toward a methodology for knowledge systems development. In: Verteilte Künstliche Intelligenz und kooperatives Arbeiten: 4. Internationaler GI-Kongress Wissensbasierte Systeme, Berlin, Springer-Verlag (1991) 53–64
28. Schamber, L., Eisenberg, M.B., Nilan, M.S.: A re-examination of relevance: Toward a dynamic, situational definition. Information Processing & Management **26** (1990) 755–776
29. Froehlich, T.: Relevance reconsidered. Introduction to the special topic issue on relevance research. Journal of the American Society for Information Science **45** (1994) 124–134
30. Ellis, D.: The dilemma of measurement in information retrieval research. Journal of the American Society for Information Science **47** (1996) 23–36
31. Simon, H.A.: The Sciences of the Artificial. MIT Press (1969)
32. Dix, A., Rodden, T., Davies, N., Trevor, J., Friday, A., Palfreyman, K.: Exploiting space and location as a design framework for interactive mobile systems. ACM Transactions on Computer-Human Interaction **7** (2000) 285–321

The Fastap Keypad and Pervasive Computing

David Levy

16 Blake Street, Cambridge, MA 02140, United States

Abstract. The need for an effective and adaptable data input method is fundamental to pervasive computing on hand held devices. The Fastap ™ keypad is a new paradigm of keyed input, providing more functionality in a smaller space than previously considered possible. Fastap technology may prove a useful advance to the widespread adoption of the mobile Internet, especially messaging, and more broadly to pervasive computing in general. The purpose of this white paper is to provide a framework for understanding the problems with existing mobile telephone interfaces, to introduce the Fastap technology and to qualify the assertion that the technology is a fundamental advance in keyed input that is well-suited as a replacement of the existing 12-button telephone interface.

1 Overview

Ironically, while few doubt the value of portable computing or of wireless communication, there is widespread speculation about the value of the mobile Internet, the first mass market effort to unite the two. Analysts and industry press devote significant attention to the search for the mobile Internet's "Killer App"… a search to validate the need for the *portable* version of what we all know to be a vitally important tool on the desktop.

Why? The search for a Killer App seems to belie an inherent belief that the desktop experience will not translate to the pocket. A central source for this disbelief, as expressed by analysts, industry press and the public, is dissatisfaction with the manner in which data is entered into mobile devices.

The object of this white paper is to:
- discuss the impact of Interface on the adoption rate of new technologies
- introduce the Fastap™ keypad
- demonstrate that the Fastap keypad is a new paradigm of keyed interface
- discuss qualitative issues comparing Fastap to its alternatives

1.1 Impact of Input

Interface technology is often under-appreciated. The fact is that systems of great utility can remain relatively unused only because they are slightly too complicated to operate. Any feature that requires reading a manual has precipitously decreased

F. Mattern and M. Naghshineh (Eds.): Pervasive 2002, LNCS 2414, pp. 58-68, 2002.

chances of being used. Perhaps a condemnation of human nature, but the nature exists.

Relatively small advances in interface technology often provide the missing element necessary to unleash phenomenal industry growth.

Some examples:

- The mouse and Graphical User Interface. (Iconic desktop). We can only speculate how the world would be different if computers still operated in a DOS/command line environment. But it is clear that Microsoft, Dell, Apple, Compaq etc. owe their overwhelming success to the relatively small ergonomic niceties that made rapid and massive computer adoption possible: the mouse and the GUI. [1]
- Graffiti Handwriting Recognition. As tenuous as Graffiti's popularity may be, it is indisputable that without it the handheld industry would basically not exist. Handspring and Palm owe their existence to this interface innovation.
- The Browser. The Internet has been around since the 1970's. It lay dormant for twenty years until an easy-to-use interface (a browser) woke it in 1991. The Internet then quickly became one of the largest economic expansions in history, and provided a fundamental societal change.

Each one of these economic and social engines was started by a relatively small ergonomic advance that made the underlying and (pre-existing) technologies easy to access.

This is because:
Ease-of-use = USE.

This fundamental idea is often overlooked in the development of products, and sometimes entire industries, such as the mobile Internet. Billions of dollars will be invested to develop amazing new technologies and infrastructure, with an attitude of "if we build it, they will come" while minimal effort will be spent on providing easy access to the technology. Developing a new technology, and not providing an appropriate interface is akin to building a baseball diamond but making it really hard for people to get inside.

A more topical example may be demonstrated in reverse, as a thought experiment.

How often would you use the desktop Internet if your desktop keyboard was replaced by a 12-button telephone keypad? Imagine writing and addressing an email, or browsing the Internet. Would the decrease be small or precipitous? Might you stop using the Internet at all?

This suggests that trying to evaluate the utility or economic potential of the mobile Internet under the constraints of an inappropriate interface will lead to vast underestimation. The existing 12-button interface was never intended to access computer functionality, and the result is a distorted negative impression of the utility, and hence the value of the first consumer implementation of pervasive computing. As with technologies before it, the adoption and use rates would be far higher if the interface was designed to make the device easy to use.

1.2   Telephone History

With early telephones, the user would ring a bell to summon the operator to make a connection. Strowger invented the rotary dial in 1888, allowing users to dial their party directly. In the 1950's Bell Labs invented the push button telephone interface we use today.

Half a century has passed since then. One could reasonably expect the underlying technology to have outgrown the ability for the interface to support it. Indeed it has. Most mobile telephones today are able to: write and receive messages; browse the Internet; store, write and retrieve personal data; trade stocks; play games; perform banking operations; and buy and sell goods. The telephone has developed computer functionality, and not surprisingly, the interface is badly outdated, and simply ill-suited to the task.

Fig. 1. A Fastap Keypad on a mobile phone (49mm x 104mm x 16 mm)

2 Fastap Technology

In general, the ideal modern telephone interface would support the phone's traditional capabilities and operate in a traditional way, but also provide access to the alphabet, punctuation and auxiliary functions in an intuitive and unrestricted manner.

This paper now turns to discussion the Fastap technology and how it satisfies the demands of a modern telephone interface. The technology provides the traditional 12-button telephone keypad, plus a full alpha keyboard with punctuation, in approximately the same area as the existing telephone interface, and with similar ease of use. This section will explain system operation, and provide a quantitative and qualitative performance analysis. While the technology may be implemented in any layout, including the traditional "QWERTY" desktop layout, this paper is limited to discussion of the telephone, with emphasis on the mobile Internet telephone.

2.1   Operation

Fastap technology is comprised of two superimposed keyboards, as shown in Figure 1. The traditional telephone keypad is on the lower level and the letters are available on the raised keys. Punctuation and control characters are interspersed between both in a logical manner. To access a number, letter or punctuation character, the user simply presses it.

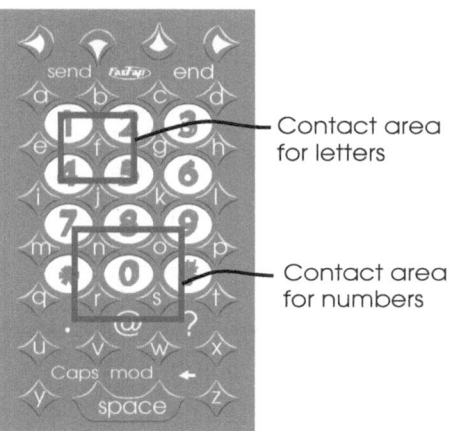

Fig. 2. Operation of letters and numbers. (Drawing is full scale)

Figure 2 demonstrates how the surface contour provides ample space for a finger or thumb to actuate any key. The raised letters are spaced sufficiently from adjacent letters to provide the contact area of a full-sized desktop key. Likewise, the user enters lowered characters simply by pressing the desired character, and ignoring the inevitable consequence of pressing adjacent letter keys. The result is that the finger (or thumb) is always provided ample space. People with extremely large hands have expressed preference for Fastap keypads to the keypads of their own mobile phones.

Technically, it is the actuation of opposite adjacent independent keys (around a lowered character) that activates one of the combination keys.

For example, pressing either the "n" and "s" *or* "r" and "o" are sufficient to uniquely define the number "0," so the user can press off-center (missing a key) and still receive reliable operation. In practice, the system demands similar accuracy to a traditional keypad of similar size. The system operates on the principle of "passive chording." Although multiple keys must be struck simultaneously to provide output, the system operates transparently to the user so that no special training or memorization of stroke sequences are necessary. Other chorded keyboards have had poor acceptance because they require training and/or memorization.

Because passive chording provides each character with a single press, the device is as intuitive as any desktop keyboard. The device requires no training or explanation, yet allows 50 independent full-size keys to fit in area one-third the size of a business card, bringing efficient, intuitive and ergonomic computer functionality to small handheld devices.

2.2   New Paradigm

The Fastap keypad is a new paradigm in keypad technology. The design increases maximum key density (number of keys per square cm) by a factor of approximately 2.4 and *simultaneously* increases the contact surface area of each key. The result is a computer-level interface far smaller than previously thought possible, and yet more comfortable to use than devices with less functionality. By providing one-touch access to each number letter, and basic punctuation, Fastap technology eliminates the need to "triple tap" or to use disambiguation software. In a study of 113 cases extracted from 57 HCI studies Nielsen & Levy [2] found high correlation between user preference and device efficiency.

3   Quantitative Analysis

3.1   Key Density – Theory

In order to demonstrate Fastap technology is truly a new paradigm key input, we must first establish a theoretical construct of the existing technology.

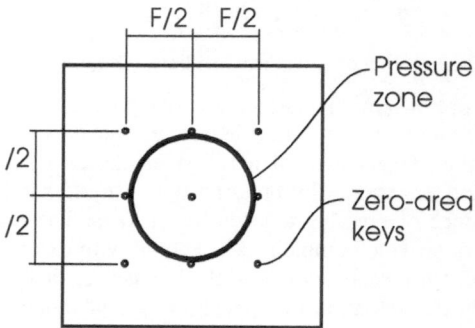

Fig. 3. Theoretical maximum key density

Figure 3 shows a keyboard reduced to its theoretic limits. The keys are reduced to regions of near-zero contact area. While such a design would certainly be uncomfortable for the user (if it could work at all) the design provides an unassailable theoretical model of the absolute maximum possible key density within the existing keypad paradigm.

The key caps of near-zero area are placed at a distance of F/2 from each other, where F is the diameter of a theoretical pressure zone at the tip of an idealized human finger, the region that would apply pressure to a surface as the finger activates a button. This model therefore uses the smallest possible keys, located as close to each other as they can possibly be without necessarily striking more than one key at a time.

Using this layout, we can calculate the theoretical maximum density (keys/sq. cm) of independent key caps in a keypad M finger units wide by N finger units high:

$$\text{Max Density}_{\text{Theoretical}} = \frac{4MN - 2N - 2M + 1}{MNF^2}. \tag{1}$$

Figure 4 shows the pressure zone of a finger as it actuates a "combination key" of a Fastap keypad. Each combination key is identified by a graphic located at the intersection of four keys.

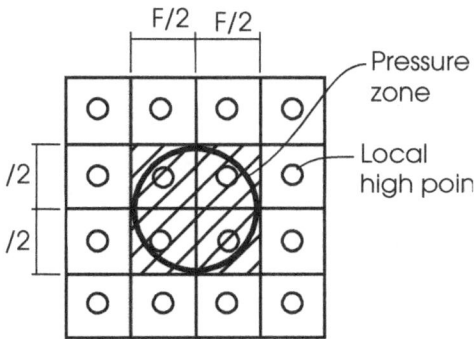

Fig. 4. Combination key of Fastap Keypad

Note two things: 1) the combination key density (below) is exactly the same as the theoretical maximum of the old style keypad. 2) The area allocated to each Fastap button is F^2 (i.e. finger-sized) instead of near-zero, as with the old style keypad of Figure 3. The density of the combination keys is given by Equation 2.

$$\text{Key Density}_{\text{Combo Keys}} = \frac{4MN - 2N - 2M + 1}{MNF^2}. \tag{2}$$

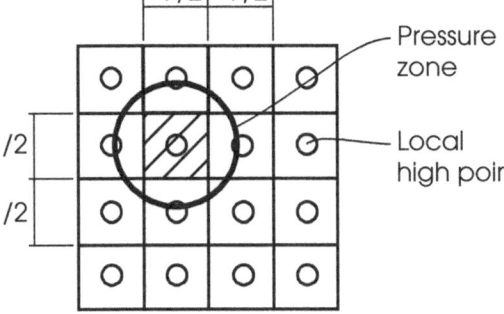

Fig. 5. "Independent key" of Fastap Keypad

By itself, this is an interesting result. However, Fastap technology offers far more. With Fastap technology, the individual keys are elevated, allowing the user to operate an *additional* set of 2M x 2N keys, as shown in Figure 5.

The combined densities of the combination and independent keys yield a key density of:

$$\text{Key Density}_{\text{Total}} = \frac{8MN - 2N - 2M + 1}{MNF^2} . \tag{3}$$

By entering values of M=3 and N=4 we can determine theoretical values for the densities of a telephone-sized keypads.

Existing button paradigm:

$$\text{Max Density}_{\text{Theoretical}} = 2.9 \text{ keys/finger unit} . \tag{4}$$

Fastap Technology:

$$\text{Density}_{\text{Fastap}} = 6.9 \text{ keys/finger unit} . \tag{5}$$

Because the "finger unit" is arbitrary, only the ratio is significant.

Taking this ratio (6.9/2.9) we see that the Fastap keypad technology provides a key density ~2.4 greater than smallest possible traditional keypad.

3.2   Key Density – Practice

How does the theory work in practice? The keys of real keypads must have some physical area.

This fact applies to both traditional and Fastap keypads, which must both scale approximately equally as they grow from theoretical abstraction to a useful size. The result is that the ratio does not differ significantly between theory and practice.

This analysis is proved correct by simply measuring the key densities of existing products and Fastap keypads. The results are shown in Figure 6.

The thick line demonstrates how products approach an asymptote of maximal key density of approximately 1.2 keys per square cm. Apparently higher densities are ergonomically infeasible for a traditional keypad. The Fastap technology, with 3.3 keys/square cm, provides an increase in key density of approximately 2.7 over these successful small products. So theory (at 2.4) is reasonably close to practice, at 2.7.

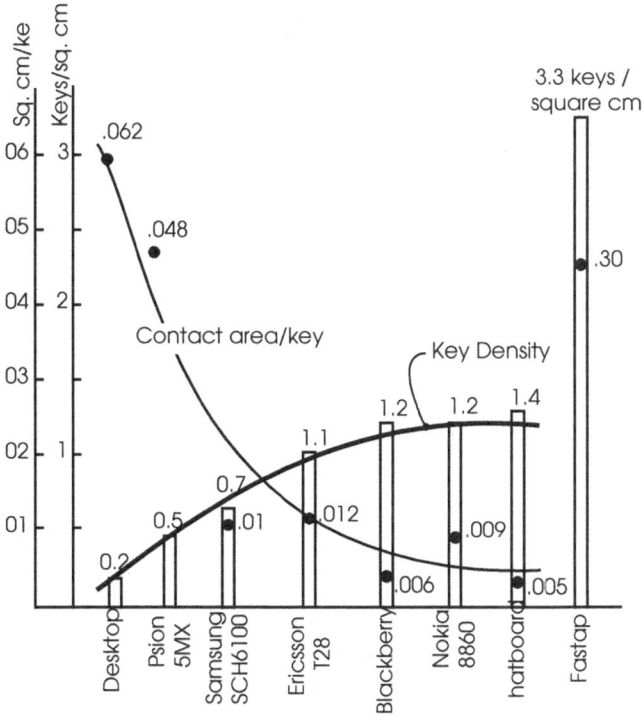

Fig. 6. Key density and contact area

3.3 Key Size

Figure 6 also tracks the key size of the same products, shown as dots on the graph. Although the "curve" tracks the data relatively poorly (demonstrating that key size and density are somewhat independent) Fastap technology is clearly independent of the previous trend, with an increase of key area of 4-6 times.

3.4 Summary

Fastap technology provides an increase in key density of approximately 2.4 - 2.7 (theoretic and empiric, respectively) while *simultaneously* providing an increase in key contact area by a factor of approximately 4-6 times (empiric).

4 Qualitative Analysis

To perform a qualitative evaluation we will identify a set of ideal criteria and use them to compare Fastap technology to its competitors. For the purposes of this paper, we will limit the analysis to: 1) the mobile Internet (the most important aspect of pervasive computing at this time) and 2) technologies that are available now or that may become viable in the next few years:

- Voice recognition
- Hand writing recognition
- Graffiti
- On-screen keyboards
- Accessory keyboards (e.g. Chatboard, iBoard, Cirque's Pocket Keyboard)
- 12-Button keypad (multi-tapping)
- 12-Button keypad (with predictive algorithms such as T9, eZtap, and iTap)
- Integrated QWERTY (e.g. Treo, Nokia 9xxx)
- Fastap keypad

Defining the Ideal.
In order to provide a more detailed framework for analysis, the following list of "ideal" mobile Internet interface characteristics are provided:

1. *No training to use.* It is been shown repeatedly that increasing the amount of training necessary to use a product will decrease the adoption rate and reduce the final penetration rate.
2. *Error-Free operation.* Some technologies inherently create errors, even if the user performs perfectly, a highly undesirable feature.
3. *Speed.* Obviously desirable.
4. *Convenience/Portability.* Obviously desirable.
5. *Transparence/no-modalities.* A "transparent" interface allows the user to focus beyond it, on what they are trying to do, instead of the interface itself: as if the interface was not there. Transparency is extremely desirable.
6. *Alphabet-centric* (instead of language-centric) vastly increases cross-cultural communication abilities, and/or reduces manufacturing and localization costs.
7. *Cost.* Preferably low.
8. *Hands-free operation.* Only voice recognition is hands-free. One-handed devices are given a "very good" rating in the chart below. Two-handed devices are given a poor rating, because analysts have identified one-handed operation as important for mobile Internet devices.
9. *Universal Access* – Access for the deaf and blind is and important criteria.
10. *Non-restrictive* – The interface should not restrict what people can say or how they say it.
11. *Private.* The interface should allow the device to be used without compromising privacy.

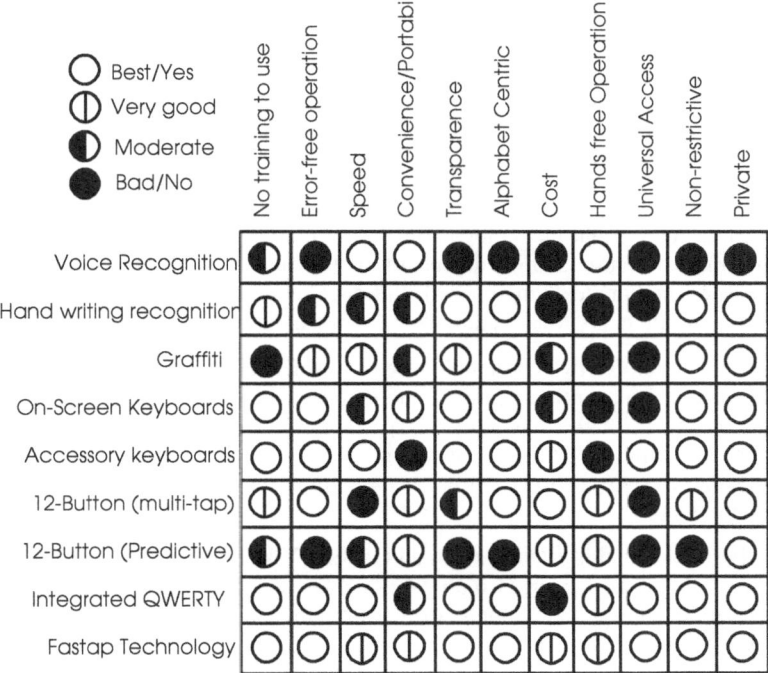

Fig. 7. Interface qualities for mobile phones

Figure 7 shows a graph of the qualities of each of the competitive alternatives available in the foreseeable future as filtered through the ideal mobile phone criteria listed above.

Discussion.
While no assumption is made each criterion is of equally importance, the amount of light color provides an interesting first-order approximation indicating the Fastap keypad is an interesting technology, worthy of further evaluation.

A few comments on the rankings:

• Graffiti was the only system given a "bad" rating for training because it is the only system in which continual practice is needed.

• While Voice Recognition is widely seen as the next generation of interface, it is worth noting that the most significant benefit of pervasive computing is to provide the ability to work any time and any place... and likewise, the most significant problem with voice recognition is that it *limits* the ability for a person to work at any time and any place. Furthermore, many of the cases in which VR would be inappropriate to use are the same cases in which pervasive computing are most compelling: in meetings, on planes, trains, buses, subways, waiting in lines, etc.

VR will one day be an important input technology, but it will not obviate the need for an alternate (private) means of entering data. Continuous VR (required for messaging) currently requires over 100MB of storage, 64 MB of RAM and a high-end processor, making it two orders of magnitude more expensive than the alternatives, and despite its availability on the (relatively quiet and controlled) desktop environment, it remains a disability solution .

- Because of the familiarity we all have with using a stylus, hand writing recognition is an intriguing input technology. But in the mobile context, two-handed input is undesirable because a high degree of mobile use is conducted while carrying something, and/or steadying oneself. Indeed, Christian Lindholm, Nokia's Director of Usability includes "one-handed device" in the definition of a mobile telephone.

- Disambiguation software, though often considered an improvement to the 12-button keypad for writing text on a phone, fairs quite poorly with respect to triple-tapping, except in the "category of speed. Disambiguation methods are slightly faster than triple-tapping as long as the dictionary contains at least 85% of the words the user wishes to type [3], also assuming the user makes no typographical or spelling errors.

- The Fastap keypad is rated lower than QWERTY keypads with respect to training, because the pervasiveness of the QWERTY keyboard provides other opportunities to learn it. The actual amount of training to achieve the same familiarity is obviously the same.

5 Conclusion

History has shown interface quality is critical to the successful adoption of new technologies. The mobile internet, one of the first mass implementations of pervasive computing, suffers from a poor interface. Nearly half a century old, the telephone's 12-button keypad is simply ill-suited to access the computer functionality of modern phones. The Fastap keypad is a new paradigm of key input technology that brings computer-level functionality to the telephone. By increasing key density by 2.4 - 2.7 times, while simultaneously increasing the size of the keys from 4 - 6 times, the technology provides a complete and easy-to-use alphanumeric keyboard, including basic punctuation, in an area of 1/3 of a credit card. The technology also excels at a set of qualitative criteria that identify an ideal mobile telephone interface.

References

1. Donoghue, K.: Built for Use: Driving Profitability Through the User Experience. McGraw Hill (2002)
2. Nielsen, J. and Levy, J. Measuring Usability: preference vs. performance. Communications of the ACM, vol. 37 no. 4, 66-75. (1994)
3. Mackenzie, S., et al: Letterwise: Prefix-based Disambiguation for Mobile Text Input. Proceedings of the ACM symposium on User Interface Software and Technology. New York (2001) 111-120

Going Back to School: Putting a Pervasive Environment into the Real World[1]

Victor Bayon, Tom Rodden, Chris Greenhalgh, and Steve Benford

Mixed Reality Lab, School of Computer Science and IT,
Nottingham University, Jubilee Campus,
Wollaton Road, Nottingham NG8 1BB
{vxb;tar;cmg;sdb}@cs.nott.ac.uk

Abstract. This paper presents the lessons learnt from the development of a ubiquitous computing environment for use within the real world. Such systems are currently purpose built demonstrators, often within research labs. This paper is based on the development of a storytelling environment for use within schools. This migration – from the lab to the school – required the redevelopment of the platform, and highlighted the importance of providing support for the maintenance and management of the environment when access to a sophisticated infrastructure and dedicated space can no longer be guaranteed.

1 Introduction

Environments that are populated by a wide range of interactive devices represent the perennial articulation of pervasive computing [2,17]. Building upon Mark Weiser's early vision of ubiquitous computing [18] a number of researchers have articulated a research agenda for pervasive computing based on a world populated by a range of embedded devices. Understanding the nature of these devices and the ways in which computationally rich environments may be used to support different digital experiences has emerged as one of the predominant challenges for pervasive computing.

In order to understand a world populated by a diverse set of devices the research community has had to strike a balance between its exploration of the use of these devices and the development of radically new technologies. Most visions of ubiquitous computing make strong assumptions about technological infrastructure, which makes achieving this balance more difficult. This problem has been compounded by the need for researchers to develop these infrastructures in tandem with the construction of the interactive environments that they support. The core of assumption is that devices will

---

[1] This work has been conducted as part of the EU KidStory project and the UK EPSRC Equator IRC. The authors acknowledge support from partners in both endeavours particular those in Kidstory whose work this builds upon. Thanks also to the teachers and children at Albany School, Nottingham for their collaboration.

F. Mattern and M. Naghshineh (Eds.): Pervasive 2002, LNCS 2414, pp. 69-83, 2002.

be built into the fabric of the spaces we inhabit and advanced computing and communication facilities will always be available.

This dependence on a relatively sophisticated infrastructure has meant that the majority of pervasive environments have had to be developed and explored within research labs. The development of these laboratory-based systems has allowed researchers to consider the long-term use of these systems [6]. As technologies have matured these test-bed environments have become increasingly sophisticated in nature. Our view of these environments has altered with the emergence of mobile devices [10], large screen displays and PDAs [11, 12] and tangible interface devices [9]. It is not unusual for these technologies to be explored in terms of the "potential" they offer users, through the construction of purpose build demonstrators in lab settings. Research-oriented visions and scenarios are often used to articulate these systems, with less attention paid to their practical application than to real world activities [5].

The construction of technologically laden pervasive environments sets a challenge in terms of how researchers migrate from the sophisticated technological surroundings of research labs to real world setting where fewer assumptions can be made about the supporting technological fabric. Essentially, within a research lab the developers of a pervasive environment have complete control of the space. Its purpose is solely to demonstrate the pervasive space. However, outside research labs space is a scarce resource and multiple competing demands are often placed upon the space. How might we consider the development and deployment of pervasive environments when these environments have to co-exist with all of the other demands on a space?.

This work considers the lessons we have learned through the development of a series of interactive storytelling environments within real world school settings, and the implications that these have for the technological infrastructure needed to support these environments. This shift toward *situating* interactive environments within the real world rather than the research lab reflects a maturing of the technology, built on the considerable body of existing research. It also represents a shift in the role of these environments. Previous work has explored them as a medium of expression, with an emphasis on articulating how people experience these environments. In contrast we must now consider how this class of interactive environment is put to work for a particular purpose in the real world with its myriad constraints.

The paper is based on the development of a technological infrastructure to support an interactive storytelling environment within a school setting. The following section briefly describes the nature of the environment and the multi-computer set-up used to support these activities. The challenges that emerged when using this environment within schools are described in the section that follows, together with guidelines and strategies that might be employed to address these issues from a technological perspective. We then describe how we have begun to explore these strategies in the development of an appropriate supporting infrastructure. Finally, we discuss some of the outstanding issues and areas for future work and exploration.

2 Developing Storytelling Environments

Although computers are becoming a more common resource in schools, they are not generally integrated within routine classroom practice. Rather, they are located in a special "IT room" somewhere in the school, or at the periphery of the classroom (both in terms of space and use). Our research has been concerned with bringing the potential benefits of storytelling-based interactive computer environments to the everyday teaching space, where it can be more integrated with other educational resources.

We have collaborated with teachers and children from a local UK school over almost 3 years. The project has made use of a number of different arrangements of technology, employed in different contexts within the classroom situation. In this paper we are concerned with relatively complex multi-machine systems with multiple (often tangible) input/output devices, which have the property of allowing groups of children (or adults) to interact and work together, i.e. Storyroom-type technologies [2], which allow children to create their own story spaces and author stories using physical props. Working in the design space of a real school environment introduces a number of technical and practical problems that such systems must take into account if they are to be usable for teaching on a day-to-day basis.

For more than 20 years, the research community has been exploring interactive computer based systems [14], and several other examples of this kind of story-telling interactive environment exist. In Immersive Environments [7], Druin demonstrated the concept of physical interaction with computer systems. In the Kidsroom [14] set-up group video tracking techniques were used to control the narrative flow of a story. While Swamped! [14] used a novel sympathetic interface to control the interactions of a virtual character within the a virtual environment. The focus of this paper is on the practical problems that arise when these kinds of technologies move from a research lab or exhibition into a school setting to be used by teachers and pupils.

2.1 The Original Room-Scale System

Our starting point is the room-scale system used for story-telling in the second year of our project. This comprised of an extended version of KidPad [8], the multiple-mouse-based story-telling system (with supporting applications), used on a network of PCs, with various peripherals supporting tangible interaction, and a large back-projected display [15]. This is illustrated in Figure 1.

The original room-scale KidPad system was used for whole-class retelling of stories. This configuration was designed to explore scalable performance and interaction. In particular, it allowed up to 6 children to simultaneously use the "Magic Carpet" (a set of large floor pressure pads) and video tracking for navigation of the stories. It also enabled a large audience to view the retelling. The right part of figure 1 shows a rear view of the retelling system, which included: 2 desktop PC's, 1 laptop, the navigation device, video camera, printer, barcode readers, projector, a projection screen and cables.

Fig. 1. Front-view and rear-view of the initial room-scale configuration

3 Moving into the Classroom Environment

This room-scale multiple machine system was developed and tested in our laboratory space, where we could make use of a supporting technical infrastructure and dedicate space to the system. During development and initial trails the assembled environment occupied dedicated space within the lab for over six months. This was an iterative design process, interleaving user sessions and development. However, moving to a school environment revealed a number of specific challenges that are described in this section. We also suggest a number of strategies that might be used to address these challenges.

3.1 Competing for Space in the School

The first challenge was the lack of available space in the school environment for establishing such a large system. As can be seen from the previous section the system required space "front stage" for the story tellers and listeners, space "back stage" for the equipment, and an unobstructed volume for the image to be back projected onto the screen. Space is typically a scarce resource in the classroom – and the school at large – and only limited space can be dedicated to a system such as this that will receive only intermittent use. The lack of space meant that (a) the room-scale system had to be deployed in the assembly room rather than a normal classroom and (b) the system could not be left in place, as that space was regularly used for other activities.

Activities within the school were routinely managed in terms of configuring the space. Children were asked to fetch the materials associated with the activities (e.g. paints and easels), configure the room to allow the activity to take place (often moving furniture) and to then pack everything away at the end of the activity. This routine allowed classroom spaces to meet the needs of different activities. Essentially, the room was the home of the class and children were seldom moved from class to class.

This routine of configuring the class to meet the activities available represented our first major challenge. Essentially the spaces were not sufficiently dedicated to a single activity so that we could build devices into those spaces. For example, in Classroom2000 [1] lecture rooms are equipped with devices dedicated to the activity of lecturing. In our classrooms children would move from lecture style teaching, to painting to drama within the one classroom space. Consequently the system had to be unpacked, set-up and packed away every time we ran a school session. For a typical 30 to 60 minute retelling session the researchers had to have at least 60 minutes to unpack and set-up and 30 minutes to pack up and leave the assembly room empty. The typical operations performed during the set-up of the retelling configuration were: (a) unpack the components in the available space, (b) place the different parts of the set-up in different pre-determined physical locations (to position the various elements relative to one another), (c) connect all the input devices, output devices, networking and power cables, (d) power up all of the computers and the network (e) start all applications. (f) Check that everything was working.

The most complex part of the set up was making all the cable connections, especially connecting and bringing up the network that connected all the computers. One computer had to be dedicated solely to KidPad. The other two ran the Magic Carpet and video tracking system respectively, and communicated over the network to the computer running KidPad. Running the Magic Carpet and the video-tracking as applications on different machines was a necessary requirement. Distributing the input and output among several machines allowed us to increase the responsiveness of KidPad while children interacted with it. However, this version of the system could not cope with any part of the system (KidPad, the Magic Carpet, or video tracking application) disconnecting or restarting during the session. So if any application or machine had to be restarted or disconnected from the network then *all* of the applications had to be restarted (in the correct order).

The pressures on space are likely to be a generic issue for this class of system. In fact, space is routinely overloaded in most environments, with spaces often being reconfigured to meet particular demands [13]. In order to fit within a real world setting – where space needs to be shared across a broad range of uses – the technologies making up interactive environments will need to make efficient (and appropriate) use of the available space. This might be realized in a number of ways including:

- Reducing the physical size of devices and minimising the number of devices required to support an activity (especially "support" devices such as PCs that are "behind the scenes").

- Reducing the complexity and presence of wiring where possible by shifting to wireless devices and technologies. It should be noted that despite the maturing of various technologies a fully wireless environment is unlikely. Power leads alone accounted for over 12 cable runs in our environment.

- Where possible exploiting existing in-situ devices that are already integrated with the normal working environment. For example, in the classroom setting ceiling-mounted projector(s) and multi-purpose screens might be available.

Developing devices and software that support the widest possible range of useful applications within that particular environment. It should be noted that there is a potential conflict between flexibility and simplicity – a device able to support more activities may well require a more complicated interface and more complex patterns of configuration and use.

3.2 Finding Time

The second challenge was the lack of 'slack' time within the teaching day. An extreme example of this challenge was the requirement that the complete system is set up and taken down before and after each use (which was caused in turn by the use of space, noted above). The pressure of teaching meant that this activity could not be undertaken by the teachers in addition to their normal responsibilities. The lack of slack time also impacts what one might expect to be the more mundane running of the system. For example, on some occasions the computers had to be moved around the room after they were set up in order to adapt the system to the unique conditions of a particular session such as the number of children, special requirements from the teacher, or the presence of visitors. Because of the way the system was implemented this often meant that the whole system had to be restarted in very little time. Similarly, the teaching day is very busy, and this technology might be used for just one part of one lesson. There is no time to spare for system configuration, trouble-shooting or the other administrative and maintenance activities that might be expected within the laboratory environment. In the school setting the researcher might easily find themselves frantically attempting to locate the system's problems while a teacher and a class full of children were waiting on the other side of the screen.

The general problem of the added work required to make technology work has been reported elsewhere [4]. A number of researchers cite this invisible work as a problem for the acceptance and use of systems. Addressing this within a school setting means that the technologies should make minimal demands for time in and of themselves, so that ideally all of the available time can be devoted to the activity being supported, rather than to the technology being used. This might be addressed in a number of ways:

- *Fast and simple physical assembly*, e.g. any peripheral might be connected to any machine via any connector and cable that fits, or even no assembly, e.g. using wireless communication.

- *Flexible software*, e.g. attempting to respond meaningfully to novel device configurations (such as moving a peripheral from one system to another) by providing – possibly limited – use of the new device, or by clear feedback and guidance to the user (e.g. that the device cannot be used in this way).

- *Robust and self-healing software*, with appropriate diagnostics and user guidance, e.g. guiding the user in locating and fixing a loss of network connectivity.

- *Timely and appropriate feedback of system status and activity*, allowing users to manage their confidence in the system, and to structure and pace system's use.

As well as directly addressing the issue of time requirements, many of these strategies also address issues of system complexity (see below). This is because one of the main negative impacts of complexity on the final activity is the time required to manage that complexity (as we have found in our school sessions to date). Consequently strategies that reduce the effective complexity of the system could (in many cases) also make it faster to use.

3.3 Configuring and Maintaining the System

The third challenge was the sheer complexity of the system, in terms of the number and diversity of physical devices in use, and in terms of the specific requirements for configuring and using them. In physical terms each device had to be (a) identified (b) positioned (c) matching cables identified (d) appropriate sockets identified and secured with the right cables and (e) powered up at the right time. The software had to be (a) located on the machine (b) configured if necessary (c) started correctly and in the right order and (d) verified. If anything went wrong then the user would have to attempt to locate and rectify the problem, and resume the start up process. Many of these tasks require skills and knowledge beyond those that may be expected from those who are not IT professionals.

For long term use and integration of the technology in the curricular activities of the school the whole system would have to be used and maintained by the teachers and students themselves. While they have some experience with information technology, many of the devices and paradigms used (e.g. distributed computation) are outside of their experience and expertise. In particular, the complex requirements for setting up the system required the help of the researchers, and so the system could not be integrated into normal classroom use.

The problem of maintenance and management of complex technological infrastructures – inherent within pervasive environments – has already led a number of researchers to suggest that we need to broaden our consideration of 'use' to embrace the area of reconfiguration [6]. Any complete system must be usable on a day-to-day basis by the teachers and children themselves, and they are not typically experts in the elements of computer science or tangible technologies embodied by the technologies used. Rather they are teachers and children who are interested in meeting the goals of education rather than technological reconfiguration. To some extent the techniques that address the pressures of time, above, might also allow the broadest possible range of users to make effective use of the system and its components. In addition to this we should also consider how the system might support people in the day to day running of the system:

-   *Minimal user configuration* of the hardware and software. For non-trivial applications and environments we do not believe that it is realistic to expect all configuration to be automatic. In particular, support is needed for the addition of new devices and applications.

-   *Persistent and sharable configuration:* where configuration is required it should be possible to re-use it as much as possible, e.g. between sessions and between users. The mechanisms for persistent and sharable configuration could allow for technology specialist(s) (whether they be teachers, students or part of an external support organisation) to make full configurations available to the rest of the system's users, and to share these configurations more widely (e.g. via email).

- *Improved Management Facilities:* existing environments have simply not focused on management mechanisms for users. Any real world environment requires some form of access control, authentication, auditing and rollback to manage the total process of customisation and configuration with shared resources and multiple levels of user expertise. For example, a particular school might restrict (potentially error prone) configuration activities to certain skilled users, and the students are likely to have only limited access to the underlying components.

In short, in the laboratory and for the developers, the system might be regarded as "sufficient" and "complete", especially because the developers find it easy and natural to relate to the system in terms of technology. The system is also located in a dedicated and stable research space where once set up the technology seldom needed to be reconfigured. However the teaching environment considered here addresses the technology only in passing, as one possible means among many to address goals of teaching, learning, involvement and collaboration. Moreover, the technology needs to exist in a crowded space where it needs to compete with multiple simultaneous demands on the space. This shift in context has highlighted these sets of generic strategies that need to be considered when we seek to make ubiquitous computing an everyday aspect of the world we inhabit. The next section describes how we have begun to address some of the issues described, and experiment with some of these strategies within the latest prototype system.

4   The Final Room-Scale System

The previous section has described the challenges that we have encountered taking experimental technologies into a working classroom environment and also suggested a range of strategies that might address these challenges. Of necessity, we have had to focus our ongoing activities on exploring just a few of these areas. In particular, we have focussed on software-related issues of configuration, failure and feedback. Our final prototype large-scale system seeks to improve on the original as follows: (a) reduced configuration demands by reducing the pre-allocation of software, machines and networks (b) improving failure recovery by allowing elements of the system to fail or be stopped and subsequently restarted independently, maintaining a working system (c) timely feedback of system state and activities. In addition, the framework introduced in the new prototype needs to be lightweight. It can make minimal assumptions about network connectivity outside the room and should not require a significant investment by teachers in managing the infrastructure.

4.1 Problems with the Original System

The original room-scale implementation had some specific shortcomings that arose from its simple Client/Server networking model. These constrained the system's operation as follows.

- The server(s) had to be present on the network and available before the client(s) were started (in our case the server was KidPad).
- Each client had to know the location of the server(s) that it was to connect to (in our case the clients were the Magic Carpet and Barcode reader interfaces).
- Once the client and server were connected this connection could not be broken until the communication between the server and the client was finished.

The simple client/server approach used was not flexible enough to be able to adapt quickly to variations in the physical environment where the system had to be deployed. In particular, the system's inability to recover from any single element being restarted or disconnected made reconfiguration slower and more prone to difficulties (all of the software had to be restarted).

The simple client/server model also required that each client be pre-configured to locate the server. This meant that the software could only be used with specific machines playing specific roles: using the system on other machines – for example for multiple sessions or in the event of a machine failure – would mean changing system configurations, which was a relatively time consuming and error-prone task that assumed expert users.

4.2 Supporting a Localized Dynamic Infrastructure

The fundamental change between the approach used in the original room-scale system and the new version may be regarded as a shift in programming and operational paradigm. The original system was essentially a one-off fixed-configuration lab-oriented prototype. It could be adapted to new contexts, but only through low-level intervention (configuring scripts and/or changing programs).

The new system, in contrast, reconsidered (and re-engineered) the system as a loosely coupled localized network of services (many of which represent tangible devices), relying on robust dynamic discovery for self-configuration and failure recovery. By exploiting the fact that these devices are physically co-located we can focus on the development of a lightweight set of services that support a dynamic infrastructure that is localised to the room set-up.

When building this second system we considered a number of potential distribution platforms, including Java RMI and Jini, CORBA, DCOM, and Web Services. RMI, CORBA and DCOM are designed around a more tightly coupled client-server paradigm than we wanted to use (after the problems noted above). They also do not directly address the goals of reduced configuration and failure recovery. DCOM is also Microsoft-specific, not supported on other platforms such as Linux.

Jini is built on the top of RMI, and provides a framework to deploy services and allows services to discover and participate in certain activities. Although Jini seemed

ideal, the functionality achieved requires, in practice, a complex set up and has a very
high overhead. A minimal Jini environment needed a significant number of generic
servers to be maintained. Jini assumes that services such as lookup are always on and
can be discovered on the network. However, in practice within the school each installation required the reconstruction of the network, exposing the users to complex service management interfaces. Furthermore, by considering discovery to be localized to
the particular assembly of devices making up the room-scale environment we can
significantly reduce the complexity of management involved.

Web services are the natural evolution of the Client/Server model of the World
Wide Web. In essence a web service is an application that exposes its functionality
and can be accessed using an URL. Applications access those services using standard
GET and POST HTTP actions. These actions have the advantage that they are both
human and machine-readable.

We chose to implement extensions to our second system using a combination of a
lightweight multicast-based service discovery (implemented in Java), with a web
services approach to service access (although using our own plain text for data transfer, rather than provided by XML/SOAP). We feel that this approach provides a simple yet powerful paradigm for network interoperability, portable across many platforms, and with loose coupling of components. Re-implementing the services ourselves (rather than directly using Jini or SOAP) allowed us to trade flexibility for
compactness and efficiency.

4.3 Implementation

We integrated the original room-scale system with a new set of components that ex-
panded the types of media that could be incorporated into KidPad (to be published).
Among these new components, we developed devices that could take still pictures
from a web camera and another one that could scan images using a scanner.[2] We have
used these new devices to gain experience with the new design outlined above.

We extended KidPad to act as a web services client, to interface to the new camera
and scanner components. These new components, the WebCamera and WebScanner,
interface to our resource discovery system, and expose their services via well-defined
HTTP interactions. Each new device implements a basic web server with scripting
capabilities. There is also a multicast packet broadcaster that announces itself to the
network. Each component maintains a list of known components (the "host list") that
is updated automatically as multicast announcements are detected on the network. In
this way, the number and location of services on the network is discovered dynamically. A device that does not announce itself for a period of time is removed from the
list. An application that is interesting in using a particular device can query that device
– via a standard URL – to obtain detailed information about the device and its capabilities. It can then further interact with the device via device-specific URLs.

---

[2] The new system included other components that integrated audio capture and PDA drawing.

4.4 Sample Use

The WebCamera and KidPad both run the multicast announcer and host list software. So, once both are running, they will discover each other on the network. When a child presses the button on the camera the WebCamera component makes a HTTP GET request to a specific URL based on the discovered network address of the KidPad application; this notifies KidPad that a picture is being taken. KidPad responds with its own request to the WebCamera component, again an HTTP GET request, that returns a new image from the digital camera. KidPad then integrates this new image (typically a GIF file) into the current story. Figure 2 shows the web camera in use in the new system.

Fig. 2. Using the web camera in the revised room-scale system

This example indicates several advantages of this new implementation: (a)KidPad and the WebCamera can be run on any machine(s) on the network without (re) configuration. (b) KidPad and the WebCamera can be started in any order and stopped and re-started independently and the interaction will still work correctly once they are both running again. (c) Any number of WebCamera components can be added to the network and they will all be able to contribute images to KidPad.

We have also given consideration to issues of user feedback, pacing and confidence monitoring in this system that are also illustrated by reference to the above interaction example. Grabbing a digital picture can take a few seconds, while scanning a digital image on our flat bed scanner can take about 40 seconds. This is a considerable length of time in terms of user interaction, and specific feedback and monitoring is required. Our revised implementation uses two levels of feedback: local (generated by the components around their physical periphery as result of children's interaction) and global (generated by KidPad to represent the status and progress of the individual components when they are active).

5 Discussion

As our project ends – of which this work is part – we have had to rely on 4 small-scale trials of the final system. None the less, we have gained some useful experience and insights in testing this revised architecture, and have provoked ongoing reflections. We summarise these reflections in this section, indicating areas for future thought.

First, the move to a lightweight multicast approach has certainly eased system configuration, at least for simple deployments. It is refreshing to be able to 'just start' a camera or scanner component on a new machine and have it work with no additional effort. It is also a relief to be able to start and stop these components without having to worry about restarting other parts of the system, and to be able to start services in any order, or add them dynamically to an ongoing session.

Second, we have found the web services model (i.e. remote operations as HTTP requests) to be simple and robust, but relatively slow, at least for our Java-based implementations. For example it is adequate for inserting a scanned image into KidPad, but would not be fast enough to operate as a remote interface to additional mice or other highly interactive components. For these latency-sensitive interactions a finer-grained and higher-performance method of communication is required (currently the system uses native TCP sockets).

Third, we have found multiple levels of feedback – both locally and globally – to be very useful in maintaining a timely understanding of the system's state and activity, especially when a high number of devices and users are involved. This highlights the potential need in pervasive computing environments to present to users the relationship between the global (user-oriented) activity and the local devices and activities that comprise it. For example, how do I know which 'nearby' printer is printing my document, and how do I find it?

Fourth, while simple dynamic discovery has been adequate for a simple configuration-free system there are problems that will arise with more complicated configurations. For example, suppose two teachers are using KidPad at the same time and a child takes a digital photo; which KidPad application does it go to? Both, one, the other or neither? In non-trivial contexts there is clearly a need for configuration and customisation in some form, e.g. dynamically associating the camera with a particular instance of KidPad, or associating a particular child (with some form of tagging that the camera can read) with a particular instance of KidPad. However, the difficulty that arises as we seek to generalise from our simple lightweight approach is the eventual need for users to understand and manage a network-based infrastructure. Certainly, a balance needs to be found between the generic "always-on" philosophy of systems such as Jini and the more targeted but much lower overhead approach of developing localised discovery based systems.

6 Conclusions

This paper has reported the issues involved in moving a storytelling environment out of a research lab and into a real world school. Migrating the environment from a technologically rich research lab to a school environment provided an illuminating insight into how the common vision of ubiquitous and pervasive computing might be realised in practice. This transition has led us to reflect on the general issues surrounding the development of ubiquitous computing environments.

Space is scarce. In the real world and dedicating space is very difficult. Space is routinely overloaded and continually reconfigured to meet demands over time. Having built an environment inside a dedicated research lab space we were then faced with the problems of making it exist in an environment of ongoing continual reconfiguration. The cost of building and taking down the environment was such that it would prohibit its widespread use. Existing considerations of ubiquitous environments seldom consider how the environment itself is configured and amended by its inhabitants.

All objects in a space can be rearranged with the artefacts within a space reconfigured to meet particular needs. Some of the devices in our environment proved difficult to rearrange and the cost of connecting devices together was high. Making devices easier to physically rearrange proved problematic particularly with power leads and cabling runs. Existing considerations of movement of devices in ubiquitous computing tends to focus on interactive devices carried by users. The general assumption is that sensors and larger scale devices are embedded into the environment. However, we found that even "fixed" devices such as sensors, floor pressure pads and displays where routinely moved in order to reconfigure the space.

Spaces are dedicated places. While it is obvious to say that we built our initial environments in a research lab, the fact that this was a place where new technology was constructed became significant when the environment moved to another place. Schools are places where children are taught and the inhabitants wish to teach rather than manage a surrounding electronic environment. The cost of management of the environment was simply very high with those inhabiting it were unwilling to accept the additional overhead of managing a potentially complex system infrastructure.

Spaces have a legacy. Changes to the physical fabric of the buildings such as schools take place over a long timescale and are balanced against a variety of demands. The slow evolutionary nature of real world spaces presents a real problem when placed alongside the revolutionary vision of ubiquitous computing. Simply put, the continual availability of certain information and services assumed by most ubiquitous environments does not exist and may never be universal. To address these issues required us to reconsider the development of our environment in practice In particular we found that the needs of our local environment did not match current infrastructure assumptions and impacted interactions with the story telling environment. The removal of the protection provided by standard interfaces suddenly made the infrastructure much more visible and the exposure of the infrastructure taught us some significant lessons.

Software services are currently not ubiquitous but need to be locally available. Current supporting infrastructures focus on developing scalable platforms. However interaction is locally contained within the space and often needs to be managed locally. Consequently we found ourselves building a self-contained discovery mechanism that focused on lightweight local availability within the space.

The environment needs to be managed as well as used. We had focused on the demands of the children and teachers as users. However, before the environment could be used we needed to reduce the management costs significantly by developing dynamic reconfiguration facilities into the infrastructure.

The environment needs to make dynamic actions visible. The complexity of the interactions involved in the environment meant that we had to extend the feedback provided by the environment to make the dynamic effects more readily available. This means that we need to consider carefully how we expose the behaviour of the environment in order to allow it to be readily interpretable by its inhabitants. We are currently developing a lightweight self-contained platform for data distribution and component-based application building that further builds upon our lessons called EQUIP [16]. EQUIP is a platform that allows ad-hoc interactions and notification across a number of locally available components without relying on extensive network services. The platform has been designed by trailing within the real world and has already been used within a town centre square and within a museum set up.

References

1. Abowd G., Mynatt , E. D.; Charting past, present, and future research in ubiquitous computing; *ACM Trans. Comput.-Hum. Interact.* 7, 1 (Mar. 2000), Pages 29 – 58.
2. Abowd., G.,D, Classroom 2000: An Experiment with the Instrumentation of a Living Educational Environment, IBM Systems Journal, Special issue on Pervasive Computing, Volume 38, Number 4, pp. 508-530, October 1999.
3. Alborzi, H., et al. (2000)., Designing StoryRooms: Interactive Storytelling Spaces for Children. In Proceedings of Designing Interactive Systems (DIS 2000) ACM Press, pp. 95-104.
4. Bowers, J., The work to make the network work, in Proceedings of CSCW'94, ACM Press, Chapel Hill, North Carolina, 22-26 October, 1994, pp 287-298.
5. Button, G. & Harper, R. (1996) The relevance of 'work-practice' to design, CSCW: The Journal of Collaborative Computing, 4 (4), 263-280.
6. Dourish, P., Adler, A. Bellotti, V. and Hendersen, A. Your Place or Mine? Learning from Long-term Use of Video Communication." Computer Supported Cooperative Work: An International Journal, July 1996.
7. Druin, A., & Perlin, K. (1994). I., Immersive Environments: A Physical Approach to the Computer Interface. In Proceedings of Extended Abstracts of Human Factors in Computing Systems (CHI 94) ACM Press, pp. 325-326.
8. Druin, A., Stewart, J., Proft, D., Bederson, B. B., & Hollan, J. D. (1997), KidPad: A Design Collaboration Between Children, Technologists, and Educators. In Proceedings of (CHI'97 ACM Press, pp. 463-470. http://www.kidpad.org.
9. Ishii, H., and B. Ullmer. Tangible bits: towards seamless interfaces between people, bits and atoms. In CHI'97, pages 234–241, 1997.

10. Kristoffersen, S., Ljungberg, F., Making place to make IT work: empirical explorations of HCI for mobile CSCW; Proceedings of Group'99: ACM SIGGROUP conference on Supporting group work , 1999, Pages 276–285

11. Moran, T.P., van Melle, W., & Chiu, P. Tailorable domain objects as meeting tools for an electronic whiteboard. Proceedings of CSCW'98. New York: ACM Press.

12. Myers, B.A, Stiel, H., Gargiulo; R., Collaboration using multiple PDAs connected to a PC; Proceedings of CSCW'98. New York: ACM Press.

13. O'Brien, J., Rodden, T., Rouncefield, M., Hughes, J. "At home with the technology: an ethnographic study of a set-top-box trial" ACM Trans. Computer Human. Interaction. 6, 3 (Sep. 1999), Pages 282–308.

14. Pinhanez, C. S. Intille, J.W.D et al. Physically interactive story environments, IBM System Journals, Vol. 39, Nos. 3 & 4 - MIT Media Laboratory (2000).

15. Stanton, D. Bayon, V. et al (2001) In CHI 2001, Vol. 3 ACM Press, Seattle, 482-489.

16. The Equip Platform. Download page http://www.crg.cs.nott.ac.uk/~cmg/Equator/.

17. Streitz, N.A et al "i-LAND: an interactive landscape for creativity and innovation" , Proceeding of CHI 99, Pages 120–127, ACM Press.

18. Weiser, M., "The Computer for the Twenty-First Century," Scientific American, pp. 94-10, September 1991.

Pervasive Web Access via Public Communication Walls

Alois Ferscha and Simon Vogl

Department of Computer Science
Johannes Kepler University of Linz
{ferscha,vogl}@soft.uni-linz.ac.at

Abstract. Multi-user communication and interaction via public displays together the pervasive and seamless access to the WWW in public areas via mobile phones or handheld devices is enabled via the WebWall system. A software framework for the operation of WebWalls has been developed, strictly separating WebWall access technologies (like HTTP, email, SMS, WAP, EMS, MMS or even simple paging protocols found on mobile phones) from the physical display technologies used and the presentation logic involved. The architecture integrates ubiquitous wireless networks (GSM, IEEE802.11b), allowing a vast community of mobile users to access the WWW via public communication displays in an ad-hoc mode. A centralized backend infra-structure hosting content posted by users in a display independent format has been developed together with rendering engines exploiting the particular features of the respective physical output devices installed in public areas like airports, trainstations, public buildings, lecture halls, fun and leisure centres and even car navigation systems. A variety of different modular service classes has been developed to support the posting or pulling of WWW media elements ranging from simple sticky notes, opinion polls, auctions, image and video galleries to mobile phone controlled web browsing.

1. Introduction

The growing availability of wireless communication technologies in the wide, local and personal area, together with the pervasive use of mobile and embedded computing devices gives strong raise for WWW services adapted to context, particularly to the person, time and location of their use. The seamless provision of services to anyone (personalized services) at any place (location based services) and at any time will presumably fertilize – besides the "desktop-WWW" – a qualitative growth of the Web towards an "embedded WWW", enabled by wirelessly networked autonomous special purpose computing devices (i.e. Internet appliances). Applications and WWW services will have to be greatly based on the notion of context and knowledge, will have to cope with highly dynamic environments and changing resources, and will need to evolve towards a more implicit and proactive interaction with users [3,16], and will have to be accessible in a more ad-hoc fashion – not only in privacy (from the desktop), but also in the public (via shared display artifacts). We therefore explore the fact that visual displays have played an important role in individual WWW usage, but very little research has been conducted to explore the potential of large, shared visual displays for group and community communication

F. Mattern and M. Naghshineh (Eds.): Pervasive 2002, LNCS 2414, pp. 84-97, 2002.

and interaction. While the use of visual displays and desktop projections is getting quite popular in group work settings (shared whiteboards, smartboards, etc.), their use in public spaces to allow for a ubiquitous WWW access for a broad, loosely related, non-determined and unstructured audience is only rudimentarily understood today.

With this work we address the potentials of ad-hoc communication in public spaces using a wall metaphor. We have developed a software framework, the WebWall framework, providing a seamless WWW access over visual displays in public spaces via a manifold of access technologies including HTTP and email, but most importantly SMS and WAP. The WebWall framework is presented as a means to enrich public places with digital communication and interaction means for people to access their personal 'multimedia memories', to share information (e.g. notes, videos, pictures) with others or to interact with others (e.g. opinion polls, auctions, games) – all over the WWW and possibly all over mobile devices like mobile phones or Internet appliances.

2. The WebWall Framework Architecture

The design principle of the WebWall framework appears to be independence with respect to potential access technologies, display technologies, and configurability and dynamicity of interactive services. The software architecture hence isolates a request handling component on the input side, a data management component in a backend system, and a presentation component on the output side. As for the input side, the integration of Internet- and mobile networking technologies demands for flexible and standardized access to a WebWall system which is granted via the representation of access requests in a standardized format, irrespective of the media (SMS, WAP, email, and HTTP). The strict separation of request handling and display rendering provides extensibility by means of being able to integrate new technologies as they evolve. As for the physical presentation of WebWalls, various display technologies exist today (projectors, plasma screens, CRT, etc.), and further technologies will evolve (laser projection). The WebWall system therefore is designed to provide flexible support for the full range of exisiting and upcoming display technologies. A presentation module is responsible for arranging the service instances on a physical screen according to service type and priority. Services are provided as instances of services classes with dedicated functionality. A so called "rendering engine" for each service class is responsible for translating the requested WebWall data into a displayable form, for example a HTML page of a given size.

Users interact with the objects (i.e. service instances) on a WebWall by passing messages and/or commands through one of the access modules. The current implementation of a WebWall provides GSM, IEEE802.11b WLAN as well as standard LAN access to receive requests, which are then passed on to the service access module that is responsible for translating the text into requests to specific service instances or classes. Personal preferences, login information as well as pre-defined objects are managed by the backend system.

Users may create service instances not only by direct interaction with a WebWall, but also by accessing the backend system via a web-interface. This way, many service classes – like picture *galleries* or personal *videos* – can be costumized and saved for display on a WebWall at any later time. Besides the user related data, the backend system hosts the code for the service classes and the renderer classes that are downloaded to Java-based clients whenever they start up. It also handles configuration sets for individual clients that define the services that should run, as well as the display areas where individual instances may appear on a visual display. This central storage of configuration sets and class code enables application providers to implement new service classes and distribute them to a defined set of clients.

WebWalls support a range of service classes that differ in presentation as well as in functionality (see Figure 1). The most basic service is the one for posting notes (service class *Note*) to a WebWall that can be viewed by everyone in the spatial proximity of the (public) display. Replys to a note may be sent to a WebWall, which, depending on the reply mode, either display on the WebWall or are routed invisibly to the author of the referred note. After a defined lifetime, notes are removed from the WebWall. While notes may be posted instantly when viewing a WebWall, there are other service classes that are better defined first over the Web-client: *Video* and picture *galleries* (service class *Gallery*) can be used to display multimedia content by composing URLs of the media to display and save them under a userdefined names. *Polls* may be used to solicit the public opinion on local issues that may arise in the geographical vicinity of a WebWall, for example. Polls display an up-to-date view of the current collective opinion, thus providing an effective means for instant democracy. To allow for ad-hoc buy&sell applications and commercial advertising the framework provides two service classes: *Auction* and *Banner*. Banners work analogously to their WWW counterparts, but could be used to send vouchers to the interested reader upon request. Auctions lets users bid for an item on sale, with the highest bid being on display on the WebWall.

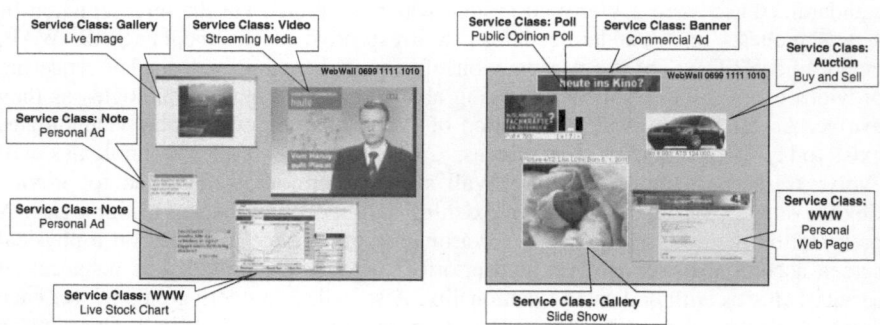

Fig. 1. WebWall and service classes

The software architecture of the WebWall framework can be separated into four major entities (Figure 2, grey rectangles) and two interfaces (Figure 2, red rectangles). The *Request Generator* (RG) module on the access side, and the *Backend System* (BS), the *Community Management System* (CMS) and the *Show Module* (SM) on the processing and display side.

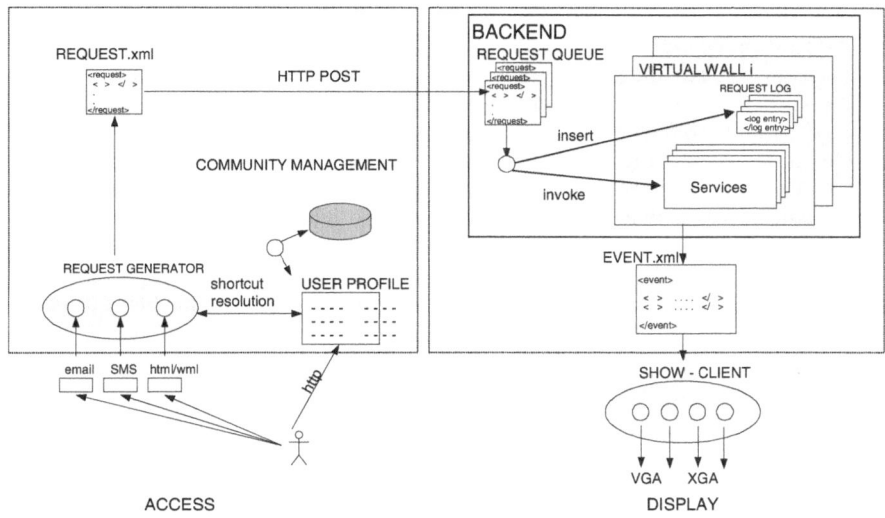

Fig. 2. WebWall software architecture overview

The *Request Generator* module accepts incoming requests from different media types (currently SMS, E-mail and HTTP) and prepares those requests for further processing. The RG is responsible for transforming each request into its XML representation (request.xml) which is the input for the BS. The RG logs all incoming requests, compares them with profiles managed by the CMS and adds data from the CMS where necessary (profile, resolves shortcuts, etc). A *request.xml* file is the interface between the RG and the BS. It is created by the RG and holds information of a specific request of a specific service class. request.xml is being passed from the RG to the BS via a HTTP-Post-command. In the BS the data of this file will be transformed into a JDOM-Object for further processing [15]. The *Backend-System* processes the request.xml files generated by the RG. It is the most complex module of the whole WebWall-system. It manages so called *Virtual WebWalls* (VWWs) and permanently stores the state of each VWW. It is responsible for creating, scheduling (including a waitlist which also can be displayed on the WebWall if needed) and positioning all service-class-instances, generating HTML-content for each visual representation of an instance and creating the file event.xml which contains display-information for the SM placed on a specific WebWall client. The BS can be configured via a WWW-Interface. The *event.xml* file is being created by the BS each time a SC is due to be displayed on a WebWall. It contains data which is needed by the SM to successfully display a service class. Like request.xml this file is sent to the SM via a HTTP-Post-command, which means that the BS contacts the SM every time a event.xml is generated and transmits that file. The *Show Module* component is responsible for displaying the results of BS processing. It receives standardized event.xml documents which are passed over from the BS via HTTP/POST. As this passing happens in a "push" oriented manner, the SM has to provide rudimentary HTTP server functionality, in order to listen to incoming requests from BS. As the content of the event.xml is classified by MIME types, the SM has to provide functionality in order to interpret the corresponding data formats. The current prototype is based on the "text/html" MIME type as a visualization format for service instances [13]. The SM provides the functionality to interpret and render the html encoded content passed over with the event.xml, creates browser (IE) instances containing the HTML-content rendered by the BS, and displays the WebWall service within an IE-instance (neither contain menu-bars, task-bars nor status-bars, only the common browser-window is being displayed). The *Community Management System* is implemented as a WWW-interface for users and administrators of the system for customisation and administration. Customisable features include the definition of user profiles, shortcuts (representation of longer strings or even whole services for easier handling via SMS), styles (colour and text properties), picture-upload and instant or scheduled posting.

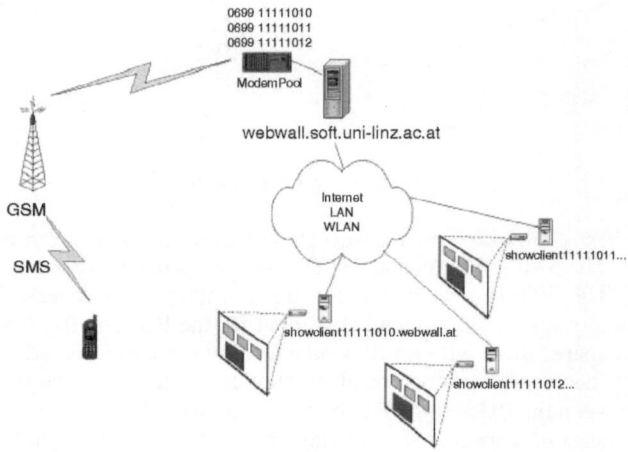

Fig. 3. Virtual WebWalls in a distributed multiple display setup

The WebWall system can either be distributed on physically different servers/clients, or be set up on one single machine to satisfy different applications of this service (for example a system managing many WebWalls used in Underground stations). All modules communicate with each other over HTTP gateways and pass data in XML format (request.xml and event.xml). As an example for a distributed WebWall setup, Figure 3 shows a configuration of a geographically dislocated WebWall system. It illustrates the possibilty to operate multiple WebWalls (Show clients), which are distributed over the Internet (LAN, wireless LAN, etc.). All clients are managed by one server with the WebWall-Backend component installed. The example depicts the centralized request reception via SMS with a modem pool attached to the WebWall server. Three individual *Virtual Web Walls* are maintained and their content is displayed at possibly three different physical displays.

3. Visual Components and Styles

A variety of different visual components have been created for the individual service classes, some of which are displayed below. For the video service class (Figure 4, upper left) the streaming video is displayed in the main frame. The service instance id is placed in the upper right header and can be used to stop, replay or removie the video. The gallery class (Figure 4, upper right) overlays image by image out of a collection of objects in img MIME type from the CMS. The auction class (Figure 4, lower left) displays an image and description text of an entity upon which an auction is set up in the public. New bids are posted by referring to the instance id, and once registered by the WebWall overwrite the current bid tag. An opionion poll like e.g. the evaluation of presentation by the audience accepts votes for one of the displayed alternatives (Figure 4, lower right), counts the votes, computes percentiles and displays the information in real time.

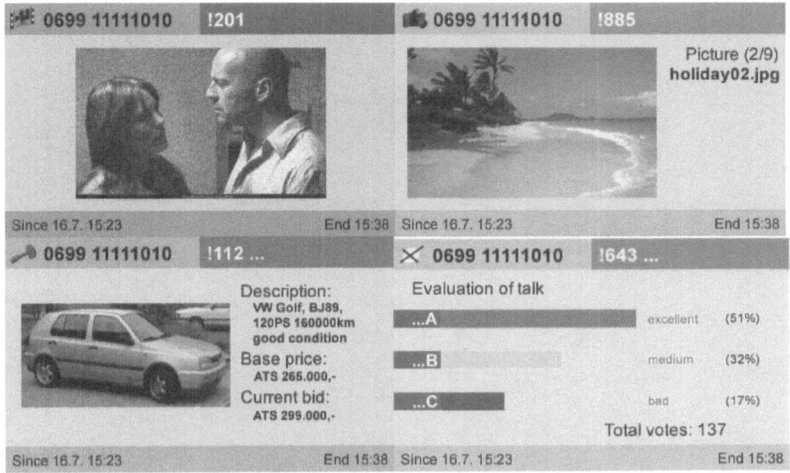

Fig. 4. Service class visuals: video, gallery, auction, poll

4. Scenario: Posting a Note to a WebWall via SMS

To illustrate the operation of the WebWall system, let us consider the creation of an instance of the service class "Note" on a WebWall identified by the MSISDN "+436991001010". Assume a person wants to post the text "Hello WebWall!" in the style "blue" with a GSM mobile phone via sending an SMS message (Figure 5). It is assumed that at the moment the WebWall does not contain any information elements (i.e. no request has already been posted to WebWall), that the requesting user is identified by the MSISDN 436645229250, and that the request to post a note "Hello WebWall!" to the WebWall has been issued at the time 10:29. In the first step the RG translates the incoming SMS information into a request document that looks like:

```xml
<?xml version="1.0" encoding="UTF-8"?>
<request id ="1" timestamp="9897668787" >
<user authType="sms" id="436645229250"/>
<service command="new" class="note" screen_id="436991001010">
        <parameter name="text" value="Hello WebWall!"/>
      <parameter name="style" value="blue"/>
      <parameter name="lifetime" value="30000"/>
</service>
</request>
```

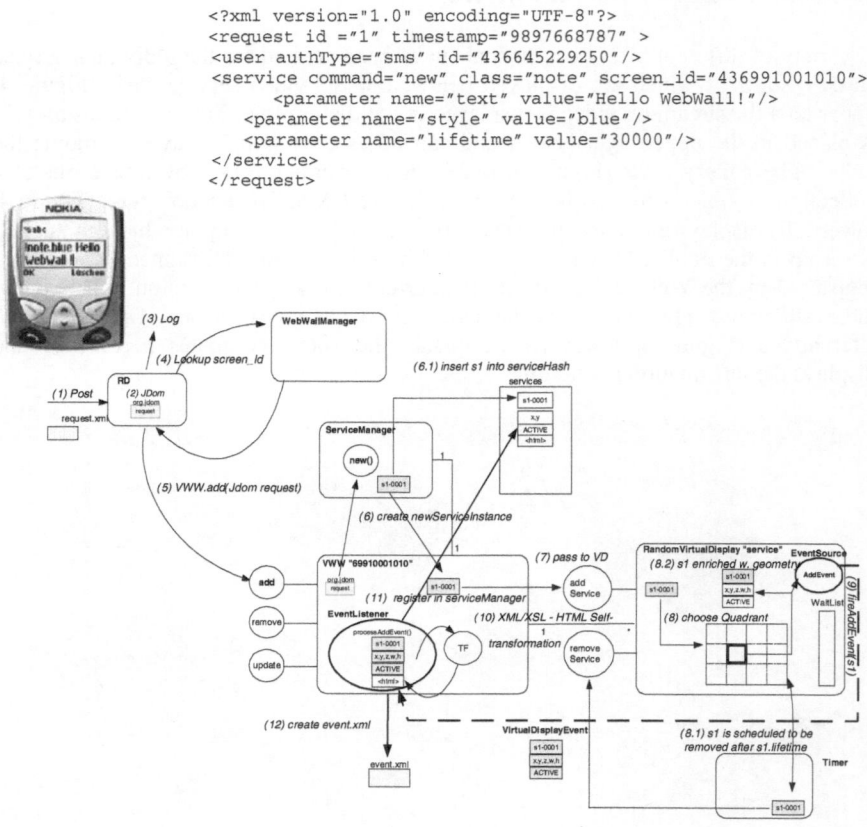

Fig. 5. SMS to WebWall 436991001010

To accomplish the request, the BS executes the following steps:

1. As a first step, after the SMS message has been received, parsed and transformed into a request file, request.xml is submitted via HTTP/POST from the RG to the WebWall BS, where the RequestDispatcher servlet receives (1) the incoming request as a *java.io.InputStream*. The RequestDispatcher creates an *org.jdom.Document* representation of the request based on the incoming string (2), and logs the request (3) appropriately.
2. With the information stored in the JDom Object (i.e. the *screen_id* attribute and the command attribute) the RequestDispatcher first checks whether the specified screen_id matches a VirtualWebWall instance within the Web-Wall Manager (4). If it exists, it delegates further processing to the specified VirtualWebWall instance. In this case it is the VirtualWebWall which is identified by the key "+436991001010". As the command attribute of the request specifies that a "new" instance of a service class note should be created, the appropriate public

add(org.jdom.Document) method of this VirtualWebWall instance is invoked (5) with the org.jdom.request object as parameter.

3. The VWW receives the *org.jdom.Document(request)* and asks its ServiceManager to create a new instance of a service class. According to the class attribute of the service Element, the ServiceManager calls its internal newNoteInstance method to create an instance of a Note (6). Assumed that this is the first request to the WebWall Backend, the initial value of RNum is "0001" (6.1). This number is the identifier of the just created Note instance, and is used as key for the storing in the ServiceManagers Hashtable (services) for later reference. The service instance is inserted into this Hashtable (6.2).

4. The Note instance is returned to the VWW(7.1). The VWW passes the Note to its default VirtualDisplay(VD) (7.2) where the layout/geometry allocation is done: As the VD is empty, a random Quadrant is chosen (8.1) and the Note is being enriched with the appropriate geometrical data (8.2) and the status field of the Note is set to "ACTIVE" describing that the ser-vice is actually being displayed on the WebWall. Also time relevant fields of the Note (ActivationDate, ExpirationDate) are set accordingly. The Note instance is scheduled within the VDs internal Timer according to its lifetime attribute. In this case a lifetime of 5 minutes (300000 milliseconds) is as-sumed as the default lifetime of the service class Note. After expiration of this lifetime, an automatical call of removeService will be fulfilled.

5. The VD invokes its fireAddEvent method, in order to indicate that a VirtualDisplayAddEvent has occured (9). The listening (VirtualDisplayListener) VirtualWebWall is notified that a service instance has been placed on the VirtualDisplay as the VD invokes the processVDAddEvent of the VWWs VirtualDisplayListener (10).

6. Based on the event that a change in the VD has occured, the VirtualDisplayListener invokes the process of rendering a HTML representation of the Note instance. First a XML representation of the service instance is generated (11.1), which in turn is the basis for the XSL transformation into the HTML representation (11.2). Subsequently, the invokation of the to-EventXml method of the service leads into the generation and pushing of the event.xml to the associated waiting Show module, which in turn uses the event.xml to display the Note instance on the physical projection display. In this scenario the corresponding event.xml looks like (HTML code omitted):

```
<?xml version="1.0" encoding="UTF-8"?>
<event requestId="1"
   id="1000456158960"
   serviceId="436991001010-0001"
   class="note"
   command="new">
    <geometry x="30" y="480" z="1" w="200" h="150"/>
    <content type="text/html"><![CDATA[]]></content>
</event>
```

The positioning within the WebWall projection display depends on the associated geometry data. In this case the Note is positioned at the coordinates (x=30,y=480) with a dimension of (width=200,height=150), and z-order=1 within the display of the WebWall.

5. Related Work

The use of the wall metaphor as a means to enable "shared artifact interaction" among humans is not new. Indeed is wall computing research closely related with artifact research enabling collaboration and awareness [10,11]. At least for awareness solutions for workgroups quite a few wall metaphor based approaches have been proposed in the literature. We briefly discuss the main efforts and relate the WebWall approach to them.

The *Notification Collage* (NC) [8,11,12,23], developed at University of Calgary, Canada, is a groupware system that provides ways to post media elements onto a real-time collaborative surface (display). It is inspired by the metaphor of a bulletin board containing a collage of randomly positioned and possibly overlapping visual elements like sticky notes, video (used for providing virtual "presence" through Webcams), slideshows, or activity indicators to provide a means of indicating the amount of activity at a persons site through collected (proximity sensoring) movement data in the appropriate users room. The emphasis of the Notification Collage lies in the consideration of support of interpersonal awareness and interaction within small communities of colleagues. It supports both co-located and distributed members of a group in the means of providing functionality to allow all group members to post, notify other users about, and use shared resources within the context of a Notification Collage. The Notification Collage is based on conventional PC and Internet technologies and incorporates MS Windows as operating system in order to provide broad support for various multimedia formats. Display technology incorporates both standard monitors for workstation oriented dual screen setups, and large projection displays for public applications.

DATA Wall is a project at MIT Media Lab [18], which aims to overcome limitations given by conventional display technologies. The DATA Wall has a resolution of 2048x3840 pixels and provides a seamless, gapless fullmotion ultra high resolution projection display. It can be used in either front or rear projection, direct or folded optics mode. The main scope of the DATA Wall is to provide a projection facility with variable and configurable displays. It therefore can be used as a physical design component for various projection applications.

LIDS (Large Interactive Display Surfaces) is a project of the University of Waikato; New Zealand [1]. The LIDS project explores applications of large display surfaces for teaching and distance learning, meeting support, and personal information management. For teaching, and particularly distance learning, the focus lies on the ability to efficiently record, retrieve and disseminate lectures, seminars and tutorials, with almost no additional effort on behalf of the teacher. Similar technology is applied to the support of meetings and tutorials conducted over multiple sites, and this technology will be adapted to support both informal and formal face-to-face discussions and meetings. The system consists of a rear projected glass screen with standard data projector and a Mimio whiteboard digitizer.

The University of Minnesota in collaboration with Silicon Graphics Inc., Ciprico Inc. and IBM Storage Products Division have developed *PowerWall*, a high performance, high resolution visualization system [21]. The purpose of the PowerWall is to visualize and display high-resolution data from large scientific simulations. In addition to this high resolution, the PowerWall provides a large 1.8 x 2.4 m display area to facilitate collaboration of small groups of researchers using the same data set. All the collaborators can see the display clearly and without obstruction, and the rear-projection technology makes it possible to walk up to the display and point to features of interest, just as one would do while discussing work at a blackboard. The PowerWall is a single 1.8m x 2.4 screen driven by a 2x2 matrix of video projectors. A successor project called *InfinityWall* has been portrayed in [5], providing a 2048x1536 pixel stereoscopic display for large audiences.

The *HoloWall* [19,20] is a wall-sized computer display that allows users to interact without special pointing devices but by gesture recognition. A rear-projection setup is used in combination with an infrared light source and an IR-camera that films the back side of the display. Since the rear-projection panel is semi-transparent, the user's hand or any other objec in front of the screen reflects IR light and thus becomes visible to the camera, if it is close enough (somewhere between 0 and 30 cm). Image processing algorithms are used to track the shapes and cause interaction with the displayed contents.

i-LAND [10,26] integrates several so-called roomware components into a combination of real, physical as well as virtual, digital work environments for creative teams. By roomware, computer-augmented objects in rooms are considered, like furniture, doors, walls, and others. The current realization of i-LAND covers an interactive electronic wall (*DynaWall*), an interactive table (InteracTable), computer-enhanced chairs (CommChair), and the Passage mechanism. The objective of the DynaWall is to represent a computer-based device that serves the need for being able to interact with virtual content in an intuitive way, relying on standard gestures known from the interaction with physical objects in the real/paper world. It is possible that information objects can be taken at one position and put somewhere else on the display or thrown from side to the opposite side. These features are realized by an advanced interaction mechanism based on the penguin concept [9]. The current realization uses three rear projection electronic whiteboards (SMART Boards [25]) with a total display size of 4.5m width and 1.1m height and a resolution of 3072x768 pixels.

The *Stanford Interactive Workspaces* (SIW) project (also called i-Room) [7] is exploring the integration of high-resolution wall-mounted and tabletop displays (Interactive Murals [14], Interactive Tables), as well as personal mobile computing devices such as laptops and PDAs connected through a wireless LAN. Specialized input and output devices such as LCD-tablets, laser pointer trackers, microphone arrays and pan-and-tilt cameras are also present in the environment. The Interactive Mural is a large, high-resolution, tiled display, constructed using 8 projectors connected either to a SGI dual-pipe IR or a cluster of 8 myrinet-connected PCs with NVIDIA graphics cards. A scalable graphics library has been designed and implented that provides a single virtual display abstraction to the programmer, regardless of physical display properties (multiple overlapping projectors, multiple independent graphics accelerators and multiple processors). The framework supports multiple people, and pointing devices in interactive spaces including dynamic configuration and deals with failure, removal, addition and reconfiguration. Another research focus lies in interaction styles and assoiated toolkits that are appropriate for large displays, multiple devices, and multiple users.

Using the Stanford Interactive Mural, Davis and Chen [6] present new input methods for people collaborating one this shared display area. They use laser-pointers as input devices and are able to discriminate between several simlutanouos input gestures to enable a natural interaction. A similar interaction technique is used at the Fraunhofer institute [27].

The *Scalable Display Wall* (SDW) at Princeton University is another exponent of large-screen tiled display architectures, delivering 6000x3000 pixels via 24 aligned projectors, driven by an array of desktop computers. Their research efforts focus on frameworks for clustered displays, especially performance and scalability issues [4] for high performance data visualization.

The Table 1 summarizes wall computing projects and compares them with the WebWall approach. Most of the research projects presented deal with problems in display technology or CSCW-related issues, especially with closely coupled multiple users (workteams, interest groups) and appropriate interaction metaphors [2].

Table 1. Comparison of research work in wall computing

	NC	DATA Wall	LIDS	POWER Wall	HOLO Wall	DYNA Wall	SIW	SDW	Web Wall
Application area	CSCW	visualiza-tion	CSCW / CSCL	scientific visualiza-tion	ubiquitous computing	CSCW	CSCW	scientific visualiza-tion	public communica-tion, CSCW
Display	Conventio-nal monitor, rear projection device	proprietary projection device	rear projection screen, whiteboard digitizer	rear projection (2x2 matrix) video projectors	rear projection (IR cut filter)	rear projection whiteboards (SMART board)	rear projection board, tiled video projectors	tiled video projectors	independent of projection system, video projectors light emitting displays
Access	direct	passive	direct	direct	direct	direct	direct	direct	direct
Interface	Windows GUI	none	pen like interaction devices	touch screen	IR LED emitter, IR filtering camera with image processing techniques	touch screen, gesture techniques	LCD tablets, laser pointer trackers, microphone arrays, pan and tilt cameras	trackers	HTTP WAP SMS EMS MMS
Internet linkage	yes	no	no	no	no	no	no	no	yes any MIMEtype
Scalability	high	single system	medium	single system	single system	single system	medium	single system	high
Extendi-bility	high	none	none	none	none	none	medium	none	high
Status	under developmt.	idle	prototype	idle	research	commercial	under developmt.	under developmt.	fully functional

The Notification Collage is more related to the WebWall system than other wall computing projects, as it supports different media types that can be shared in a distributed setup. In contrast to WebWalls, it is closely coupled to a specific operating system for its providing its services, and lacks the possibility to access it via wireless phones (GSM) or other mobile devices – keyboard access is necessary to interact with the system. Furthermore, contents looks different on every screen, as users may arrange items at will, whereas a WebWall may be exported to different clients resulting in the same view on the data. The DataWall focuses on questions of abstracting logical from physical displays to construct larger interactive areas, and does not take into account any networking aspects, but could be used as an ouput medium for WebWalls. The POWER Wall as well as the InfinityWall deals only with high-performance data visualization problems and local multi-user interaction, networking is not taken into account. Likewise, the Scalable Display Wall focuses on clustered rendering of 3D content and uses networking only to distribute internal data sets. LIDS uses a whiteboard metaphor for user interaction so users need to have physical access to the wall and a pointing device, while WebWalls can be used from anywhere with a mobile phone or the Internet. Similarily, DYNA Wall allows access to networked data but needs direct interaction with the physical device, with binds the user to a specific location. SIW supports different pointing, input and output devices, but makes also heavy use of the room metaphor – only taking into consideration objects that are in a room. HOLOWall is a singular system that explores an alternative input technology and does not deal with networking, different service classes or other multi-user considerations.

There exist several notification services that transport information from the Internet or other data sources to mobile phones [28], like the various info services of GSM network operators. Another example of Internet/phone integration is iValet [17]. It informs users of incoming emails and lets the user react to individual mails. These examples are a strict one to one type of communication, there is no ability to share information with others or even publically.

Current research efforts can therefore be summarized as concentrating on three major areas: *Display technology* research covers advanced uses of projection systems - often in combination with cameras for system feedback - to provide seamless output of multiple beamers on arbitrary surfaces, even deliberately integrating physical objects into the digital realm [22]. Several architectures for the configuration, calibration and transparent access of a multi-display Wall have been proposed [4,14]. Projects focusing on *groupware* issues deal with the interaction of a known group of users on a shared display, using a variety of input devices. These efforts deal with the cooperative manipulation of artifacts on a shared display, dealing with privacy [24]. The size of the displays creates new problems for *human computer interaction,* as normal keyboard and mouse input becomes impractical (if not impossible). Therefore, new input devices have been proposed like laser-based pointing devices [6,27].

The WebWall project, in contrast, makes use of a variety of dislocated displays to enable the ad-hoc communication and interaction of people with on another as with Internet-originated artifacts.It makes use of large displays as one possible output technology, but does not limit itself to this presentation medium. Instead it can be adapted to a wide variety of interfaces.

6. Conlusions

In this paper we have presented a framework architecture enabling multi-modal access to multimedia information sources over wireless as well as fixed networks for the purpose of communication and interaction in the public – employing the wall metaphor. By separating data from access and display technology, WebWalls provide an open, flexible, extensible architecture that offers instant access to Internet information sources on an ad-hoc basis. The access to information and the direct interaction among possibly anonymous users in public spaces is novel and unique. Furthermore, the WebWall framework extends GSM network services from simple synchronous voice-streams to an interactive WWW services with advanced multimedia and broadcast capabilities.

Fig. 7. WebWall location scenarios

Possible locations for the use of WebWalls as considered by GSM network operators are envisioned in Figure 7: in public waiting areas for general public communication and access to multimedia information; for large assemblies as public opinion polls or for democratic interaction with digital systems (such as light shows), mobile WebWalls for location-based services enhanced by its notification capabilities, or as a source of up to date event information.

Acknowledgements. This work was supported by Connect Austria under grant WebWall. The valuable design and implementation contributions of Gerold Kathan, Maxl Miesbauer, Josef Blüml, and Daniel Simma are acknowledged.

References

1.   M. Apperley, M. Masoodian: "Supporting Collaboration and Engagement using a Whiteboard-like Display." Techn. Report, Department of Computer Science, University of Waikato, Hamilton, New Zealand, 2000.
       http://www.edgelab.sfu.ca/CSCW/cscw_masoodian.pdf
2.   W. Buxton, J. R. Cooperstock, S. S. Fels, K. C. Smith: "Reactive Environments: Throwing Away Your Keyboard and Mouse." Techn. Report, Input Research Group, Computer Science, University of Toronto, 1998.
3.   M. Coen: "The Future of Human-Computer Interaction or How I learned to stop worrying and love my Intelligent Room." Techn. Report, MIT Artifcial Intelligence Lab., 1999.
4.   H. Chen, G. Wallace, A. Gupta, K. Li, et.al., "Experiences with Scalability of Display Walls", To appear in Immersive Projection Technology Symposium (IPT), March 2002.

5. M. Czernuszenko, D. Pape, D. Sandin, T. DeFanti, et.al., "The ImmersaDesk and Infinity Wall Projection-Based Virtual Reality Displays", Computer Graphics, Vol 31, No 2, 05/01/97-05/01/97, pp. 46-49, 1997.
6. J. Davis, X. Chen, "LumiPoint: Multi-User Laser-Based Interaction on Large Tiled Displays", to appear in Displays, Mar 2002.
7. A. Fox, P. Hanrahan, B. Johanson, T. Winograd: "Integrating Information Appliances into an Interactive Workspace." Techn. Report, Stanford University, 2000.
8. S. Greenberg, M. Boyle and J. LaBerge: PDAs and Shared Public Displays: Making Personal Information Public, and Public Information Personal. *Personal Technologies*, Vol.3, No.1, March 1999.
9. J. Geissler: „Shue, throw or take it! Working efficiently with an interactive wall." Techn. Report, German National Research Center for Information Technology, Darmstadt, Germany, 1999.
10. J. Geissler, T.Holmer, S. Konomi, C. Müller-Tomfelde, W. Reischl, P.Rexroth, P. Seitz, R. Steinmetz, N. A. Streitz: „i-LAND: An interactive Landscape for Creativity and Innovation." Proceedings of the ACM Conference on Human Factors in Computing Systems, pp. 120 – 127. ACM Press, Pittsburgh, Pennsylvania, USA, Mai 1999.
11. S. Greenberg, M. Rounding: "Using the Notification Collage for Casual Interaction." Techn. Report, Department of Computer Science, University of Calgary, Calgary, Alberta, Canada, 2000.
12. S. Greenberg, M. Rounding: "The Notification Collage: Posting Information to Public and Personal Displays." Techn. Report, Department of Computer Science, University of Calgary, 2001.
13. M. Grand: "MIME Overview". 1993. http://www.mindspring.com/~mgrand/mime.html
14. G. Humphreys, P. Hanrahan, "A Distributed Graphics System for Large Tiled Displays", Proceedings of IEEE Visualization 99, San Fransisco, CA, October 24-29, 1999.
15. S. L. Jones: "An Introduction to JDOM." XML Journal, 7. Sep. 2001.
16. U. Leonhardt: "Ubiquitous Location-Awareness." Techn. Report, Imperial College UK, 1998.
17. S. Macskassy, A. Dayanik, H. Hirsh, "Information Valets for Intelligent Information Access", to appear in AAAI Spring Symposia Series on Adaptive User Interfaces, 2000.
18. R. E. McGrath, M. D. Mickunas: "An Object-Oriented Framework for Smart Spaces."
19. M. Nobuyuki, J. Rekimoto: "HoloWall: Designing a Finger, Hand, Body, and Object Sensitive Wall." Proceedings of UIST'97, Sony Computer Science Laboratory Inc., 1997.
20. M. Nobuyuki, J. Rekimoto: "Perceptual Surfaces: Towards a Human and Object Sensi-tive Interactive Display." Tech. Report, Sony Computer Science Laboratory Inc., 1997.
21 POWER WALL: http://www.lcse.umn.edu/research/powerwall/powerwall.html
22. R. Raskar, G. Welch, M. Cutts et.al., "The Office of the Future: A Unified Approach to Image-Based Modeling and Spatially Immersive Displays", ACM SIGGRAPH 1998, Orlando FL, 1998.
23. J. Rekimoto,"A multiple device approach for supporting whiteboard-based interactions". CHI '98 Conference Proceedings, p. 344 -351, 1998.
24. Garth B. D. Shoemaker and Kori Inkpen, "Single display privacyware: augmenting public displays with private information", CHI 2001, pp 522-529, 2001.
25. SMART Technologies, http://www.smarttech.com/products/smartboard/
26. P. Tandler, "Software Infrastructure for Ubiquitous Computing Environments: Supporting Synchronous Collaboration with Heterogeneous Devices", In: Proceedings of UbiComp 2001: Ubiquitous Computing. Heidelberg: Springer LNCS 2201, pp. 96-115, 2001.
27. M. Wissen, " Implementation of a Laser-based Interaction Technique for Projection Screens", ERCIM News No.46, July 2001.
28. T. Woo, T. La Porta, „*Pigeon: A Wireless Two-Way Messaging System* ", IEEE Journal on Selected Areas in Communications, Vol.15, No. 8, 1997.

Efficient Object Identification with Passive RFID Tags

Harald Vogt

Department of Computer Science
Swiss Federal Institute of Technology (ETH)
8092 Zürich, Switzerland
vogt@inf.ethz.ch

Abstract. Radio frequency identification systems with passive tags are power-ful tools for object identification. However, if multiple tags are to be identified simultaneously, messages from the tags can collide and cancel each other out. Therefore, multiple read cycles have to be performed in order to achieve a high recognition rate. For a typical stochastic anti-collision scheme, we show how to determine the optimal number of read cycles to perform under a given assurance level determining the acceptable rate of missed tags. This yields an efficient pro-cedure for object identification. We also present results on the performance of an implementation.

1 Introduction

Identification is a central concept in user-oriented and ubiquitous computing. Human users are usually identified (and authenticated) by passwords or biometric data. Most applications require some kind of identification in order to deliver personalized infor-mation or restrict access to sensitive data and procedures. Object identification, on the other hand, is most useful for applications such as asset tracking (e.g. libraries, animals), automated inventory and stock-keeping, toll collecting, and similar tasks where physical objects are involved and the gap between the physical and the "virtual" world must be bridged. In a world of ubiquitous computing, unobtrusive object identification enables the seamless connection between real-world artifacts and their virtual representations.

Reliable identification of multiple objects is especially challenging if many objects are present at the same time. Several technologies are available, but they all have limita-tions. Bar codes are the most pervasive technology used today, but reading them requires a line of sight between the reader device and the tag, and often doesn't work without human intervention. Other visual recognition techniques that identify shape, color, or size, may not be able to identify single instances, but only object classes.

Radio frequency identification (RFID) promises to be an unobtrusive, practical, cheap, yet flexible technology for identification of individual instances. There is a wide variety of products and technologies available; the book [1] provides a good overview.

Research efforts are under way to develop radio frequency tags that are either small enough to be embedded even into paper in an unobtrusive way, or cheap enough to be attached to large quantities of inexpensive goods. The μ-chip by Hitachi [14] is an example of a tiny RFID chip that can be worked into thin materials; sample applications would be paper-based document management and additional security features for bank

F. Mattern and M. Naghshineh (Eds.): Pervasive 2002, LNCS 2414, pp. 98–113, 2002.

notes. Another research project, at MIT's Auto-ID Center, aims at developing a very cheap RFID chip, primarily for enhancing supply chain management processes.[1]

In most applications today, typically a single RFID tag is recognized at a time. In electronic article surveillance, for example, it is sufficient to recognize only one unpaid item in order to take appropriate measures. In many other applications, objects are presented sequentially to the reader device, e.g. on a conveyor belt, thus making it unnecessary to recognize more than one item at a time. This allows for very fast object identification.

The ability to recognize many tags simultaneously is crucial for more advanced applications, however. As examples, consider laundry services, warehouses, or the supermarket checkout. We have implemented two applications that use multiple tag identification. The first one is a monitor for card games where cards put on a table are identified in order to keep track of the game's course [10]. The second application is the "RFID Chef"[2], a kitchen assistant that recognizes food items, e.g. in a shopping bag, and makes suggestions about possible dishes involving these items [6]. It is designed to also take the abilities of the cook into consideration when the cook identifies himself with his own RFID tag. Many RFID tags are presented simultaneously to the reader device in both applications and it is crucial to reliably identify all of them.

We have used a commercially available RFID system, "I-Code" by Philips Semiconductors[3]. This system provides the feature of scanning multiple tags simultaneously employing a stochastic anti-collision scheme. (There are other products that advertise multiple tag identification, e.g. TIRIS by Texas Instruments[4].) The communications protocol between the reader and the tags uses a scheme similar to slotted Aloha where slots are provided for the tags to send their messages. Due to physical constraints, tags are unaware of each other and thus, collisions may occur when multiple tags use the same slot for sending. Since tags choose their slots randomly, collisions may be resolved in subsequent read iterations, and after a number of iterations, identification data from all tags can be retrieved.

In this paper we show how to identify a set of tags in such a system if their number is not known in advance. Multiple read cycles are performed until all tags are identified with a given level of assurance. The number of present tags is estimated in each step and the reading parameters are adjusted accordingly. This yields an efficient and reliable procedure for object identification.

The next section summarizes and discusses features of the I-Code system. Section 3 reviews the mathematical tools used for the analysis of the system and the design of the identification procedure. Section 4 presents the results of the analysis, which are exploited in Section 5 that describes an adaptive technique for multiple tag identification and presents experimental results. The remaining sections deal with related work and present a summary.

---

[1] http://www.autoidcenter.org/
[2] http://www.inf.ethz.ch/vs/res/proj/rfidchef/
[3] http://www.semiconductors.philips.com/markets/identification/products/icode/
[4] http://www.ti.com/tiris/

2   The I-Code RFID System

A working configuration of an I-Code system consists of the following parts. A so-called "reader" unit is attached to the serial interface of a host (usually a PC), and an antenna is attached to the reader unit by a coaxial cable. The reader unit controls the power that is transmitted by the antenna and encodes the data exchanged between the host and the tags. The programming interface of the reader consists of a set of commands that are described below.

Communication and power transmission between the reader and the tags takes place by inductive coupling between the coil of the reader and the coils of the tags. The channel from the reader to the tags is suitable only for broadcasting and normally all tags within reading range will answer requests from the reader. There are, however, means to address specific tags by "muting" all others. On the other hand, messages sent by tags will only reach the reader device; tags are not aware of each other and cannot exchange messages directly.

The sizes of the coils (of both the reader and the tags) determine the range in which communication can take place. Mid-range antennas with a diameter of approx. 50 cm, as used in our experimental settings, can recognize tags within approx. 50 cm range. The tags we were using were about 5×8 cm^2, so-called "smart labels" attached to an adhesive foil. The tags consist of a chip attached to a printed coil antenna.

The full technical documentation of the system is available from the Philips Semiconductors Web site. Some aspects are reviewed here since they are important for the rest of the paper.

2.1   Tag Memory

An I-Code tag provides 64 bytes memory that are addressable in blocks of 4 bytes. All blocks can be read from, but writing to some blocks is inhibited, indicated by a set of write protection bits. This prevents changes to the serial number and similar data. The write protection bits themselves cannot be deactivated after activation.

Of the 64 bytes, 46 are available for application data. The rest is reserved for a 8 byte serial number and the following functionality: write protection; one bit for indicating electronic article surveillance; one bit indicating the "quiet" state of the tag. If the latter bit is set, the tag will not engage in communication with the reader unless a "reset quiet bit" procedure is executed.

2.2   Programming Interface

The programmatic interface of the system is provided by the reader device. It comprises commands for setting configuration parameters of the reader device itself, e.g. the speed of the serial connection, and commands for handling communication with tags that are in range. Communication commands include the following:

- *Anti-collision/select (ACS)*. This command causes all tags that are in range to send their serial numbers. Afterwards, these tags become "selected" and keep quiet in following ACS cycles as long as they are in range. After a tag moves out of the field it becomes "unselected". When it comes back again, it re-sends its serial number. This command can be used to detect tags that are in range, since a list of serial numbers is returned. It is also a prerequisite for writing to tags, since the write command affects only selected tags. However, we are not going to use this command since one ACS cycle takes significantly longer than a *Read unselected* command.

– *Write.* This command is used to write data to a number of tags. One data block (4 bytes) can be written to at a time, but multiple tags may be affected. The tags are selected by the time slot (discussed in the next subsection) they have used while the *ACS* command. This requires that tags don't move in and out of range while writing is in progress.

– *Read.* This command causes only "selected" tags to send their data. It is performed after an *ACS* command.

– *Read unselected.* This is similar to *read* but all tags are triggered regardless of their selection status. By specifying the blocks 0 and 1 to be read, this command can be used to read the serial numbers as well. This is our preferred reading command.

2.3 Framed Slotted Aloha Medium Access

The I-Code system employs a variant of slotted Aloha for access to the shared communication medium, known as *framed* Aloha [11]. After the reader has sent its request to the tags, it waits a certain amount of time for their answers. This time frame is divided into a number of slots that can be occupied by tags and used for sending their answers. When multiple tags use the same slot, a collision occurs and data gets lost (Fig. 1). The reader can vary the frame size, e.g. for maximizing throughput; the actual size of a slot is chosen according to the amount of data requested.

A tag reading cycle consists of two steps:

1. Reader $\Lleftarrow$: *I, rnd, N*
2. $T \to_{s_T(N, rnd)}$ Reader: *data$_{T,I}$* for all tags T

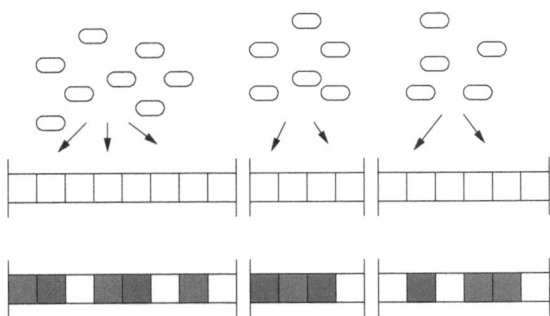

Fig. 1. Tags are randomly allocated to slots within a frame (above). This results in some slots remaining empty, and others containing one or more tags (below). The latter case results in a collision, and no data can be retrieved from these tags

In the first step, the reader device broadcasts a request for data. I denotes what data is requested by specifying an interval of the available 64 bytes of tag memory; $rnd \in [0, 31]$ is a random value whose use is explained below; $N \in \{1, 4, 8, 16, \ldots, 256\}$ is the frame size and denotes the number of available slots for responses.

In the second step, tags that are in the proximity of the antenna respond ($\rightarrow_s$ denotes a tag sending in slot s, $0 \leq s < N$). A tag T uses a tag-specific function s_T to compute its response slot number, using the frame size and the random value as parameters; the random value is supposed to avoid the same collisions occurring repeatedly. However, we found that this function is to some degree indeterministic. Generally, collision patterns will differ even if the same parameters are provided.

For the purpose of analysis, we are not interested in the actual data returned by the tags. We therefore view the result of a read cycle as a triple of numbers $\langle c_0, c_1, c_\kappa \rangle$ that quantify the empty slots, slots filled with one tag, and slots with collisions, respectively.

We will not take into consideration the *capture effect* by which a tag's data may be able to be recognized by the antenna despite of a collision. The capture effect is quite common if tags are placed close to each other. This means practically that data, which would normally be lost due to the occurring collision, can be read, and thus the system performance rises. However, it seems that whenever such a "weak" collision between the same two tags occurs, one of them always "wins" and the data from the other one is lost. Therefore, the influence of the capture effect is only minimal and seems not to have great impact on the performance.

3   Mathematical Preliminaries

This section reviews some mathematical tools we will use in subsequent sections. The number of slots in a time frame available for tag messages is called "frame size" and will be denoted by N. The number of tags is usually denoted by n.

3.1   Occupancy Problems

The allocation of tags to slots within a time frame belongs to a class of problems that are known as occupancy problems, which are well-studied in the literature [4,5] and widely applied [8,9]. These problems deal with the random allocation of balls to a number of bins where one is, e.g., interested in the number of filled bins. In the following, we will speak of "tags" and "slots" instead of "balls" and "bins".

Given N slots and n tags, the number r of tags in one slot is binomially distributed with parameters n and $\frac{1}{N}$:

$$B_{n,\frac{1}{N}}(r) = \binom{n}{r} \left(\frac{1}{N}\right)^r \left(1 - \frac{1}{N}\right)^{n-r}. \tag{1}$$

The number r of tags in a particular slot is called the *occupancy number* of the slot. The distribution (1) applies to all N slots, thus the expected value of the number of slots with occupancy number r is given by $a_r^{N,n}$ (see also [4, p. 114]):

$$a_r^{N,n} = N B_{n,\frac{1}{N}}(r) = N \binom{n}{r} \left(\frac{1}{N}\right)^r \left(1 - \frac{1}{N}\right)^{n-r}. \tag{2}$$

Let us denote by μ_r the random variable that equals the number of slots being filled with exactly r tags, $r = 0, 1, 2, \dots, n$. The distribution of μ_r depends on the probabilities

$$P(\mu_r = m_r) = \frac{\binom{N}{m_r} \prod_{k=0}^{m_r-1} \binom{n-kr}{r} \; G(N - m_r, n - rm_r)}{N^n}, \tag{3}$$

where

$$G(M, m) = M^m + \sum_{k=1}^{\lfloor \frac{m}{r} \rfloor} \left\{ (-1)^k \prod_{j=0}^{k-1} \left\{ \binom{m - jr}{r} (M - j) \right\} (M - k)^{m-kr} \frac{1}{k!} \right\}. \tag{4}$$

The rationale behind these formulas is the following. Imagine a matrix ν_{ij} with n rows (one for each tag) and N columns (one for each slot). Each allocation of tags to slots corresponds to such a matrix where $\nu_{ij} = 1$ if tag i falls into slot j, and $\nu_{ij} = 0$ otherwise; there are N^n such matrices.

The matrices we are interested in represent allocations where there are exactly m_r slots with r tags in each of them. These are the matrices for which the following condition holds. For m_r columns, $\nu_j = r$ holds, and for the remaining $N - m_r$ columns, $\nu_j \neq r$. There are $\binom{N}{m_r}$ ways of arranging these m_r columns. Each of the m_r columns defines a group of r indistinguishable rows. The first group can be arranged in $\binom{n}{r}$ ways, the second group must be drawn from the remaining columns, etc.

The remaining columns and rows can be arranged in $(N - m_r)^{n-rm_r}$ ways, but we have to be careful not to count the arrangements that include allocations of exactly r tags into a slot, i.e. containing columns for which $\nu_j = r$. Function G computes the number of arrangements we are looking for. It determines the correct value by the principle of inclusion-exclusion. The faculty accounts for the arrangements of the columns for which $\nu_j = r$. From this, the formulas (3) and (4) follow.

3.2   Tag Reading as a Markov Process

Suppose one starts identifying a fixed set of tags, having recognized none yet. In each read cycle, one is able to read data from a number of tags (the ones that don't cancel each other out due to collisions). The probability distribution of this number is given by P, defined above. In each step, a certain fraction of the recognized tags is new, i.e. they haven't been recognized in previous cycles. The number of new tags depends on the number of tags already known.

This process of reading tags can be modeled as a (homogeneous) Markov process $\{X_t\}$, where X_t denotes the number of known (i.e., already identified) tags in step t. The number of tags known in the next step solely depends on the number of known tags in the current step. The discrete, finite state space of the Markov process is $\{0, 1, \dots, n\}$. The transition probabilities are given by

$$q_{ij} = \begin{cases} 0 & \text{if } j < i \\ \sum_{r=0}^{i} P(\mu_1 = r) \frac{\binom{i}{r}}{\binom{n}{r}} & \text{if } j = i \\ \sum_{r=j-i}^{n} P(\mu_1 = r) \frac{\binom{n-i}{j-i}\binom{i}{r-j+i}}{\binom{n}{r}} & \text{if } j > i \end{cases} \tag{5}$$

The first case is simple: we cannot go to a state where less tags are known than before, therefore the probability for such transitions is zero. The second case accounts for the possibility of recognizing only such tags that are already known, i.e. the recognized tags in that step are all drawn from the set of already known tags.

In the third case, we recognize exactly $j-i$ tags we do not already know (and possibly some we already do know). That means, we draw $j-i$ tags from the yet unknown $n-i$ tags (and possibly some from the i already known tags).

We will use the matrix $Q = (q_{ij})$ to compute a lower bound of the number of reading steps necessary to identify all tags with a given probability. The initial distribution $q(0)$ reflects the fact that in the beginning we are sure that no tags are known, and is given by

$$q(0) = (P(X_0 = 0), P(X_0 = 1), \dots , P(X_0 = n)) = (1, 0, \dots , 0) . \qquad (6)$$

3.3 Parameter Estimation

In order to pick the appropriate frame size N for the (a priori unknown) number of tags n in the field, we have to estimate n. Based on the the results of read cycles $c = \langle c_0, c_1, c_\kappa \rangle$ (denoting the number of empty, filled, and collision slots), and the current value of N, we will define functions that compute estimations of n:

$$\mathcal{E} : N, c \mapsto n_\varepsilon . \qquad (7)$$

The expected error ε of an estimation function $\mathcal{E}$ tells us something about the quality of the estimate. We use the following error function, which sums up the weighted errors over all possible outcomes of the read cycle:

$$\varepsilon = \sum_c |\mathcal{E}(N, c) - n| P(\mu = c) . \qquad (8)$$

4   Identification Performance

For applications like a supermarket checkout or the RFID Chef scenario described in the introduction, where the number of tags is not known in advance, it is not clear how many read cycles have to be performed until the tags are identified with sufficient accuracy. If too many cycles are performed, the delay will be high, inducing cost and worsening the user's experience. On the other hand, some tags might be missed if too few cycles are performed. Therefore, an "optimal" value for the number of cycles should be used, minimizing the required time while maintaining high accuracy. This value, however, varies with the frame size N and the actual number of tags n.

If the frame size is small but the number of tags is large, many collisions will occur and the fraction of identified tags will degrade. Therefore, one might choose to use large frames, but then, response time is always high, even if there are only few tags in range. The choice of large frames also poses a problem in highly dynamic applications where tags leave the range of the reader quickly after entering it: tags that enter the field after the initial request has been sent by the reader will not send an answer.

In this section, we show how to compute the parameters (frame size, number of read cycles) for optimal tag identification w.r.t. to the time required to identify all tags under

Table 1. Execution time for *read unselected*. Figures shown are for a 57600 baud connection over RS-232

N slots	1	4	8	16	32	64	128	256
t_N (ms)	56	71	90	128	207	364	676	1304
σ	4.96	2.19	2.26	3.80	4.79	4.82	5.05	4.36
t_N/t_{N-1}	–	1.3	1.3	1.4	1.6	1.8	1.9	1.9

a given assurance level. The assurance level is given as a probability of identifying all present tags.

4.1 Full Tag Set Identification

Due to the stochastic nature of the reading process, we cannot expect to identify all tags with complete certainty, but we can reach for higher assurance if we are willing to perform more read cycles and wait for their completion. We define the *assurance level* as the probability α of identifying all tags in the field. The desired level of assurance depends on the requirements of our application. We will give figures mainly for $\alpha = 0.99$, which allows for one or more tags missing in less than 1 % of all runs. Note that, the number of tags missed, if any, will be typically very small, thus leading to a high overall recognition rate.

In order to compute the time required to achieve a given assurance level, we need to take into consideration the time requirements for single read cycles. Table 1 shows the cycle time t_N for all possible settings of N in the considered RFID system. The values were obtained by performing read cycles for one minute and computing the average consumed time. The variation of t_N is rather low, as the low standard deviation σ shows. Note that t_N is nearly linear in N as we would expect. Note also that t_N depends on the connection speed between the reader device and the host.

For a fixed frame size N, the time T_α required to achieve an assurance level α is given by

$$T_\alpha = s_0 \cdot t_N ,\tag{9}$$

where s_0 is the minimum number of read cycles required to identify all n tags in the field with probability α. s_0 is therefore the minimum value of s for which the following condition holds:

$$Q^s q(0)[n] \geq \alpha .\tag{10}$$

Note that the resulting vector of the product $Q^s q(0)$ contains the probabilities of identifying k tags after s read cycles, $k = 0, 1, \ldots, n$. We choose its nth component and compare it to α.

We can now compute the optimal frame size N for a given number of tags n under a desired assurance level. We have performed this computation for $\alpha = 0.95$ and $\alpha = 0.99$. Figure 2 shows the resulting time requirements for optimal choices of N for up to 80

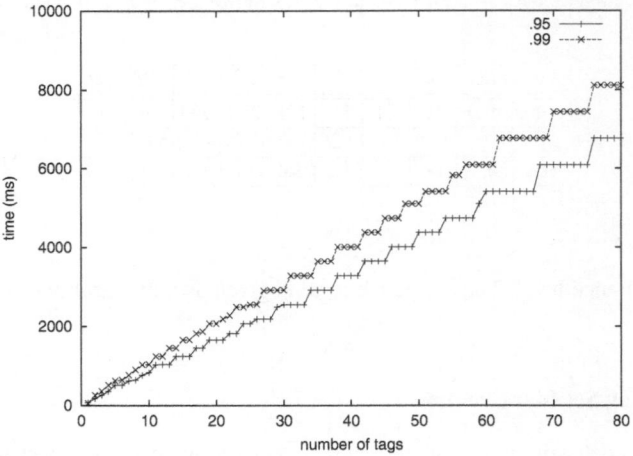

Fig. 2. Time requirement T_α (for optimal N) for $\alpha = 0.95$ and $\alpha = 0.99$

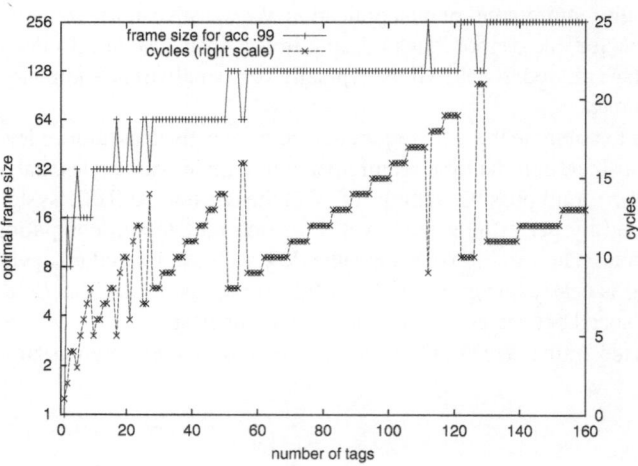

Fig. 3. Optimal frame sizes and numbers of read cycles to perform

tags. What can be seen is that, the time required increases linearly with the number of tags, and it takes approximately 3 seconds to identify a full set of 30 tags with high probability—if the optimal frame size is known, e.g. if n can be estimated correctly.

4.2 Dynamic Slot Allocation

Figure 3 shows optimal frame sizes and the respective numbers of cycles to perform in order to achieve $\alpha = 0.99$ for up to 160 tags. From this graph we can obtain the optimal value for N for a given number of tags n. (The number of cycles multiplied by t_N from Table 1 yields the graph in Figure 2.) However, in practice it is reasonable to assume that n is not known and has to be estimated based on observed read results. For a read result $c = \langle c_0, c_1, c_\kappa \rangle$ (and the current setting of N), we give two estimation functions that yield approximations for n.

The first estimation function is obtained through the observation that a collision involves at least two different tags. Therefore a lower bound on the value of n can be obtained by the simple estimation function $\mathcal{E}_{lb}$, which is defined according to the template (7) as

$$\mathcal{E}_{lb}(N, c_0, c_1, c_\kappa) = c_1 + 2c_\kappa . \tag{11}$$

A different estimation function is obtained as follows. Chebyshev's inequality tells us that the outcome of a random experiment involving a random variable X is most likely somewhere near the expected value of X. Thus, an alternative estimation function uses the distance between the read result c and the expected value vector to determine the value of n for which the distance becomes minimal. We denote this estimation function by $\mathcal{E}_{vd}$; it is defined as

$$\mathcal{E}_{vd}(N, c_0, c_1, c_\kappa) = \min_n \left| \begin{pmatrix} a_0^{N,n} \\ a_1^{N,n} \\ a_{\geq 2}^{N,n} \end{pmatrix} - \begin{pmatrix} c_0 \\ c_1 \\ c_\kappa \end{pmatrix} \right| . \tag{12}$$

In order to be able to assess an estimation function, we would like to know the expected error of the estimate. This error can be computed by applying the error function given in equation (8). Note that for $\mathcal{E}_{lb}$, the error always states an *underestimation* of the real value of n, which is not the case for $\mathcal{E}_{vd}$.

Approximations of the error of $\mathcal{E}_{lb}$ and $\mathcal{E}_{vd}$, obtained by exhaustive search of the space of possible read results, are shown in Figure 4. What can be drawn from the graph is the fact that $\mathcal{E}_{lb}$ is more accurate for low values of n, while $\mathcal{E}_{vd}$ is more steady for a wider range of n.

What is critical about $\mathcal{E}_{lb}$ is the fact that the error starts to exceed the error of $\mathcal{E}_{vd}$ and becomes quite large just in the range of the transition of $N = 32$ to $N = 64$ as the optimal frame size. Since $\mathcal{E}_{lb}$ always yields an underestimation, this makes the transition more likely to be missed for values $n \approx 30$. Thus, for such n, the accuracy will likely decline.

5 Adaptive Tag Reading

5.1 Choosing an Optimal Frame Size

Due to the inaccuracy of the estimation functions and the jitter as shown in Figure 3, we are—to some degree—free to choose the actual frame size for a given estimate. For example, if $n \in [17, 27]$, both 32 and 64 are appropriate choices for N, since both settings yield similar times. The intervals [*low,high*] for which a certain choice of N is applicable are summarized in Table 2. The table lookup can be implemented in a way that is shown in Fig. 5. This implementation avoids jitter in the result by making conservative transitions between interval borders.

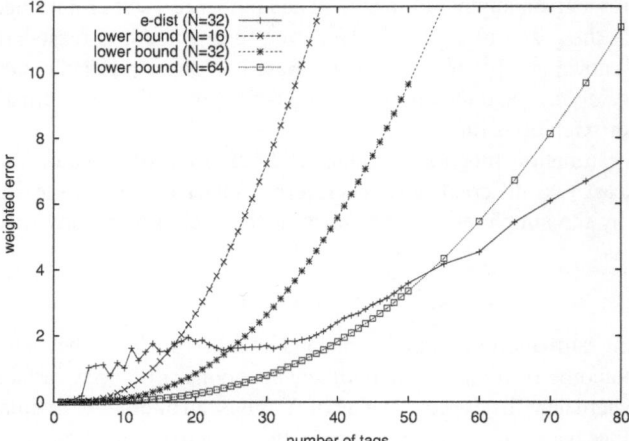

Fig. 4. Error of parameter estimation. The weighted error is the variance of the tag number estimate. $\mathcal{E}_{lb}$ ("lower-bound") is quite accurate for small n but grows fast with larger n, while $\mathcal{E}_{vd}$ ("e-dist") is more steady

Table 2. Optimality intervals for frame sizes

N slots	1	4	8	16	32	64	128	256
low	–	–	–	1	10	17	51	112
high	–	–	–	9	27	56	129	∞

```
int adaptFrameSize(N, n_est) {
    while (n_est < low(I(N))) { N = N/2; }
    while (n_est > high(I(N))) { N = 2*N; }
}
```

Fig. 5. Choosing a frame size

5.2 Continuous and Static Tag Reading

We can differentiate between two basic scenarios for tag identification. One scenario is static, i.e. a set of tags enter the field and stay there until all tags are identified (with high probability). An example is the checkout counter where a shopping bag is put and stays there until the terminal has identified all items. The other scenario is rather dynamic, with tags entering and leaving the field continuously.

In the dynamic case, tag reading proceeds without terminating, and the currently identified tag set is reported continuously. Estimating the number of tags and adapting the frame size is nevertheless necessary in order to maximize the identification rate. However, we will concentrate on the static case where the process of tag reading must come to a halt eventually.

5.3   A Procedure for Static Tag Set Identification

In the static case, the tag reading process is started when the first read result $\langle c_0, c_1, c_\kappa \rangle$ with $c_1 + c_\kappa > 0$ is obtained, i.e. at least one tag has entered the field. The process continues until all tags are identified with assurance level α. Once a good estimate for the number of tags is known, and therefore the optimal frame size N is given (by Table 2), the number of read cycles to perform in order to achieve α can be computed using equation (10).

When the process is started, a value for N has to be chosen, but this starting value will not be optimal in most cases. This also yields a first estimate of n that will be inaccurate and we can only hope that adapting N to the estimate will bring us closer to the true value of n in subsequent read cycles.

Therefore, the first few read cycles will contribute only little information about the tag set. They will only bring us closer to the optimal frame size. Once we have reached the optimal N, we can perform the prescribed number of cycles to achieve the desired level of accuracy. Of course, this introduces a penalty due to the first read cycles consuming time without contributing much.

This idea is captured in the algorithm sketched in Figure 6. The algorithm assumes that the tag set in the field is static. This allows to adapt the frame size N and the estimated number of tags n_est such that these values are monotonically increasing. This not only guarantees termination of the procedure, but also makes the process robust against too low estimates that might occur due to erroneous read cycles.

The starting value for N is set to 16, which is the lowest reasonable value according to Table 2. Note that tags only send their data if they were present in the field at the beginning of a read cycle. Since it is unlikely that the tags enter the field right at the start of a new cycle, the time for the first cycle will be wasted. By choosing a low starting value for N, we minimize this initial idle time. Another reason for this choice is that we are adapting N only by increasing it. Thus, in order to take full advantage of all possible values for N, we have to start with the lowest value.

The variable stepN holds the counter for the cycles performed with the (currently estimated) optimal setting for frame size N. When this counter reaches its maximum value, the procedure terminates. The counter is reset to zero whenever a new estimate of N is made. Since a new estimate is only accepted if it excels the old one, N will eventually reach its maximum and the counter stepN will not be reset anymore. The variable n_est is more volatile than N, but bounded by the actual number n of tags in the field (assuming we employ the estimation function $\mathcal{E}_{lb}$ that always yields a lower bound of n). Thus, termination is guaranteed.

5.4   Experimental Results

We have implemented the tag reading scheme presented here and executed the procedure identifyStatic for tag sets of up to 60 tags. The tags were arranged around the antenna of the reader device in a way that we hoped would optimize field coverage. During the test, the tags were not moved. For each tag set, we carried out 100 runs of the identification procedure (without changing the arrangement in between).

```
identifyStatic() {
    N = 16; n_est = 0; stepN = 0;
    do {
        stepN++;
        c = performReadCycle(N);
        t = estimateTags(N, c);
        if (t > n_est) {
            n_est = t;
            NO = adaptFrameSize(N, n_est);
            if (NO > N) {
                stepN = 0; // restart with new frame size
                N = NO;
            }
        }
    } while (stepN < maxStep(N, n_est));
}
```

Fig. 6. Procedure to adaptively read a static set of tags

The aspired accuracy level of 0.99 (meaning that, out of 100 runs only one run should miss any tags) could not be reached in all tests—the worst level being with a set of 34 tags, where only 92 of the 100 runs yielded the full tag set (see Fig. 7). Not surprisingly, accuracy suffers with increasing tag numbers, due to the fact that it becomes increasingly difficult to arrange a larger number of tags around the antenna while maintaining good coverage. As was mentioned above, we expect the accuracy to drop at around 30 tags due to the increasing error of our estimation function. This would explain the drop that can be actually observed in Figure 7. Another source of inaccuracy are objects located around the testbed (walls, chairs, etc.), which could have a negative influence on the tests. However, the procedure consumed only little time more than would be required if the number of tags were known in advance (see Fig. 8). There are peaks in the zones where the error of the estimate becomes large, which causes good estimates often to be made only after some time has already been spent on a suboptimal frame size. Then, the frame size is adapted, and all cycles are re-performed.

Another figure describing the accuracy of our procedure is the fraction of tags recognized over a large number of runs. In our tests, this figure (not shown graphically) never dropped below 0.99. This means practically that although we might miss some tags in a small number of runs, these misses add up to only a small percentage in the long perspective.

One has to bear in mind that these figures are very sensitive to environmental conditions and the relative positions of the tags to the antenna. We performed our tests in an office environment where we could carefully arrange the tags and were able to somewhat optimize the conditions under which the tests took place. We expect the accuracy that can be reached under real-world conditions to be much lower.

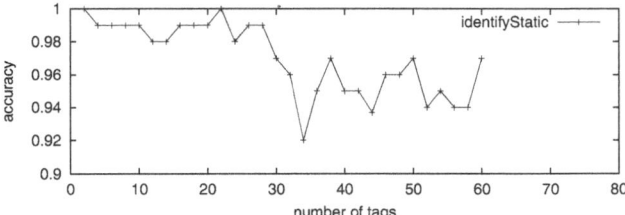

Fig. 7. Accuracy of tag identification in practice. The graph shows the percentage of runs that yielded the full tag set. Not shown is the overall recognition rate (percentage of identified tags over 100 runs), which never dropped below 0.99

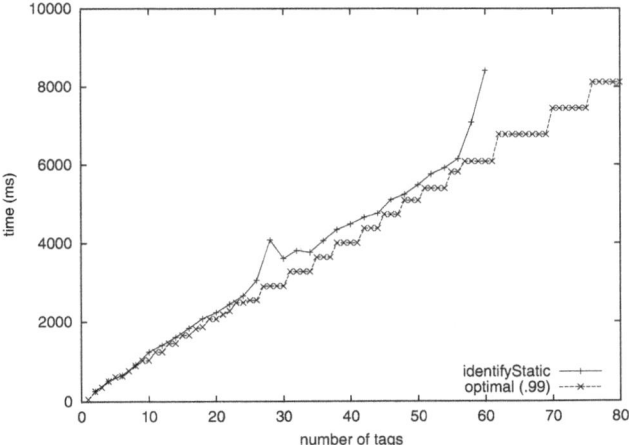

Fig. 8. Actual run time for tag identification. Within the transition areas from one frame size to the next (just below 30 and 60 tags), run time significantly increases. This is due to good estimates arriving late, causing the full number of read cycles to be performed again with a new frame size

6   Related Work

Apart from mundane applications like inventory and cattle tracking, or improving product management, tagging is often used for attaching information to real objects and for building a bridge between them and their virtual counterparts [2,15]. Such approaches are based on the assumption that it is often more appropriate to integrate the world of information into our physical environment instead of moving human beings into virtual worlds. This can help to facilitate access to information by means of familiar objects acting as interfaces.

Aloha is a classical communication protocol that is described and analysed in many introductory textbooks. Framed Aloha was introduced in [11,12]. The underlying model in that work differs from ours in the assumption that nodes are able to detect if a message could be sent successfully and will only re-send it if this was not the case. A procedure for frame size adaptation is given that depends on the outcome of a read cycle and tries to maximize throughput. Frame sizes are, however, not constrained to powers of two. The performance of framed Aloha systems is extensively studied in [16]. The analysis takes also into consideration the capture effect, by which a message is received despite of a collision.

Similar to framed Aloha is slotted Aloha with subchannels (S/V-Aloha), which is introduced by and studied with regard to "stability" in [13]. (The system is defined to be stable if the number of blocked users does not exceed a certain point beyond which throughput decreases.) This work uses a combinatorical system model similar to ours.

Some systems employ deterministic anti-collision schemes, e.g. tree-based protocols [3,7]. Trees are constructed "on the fly" as collisions occur. Each branch corresponds to a partition of the tag set, leafs represent singled-out tags that are identified without a collision. The approaches differ in how the branch of a tag is determined. In [7], ID prefixes determine the partitioning, while in [3], random coin flipping is used. The latter work also considers trees with arity ≥ 2. Both papers investigate how much effort is required for full, reliable tag identification. In contrast to stochastic schemes such as examined in our work, deterministic ones allow for a recognition rate of 100%, but note that this rate is only achievable under optimal conditions; if tags are allowed to enter or leave while the protocol is in progress, accuracy may suffer.

7 Conclusions

We have demonstrated how to efficiently identify a set of RFID tags if the number of tags is not known in advance. We have shown how to determine the parameters for tag reading in order to achieve optimal running time under a given assurance level. The practical implementation we did does not achieve that level, which could be attributed to environmental influences on the experiments. There is also room for improvement of the frame size adaptation in order to eliminate runtime peaks. It remains challenging, however, to find a procedure that would take environmental effects into account in order to adapt to them. We hope that the work done so far helps implementing pervasive computing environments that employ RFID systems.

References

1. Klaus Finkenzeller. *RFID–Handbuch*. Hanser Fachbuch, 1999. Also available in English as *RFID Handbook: Radio-Frequency Identification Fundamentals and Applications*, John Wiley & Sons, 2000.
2. L. E. Holmquist, J. Redström, and P. Ljungstrand. Token-Based Access to Digital Information. In Hans-W. Gellersen, editor, *Handheld and Ubiquitous Computing*, volume 1707 of *LNCS*, pages 234–245. Springer-Verlag, 1999.
3. Don R. Hush and Cliff Wood. Analysis of Tree Algorithms for RFID Arbitration. In *IEEE International Symposium on Information Theory*, pages 107–. IEEE, 1998.
4. Normal Lloyd Johnson and Samuel Kotz. *Urn Models and Their Applications*. Wiley, 1977.

5. Valentin F. Kolchin, Boris A. Svast'yanov, and Vladimir P. Christyakov. *Random Allocations*. V. H. Winston & Sons, 1978.
6. Marc Langheinrich, Friedemann Mattern, Kay Römer, and Harald Vogt. First Steps Towards an Event-Based Infrastructure for Smart Things. Ubiquitous Computing Workshop (PACT 2000), October 2000.
7. Ching Law, Kayi Lee, and Kai-Yeung Siu. Efficient Memoryless Protocol for Tag Identification. In *Proceedings of the 4th International Workshop on Discrete Algorithms and Methods for Mobile Computing and Communications*, pages 75–84. ACM, August 2000.
8. Rajeev Motwani and Prabhakar Raghavan. *Randomized Algorithms*. Cambridge University Press, 1995.
9. Fred S. Roberts. *Applied Combinatorics*. Prentice-Hall, 1984.
10. Kay Römer. Smart Playing Cards – A Ubiquitous Computing Game. Workshop on Designing Ubiquitous Computing Games, Ubicomp, 2001.
11. Frits C. Schoute. Control of ALOHA Signalling in a Mobile Radio Trunking System. In *International Conference on Radio Spectrum Conservation Techniques*, pages 38–42. IEE, 1980.
12. Frits C. Schoute. Dynamic Frame Length ALOHA. *IEEE Transactions on Communications*, COM-31(4):565–568, April 1983.
13. Wojciech Szpankowski. Packet Switching in Multiple Radio Channels: Analysis and Stability of a Random Access System. *Computer Networks: The International Journal of Distributed Informatique*, 7(1):17–26, February 1983.
14. K. Takaragi, M. Usami, R. Imura, R. Itsuki, and T. Satoh. An Ultra Small Individual Recognition Security Chip. *IEEE Micro*, 21(6):43–49, 2001.
15. Roy Want, Kenneth P. Fishkin, Anuj Gujar, and Beverly L. Harrison. Bridging Physical and Virtual Worlds with Electronic Tags. In *Proceeding of the CHI 99 Conference on Human Factors in Computing Systems: the CHI is the Limit*, pages 370–377. ACM Press, 1999.
16. Jeffrey E. Wieselthier, Anthony Ephremides, and Larry A. Michaels. An Exact Analysis and Performance Evaluation of Framed ALOHA with Capture. *IEEE Transactions on Communications*, COM-37, 2:125–137, 1989.

The Untrusted Computer Problem and Camera-Based Authentication

Dwaine Clarke[1], Blaise Gassend[1], Thomas Kotwal[1], Matt Burnside[1],
Marten van Dijk[2], Srinivas Devadas[1], and Ronald Rivest[1]

[1] Massachusetts Institute of Technology
{declarke, gassend, tkotwal, event, devadas, rivest}@mit.edu
[2] Philips Research
vandijk@caa.lcs.mit.edu

Abstract. The use of computers in public places is increasingly common in everyday life. In using one of these computers, a user is trusting it to correctly carry out her orders. For many transactions, particularly banking operations, blind trust in a public terminal will not satisfy most users. In this paper the aim is therefore to provide the user with authenticated communication between herself and a remote trusted computer, via the untrusted computer.

After defining the authentication problem that is to be solved, this paper reduces it to a simpler problem. Solutions to the simpler problem are explored in which the user carries a trusted device with her. Finally, a description is given of two camera-based devices that are being developed.

1 Introduction

In this paper we discuss methods for the verification of the trustworthiness of a public computer (e.g., in an Internet cafe or airport lounge). Consider Ursula, on holiday in Peru, who wishes to manage her stocks. She visits the local Internet cafe where she uses a computer to contact her bank's web-site. She uses it to review the stock market and place orders. In doing so she is exposing herself to a host of possible attacks.

Indeed, Ursula has no idea of what is going on inside the computer she is using. Even if the interface looks exactly as she expects, she could in fact be interacting with a Trojan horse. The Trojan horse has many attacks to choose from. It can store Ursula's password for later use. It can tamper with what Ursula is seeing, giving her a misleading idea of the market. It can tamper with Ursula's transaction, changing the amounts, or the stocks that are being bought or sold. A skillfully designed Trojan can completely simulate Ursula's session with her bank, while in fact doing transactions of its own choosing. It could wire Ursula's money to an account in Switzerland or change Ursula's password, thereby preventing her from contacting her bank in the future. Unless suitable measures are taken, Ursula will not even realize that she is being tricked until she checks her account through a trustworthy source. Even if Ursula knows the

F. Mattern and M. Naghshineh (Eds.): Pervasive 2002, LNCS 2414, pp. 114–124, 2002.

administrator of the Internet cafeé's intentions, the administrator's technical skill may be in doubt. Indeed, a hacker might have overcome the Internet cafe's security and installed malicious software on its computers.

This leaves Ursula in a quandary: She can use the high bandwidth, ergonomic keyboard and mouse, large screen and powerful computing of the Internet cafe and run the risk of being tricked, or she can use her simple mobile device with its low bandwidth and uncomfortable interface to do all her work.

In this paper, we will assume that Ursula is using an untrusted computer to contact a trusted computer (e.g., an Internet banking server) over a network. The goal of this paper is to present methods to provide an authenticated bidirectional channel from the trusted computer to Ursula (the user), through the untrusted computer. Authenticated means that messages received through the channel are guaranteed to be unmodified copies of messages that were sent by the party on the other side of the channel.

It is worthwhile to note that we do not attempt to insure the privacy of of any information sent through the untrusted computer. The untrusted computer has access to any information that the user enters directly into the computer, as well as any information that it is displaying to the user. Referring to the earlier example, the untrusted computer will know exactly which stocks Ursula reviewed, which ones she purchased or sold, and how much money the stocks were worth. This is unfortunately unavoidable, and must be kept in mind when considering the application and the user. Ursula may not be concerned about the disclosure of her stock portfolio, in which case the lack of privacy is not an issue. However, she will surely want her transactions to be carried out correctly, which this paper insures by providing a means of authenticating both Ursula and her bank to each other.

After a brief review of related work, section 3 models the untrusted computer problem and shows a reduction to a simpler problem. Section 4 will present several implementations that could solve the simplified problem. Section 5 will focus on the camera based solutions that we have implemented. Finally, section 6 will outline a few areas for further research.

2   Related Work

The difficulties related to using an untrusted terminal are not new. ATMs are one of the most prominent examples. Since the user gives both her card and her PIN to the ATM, she is placing total trust in it. Fake ATMs have been known to simulate a breakdown, simply keeping the user's card after the PIN was entered. Crooks can then use the stolen card and PIN until the user realizes that she has been tricked. To address this attack, many attempts [8,3,4,7] have been made to replace the PIN with a challenge-response mechanism, in which the user knows a secret that allows her to answer the challenge that she is given, by solving a little problem in her head. These systems allow the user to identify herself securely, but they fail to authenticate the rest of the session. Our system guarantees the authenticity of the whole session.

In [1] a smart-card based system is described that allows secure identification of the user, and that allows her to limit the power that is delegated to the untrusted host, as well as the duration of the delegation. We authenticate each piece of information that is transmitted, without placing any trust at all on the untrusted terminal.

The nearest system to ours that we have found is described in [9]. The protocol that they present is similar to ours, but they implement it using simple transparencies, which makes it harder to use than our camera-based authentication system.

3   Model

We now formalize the layout of the untrusted computer problem and illustrate it in figure 1.

- There is a user U. Her abilities are limited to those of a typical person.
- U might be in possession of a personal device D. Its abilities are those of a computer. We will try to restrict them as much as possible to minimize the size, cost, and power requirements of D.
- The combination of D and U will be referred to as DU.
- U will try to communicate with her proxy P, a computer that she trusts.[1] P's abilities are those of a computer, but we will not try to restrict them as strongly as those of D.
- DU and P are connected by a channel C, that embodies an untrusted computer that DU has physical access to, connected to P via an untrusted network. Most of the time C will simply convey messages between DU and P in which case we will say that it is faithful. However, in some cases C can exhibit any behavior that is possible for an arbitrary combination of humans and computers (we will assume that C cannot break cryptographic primitives).
- DU and P can send and receive messages over C, and they can sometimes decide to accept messages that they receive. Messages from DU to P will be called upwards messages, while messages from P to DU will be called downwards messages (as in uploading and downloading).
- In general, DU and P will have a shared secret to perform cryptographic operations. If the cryptographic secret is held by D then it becomes possible for an attacker to steal D and pretend to be U. In the rest of this paper, we will assume that some form of direct identification of U to D takes place, with a PIN number, or a biometric measurement.

Figure 1 illustrates the layout.

Definition 1. *We define a* Unidirectional Authentication with Secure Approval Channel *(UASAC) to be a channel that provides the following primitive operations:*

---

[1] More information about proxy-based protocols can be found in [2].

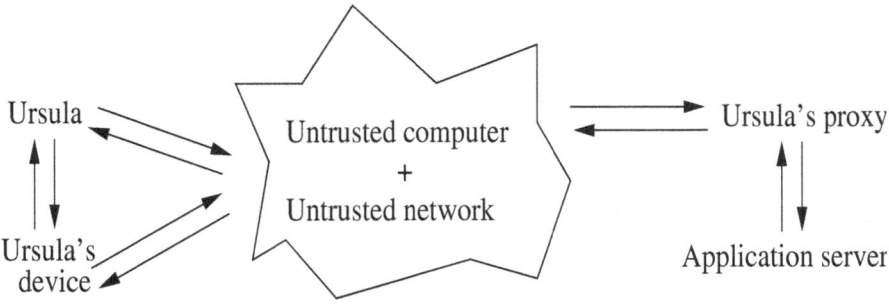

Fig. 1. The untrusted computer model: Ursula, equipped with her device, wishes to establish authenticated communication with her proxy over an untrusted channel. Once this is done, she can safely use any application on her proxy, or on some remote application server (such as her bank)

Downwards Authentication: *P can send a message to U in such a way that U always accepts the message if C is faithful. U will never accept a message that was not sent by P, or that was tampered with.*

Secure Approval: *If C is faithful then U can always inform P that it approves a specific authenticated message from P. P will never consider that a message is approved if U did not actually approve it.*

Upwards Transmission: *U can send a message to P that will be received unmodified if C is faithful.*

Definition 2. *We define a* Bidirectional Authentication Channel *(BAC) to be a channel that provides the following primitive operations:*

Downwards Authentication: *(same as above)*

Upwards Authentication: *U can send a message to P in such a way that P always accepts the message if C is faithful. P will never accept a message that was not sent by U, or that was tampered with.*

It is noteworthy, that there is no need to send user identifiers or passwords over a BAC. When P receives a message message from DU over a BAC, it knows that it is talking to DU because of the specification of the BAC. All the work needed to identify the parties that are communicating has been done by the algorithm that provides a BAC from an untrusted channel. Adding a password would be redundant. Moreover, since there is no mention of privacy in a BAC, any password sent over C would be compromised.

Theorem 1. *Any UASAC can be used to make a BAC.*

We prove this by showing an algorithm that produces upwards authentication.

1. *U sends a message M to P.*
2. *P receives M' from U, M' might be different from M if C was unfaithful.*
3. *P does an authenticated transmission of M' to U.*

4. *U approves the message M' it received from P if it is identical to M and if U did send M to P.*

   U only accepts M' in step 4 if it sent a message M that was identical to M'. Therefore, when P receives approval on M', it can accept it according to the requirements of upwards authentication.

It is of note that a similar result could be obtained by reversing U and P in the definition of the UASAC. However, this seems to lead to more complex and less interesting implementations that have not been studied.

4   The Device

We will now look at implementations of a UASAC. In the definition we have made of a UASAC, no mention was made of D. Though it might be possible to avoid the use of any device at all, we do not know of any convenient way for U alone to authenticate a message that she is receiving.

In the case where D is present, there are two basic designs: U can receive incoming messages through D, or directly from C. We will not consider methods in which U has to compare a non-secure copy of the message obtained directly from C with a secure copy displayed by D, as the non-secure copy is then redundant.

4.1   Relaying Device

In the case where U is receiving incoming messages through the device, there is a simple protocol to obtain a UASAC: Messages are sent from P in encrypted form, along with an encrypted one time password. They are protected by a nonce[2] and a MAC.[3] D receives the message from C, checks its authenticity, and passes it on to U if the checks succeed. This gives us downward authentication. Secure approval is obtained by sending the one time password back to P if U wishes to approve the message. C cannot fake approval because until U decides to approve the message, the one time password is never sent over C in the clear.

In a very straightforward implementation of the relaying device, Ursula comes to the Internet cafe with her PDA. She connects it to the untrusted computer's USB port, and connects to her proxy using her PDA's SSL-capable web browser. She can then do all her interaction with her proxy through her trusted PDA.

In a more complicated implementation Ursula would still have to use her device's screen but she could type her replies on the Internet cafe computer's keyboard, since the up-link of a UASAC need not be secure.

However, all of these solutions are dissatisfying, because Ursula is barely making any use of the untrusted computer's comfortable screen. The Internet

---

[2] A nonce is a sequence number that helps prevent replay attacks. See [5] for details.

[3] A Message Authentication Code is a cryptographic hash of a message concatenated with a key. It can only be generated and checked by someone bearing the key. See [6] for details.

cafe's computer is just being used as a network access point. Meanwhile, Ursula has to study the stock market through the tiny screen of her hand-held device.

An original implementation of the relaying device approach is presented in [9] where transparencies are used to obtain a secure channel. Here, Ursula has a secret transparency with a random distribution of black and transparent pixels. If Ursula's proxy wants to communicate a message it sends a random-looking black and white pixel pattern, such that the message emerges if Ursula looks at the untrusted computer's screen through her transparency. This is an elegant low-tech solution to our problem. However, it is impractical for a number of technical reasons.[4]

4.2   Monitoring Device

With monitoring devices, U gets information directly from C. Some extra information is also sent through C, that U need not concern herself with, but that D uses to verify the authenticity of the information U is getting. If D detects tampering, or if for some technical reason D is unable to authenticate the image (too much noise, the connection from C to D is bad, etc.), it warns U. Monitoring approaches are more convenient as U fully uses the untrusted computer's interface. However, a number of difficulties arise.

First, it is important to realize that D must be authenticating the information that U is getting. It would not be acceptable for D to receive information about what U is seeing on the screen through a USB link, as a malicious computer could send one thing through the USB port, and something completely different to the screen. This means that to authenticate screen content, D must be equipped with a camera.

Even if D and U are both getting their data from the same source (we will consider a screen), caution is still required because of noise. Indeed, it is impossible for D to reconstruct what is being displayed on the screen down to the exact RGB components of each pixel. Variations in screen brightness, camera noise and reflections off the screen all contribute to imprecision in what D can reconstruct. If a rogue message is displayed in grey on a slightly lighter grey, U will be able to read it, while D might see it as uniform grey. It is because of this difficulty that our implementations use black and white (no grey) images.

Thus we see that if D does not perceive as much information as U, then U must be aware of that limitation, and be able to tell when an image might be ambiguous to D. For example, if D can only distinguish between black and white,

---

[4] The main problem is that transparencies cannot be reused, so in order to achieve security, Ursula needs a whole sequence of one-time secret transparencies during each electronic transaction. In [9] a method is proposed to use a transparency many times, but it assumes that transparencies can be easily placed at a precise position on the screen that is unknown to the untrusted computer. Our experiments with transparencies on a screen suggest that the on-screen image must be finely scaled and moved before it can be made to line up with the transparency precisely enough; this gives away the transparency's position.

then a shade of grey should reveal to U that tampering has taken place, even if D does not detect that tampering.

The next section will give details of two camera-based monitoring device implementations.

5   Camera Based Solution

In this section, we present two implementations of camera-based authentication that are under development. In both cases the user is expected to carry a camera-equipped device that monitors the screen of the untrusted computer she is using. The visual processing involved in extracting on-screen information can be costly in computation resources. The first method we propose tries to minimize this cost, while the second one uses a high bandwidth network connection to move the computation to the proxy.

In both of the following implementations, the content displayed on the screen is in the form of an image. The transmission of an entire image from the proxy to the untrusted computer to convey generally textual information is inefficient, and instead the proxy can send formatted text. The semantics of the formatting must be exactly specified to insure exact representation of the content on the untrusted computer's screen. However, for simplicity, the following implementations describe systems in which the proxy transmits an image to the untrusted computer.

5.1   Pixel Mapping

In the pixel mapping method, the camera-equipped device is assumed not to move relative to the screen during the authenticated session. An initial calibration phase is used to construct a mapping between screen pixels and camera pixels. The mapping is then used by the device to exactly reconstruct the screen content. A small area at the bottom of the screen is used to transmit a nonce, a one-time password and a MAC.

Protocol:

– Downwards Authentication:
  - The Proxy sends (information, encrypted nonce, encrypted one-time password, MAC(information, encrypted nonce, encrypted one-time password)) over the channel. To the user, this appears as an image, with a strip of random-looking data at the bottom.
  - The device exactly reconstructs the screen content from what it sees through its camera.
  - The device checks the nonce, calculates the expected MAC and compares it with the on-screen MAC. If all checks succeed it lights a green light, and displays the decrypted one-time password on a small LCD display.
  - The user reads the screen content.

- Upwards Transmission:
  - The user types commands on the untrusted computer's keyboard. They are sent directly to the proxy.
- Secure Approval:
  - The user reads the one-time password from the device's LCD screen and uses upwards transmission to send it to the proxy.
- Calibration:
  - The untrusted computer displays a predefined sequence of images on its display. The basic property of these images is that each screen pixel flashes its coordinates in binary (with a suitable amount of redundancy to improve robustness).
  - The device records the number that was recorded at each camera-pixel. That number allows the device to know which screen-pixel is seen by each camera-pixel. Camera pixels with invalid numbers are assumed to see multiple screen pixels or to be off-screen; they are excluded from further use. If each screen-pixel was seen by at least one of the valid camera-pixels, a mapping between camera-pixels and screen-pixels can be established. It is later used to reconstruct the on-screen image. The device should check that the relative positions of screen pixels relative to camera pixels are reasonable, in order to detect if the untrusted computer attempts to tamper with the calibration process. Without this check the untrusted computer could perform an arbitrary permutation of the screen pixels.

Evaluation. A preliminary implementation of this protocol has been written. It works with black and white images, and currently requires camera resolution to be about twenty times greater than screen resolution. Current results suggest that this ratio can be reduced by a factor of 5 only by improving the calibration phase. Extensions to limited numbers of colors are also possible. At present the calibration phase displays 27 frames to calibrate a 80 by 50 pixel screen. This takes about 10 seconds. Rethinking the calibration method to take into account simple geometrical constraints should be able to reduce this to under 10 frames for any size of screen.

This approach uses simple algorithms that would be easy to implement in hardware, and does not require large amounts of computation power. Because of its calibration method, no knowledge is needed of the exact screen and camera geometry (curved screens, distorting cameras, etc.). However the geometry is expected not to change with time, which means that the device must be set on a stable surface, instead of being warn by the user (as a badge, for example).

5.2 Optical Character Recognition

In the optical character recognition (OCR) method it is assumed that the user's device is equipped with a camera and an infrared link to the untrusted computer. This link is used to exchange data with the proxy, via the untrusted computer, to take advantage of the large amount of computation available at the proxy.

Protocol:

− Downwards Authentication:
  - The Proxy sends information, in the form of an image containing text, to the untrusted computer. This image is displayed on the screen.
  - The device takes a picture of the screen, and sends ("verify", picture, encrypted nonce, MAC("verify", picture, encrypted nonce)) to the untrusted computer using the infrared link; the message is then forwarded to the proxy.
  - The proxy checks the nonce, calculates the expected MAC and compares it with the received MAC. If all checks pass it verifies that the text displayed on the screen was genuine by performing OCR on the received picture.
  - The result of the OCR is compared with the information from which the image was formed in the first step. If this check passes, the proxy sends (yes, encrypted nonce, MAC(yes, encrypted nonce)) to the device. The device checks the nonce and MAC. If all checks pass the device lights a green light.
− Upwards Transmission:
  - The user types commands on the untrusted computer's keyboard. They are sent directly to the proxy.
− Secure Approval:
  - The user accepts the data on the screen using her device. The device takes a picture of the screen, and sends ("accept", picture, encrypted nonce, MAC("accept", picture, encrypted nonce)) to the proxy. The proxy verifies the message as in downwards authentication. If the tests pass then it considers the picture approved, as only the user's device could have produced the MAC. Note that a race condition exists in which the untrusted computer changes the content of the screen after the user has pressed the accept button on her device, but before the camera has actually taken the picture. To avoid this condition, a proper implementation should consist of two separate buttons: one for capturing the image and one for sending the image to the proxy.
− Image Verification:
  An integral step in the protocol is the proxy's verification of the picture that was taken off the screen. There are two basic steps in this process. The first is to correct the distortions of the original image that are caused by the camera angle, curvature of the screen, lens deformation, and low camera resolution. An easily identifiable border is placed around the text to facilitate the correction of these distortions.

  The second step in the image verification process is to perform OCR on the processed image. This application differs from most OCR applications because in this case the intended message is known, which allows for a large speed optimization. The OCR algorithm must detect any change in the image that would result in the user seeing a different character from the one that was originally sent by the proxy.

  In order to accomplish this goal, a specific form of template matching was designed. Template matching is a method of OCR in which the detected character is compared with a known template for that character. This idea was adapted so that there are very strong penalties for areas in which the character in question does not match the template. This greatly decreases the likelihood that a false character will be accepted. The drawback to this design is that valid characters will be rejected if there is a slight distortion in the image. This drawback is necessary, given the goal of providing reliable authentication. Also, with reasonable picture quality, characters of adequate size, and the distortion reduction techniques used, rejections of a valid characters can be virtually eliminated.

Evaluation. This method's advantages over the pixel-mapping method are that it does not require any calibration, and the camera does not have to be immobile during the session. However it will take longer to verify a given screen, so authentication of every screen during a session would be cumbersome. Instead, only screens containing vital information will be verified.

The current implementation of this scheme allows the verification of a text block consisting of slightly under 100 characters in under 5 seconds. As cameras improve in quality, and processor speed increases, both of these areas will improve.

6   Possible Extensions

In this paper, we have been most concerned with authenticating communication in which the user is receiving visual information. Our protocol could be applied just as well to audio information. Though this is probably not very useful to the average user, it would certainly benefit the visually impaired.

The camera-based system does not provide any privacy, as queries and responses are transmitted in the clear. A limited amount of privacy could be added by allowing the user to point at areas of the screen. Selections made in this way would be visible to the device but not to the UC. The possibilities of such a system are yet to be explored.

For now, all our attention has been restricted to authenticating black and white images (no shades of grey). This is because of the noise-related security concerns that we discussed in section 4.2. Extending our system to a small set of easily distinguishable colors would be easy, as long as the user can be expected to notice if invalid colors are used. This ability of the user is a direction for further study. Ideally, though, it would be nice to extend the system to arbitrary colors, but then we must be sure that the device will perceive the same color areas as the user, which implies a good understanding of human visual perceptions.

7   Concluding Remarks

In summary, we have studied how a person using an untrusted terminal can communicate with a trusted computer in an authenticated way with the aid of a trusted device. To do this, we have presented a protocol that allowed us to simplify the initial problem. We then explored ways to implement the reduced problem, considering how convenient each one would be for the user. Finally, we presented our implementations that are based on a camera-equipped device.

Acknowledgements. This work was funded by Acer Inc., Delta Electronics Inc., HP Corp., NTT Inc., Nokia Research Center, and Philips Research under the MIT Project Oxygen partnership, and by DARPA through the Office of Naval Research under contract number N66001-99-2-891702.

References

1. Martin Abadi, Michael Burrows, C. Kaufman, and Butler W. Lampson. Authentication and delegation with smart-cards. In *Theoretical Aspects of Computer Software*, pages 326–345, 1991.
2. M. Burnside, D. Clarke, T. Mills, A. Maywah, S. Devadas, and R. Rivest. Proxy-based security protocols in networked mobile devices. In *Proceedings SAC*, 2002.
3. Rachna Dhamija and Adrian Perrig. Dejà vu: A user study using images for authentication. In *Proceedings of the 9th USENIX Security Symposium*, 2000.
4. Nicholas J. Hopper and Manuel Blum. A secure human-computer authentication scheme.
5. Charlie Kaufman, Radia Perlman, and Mike Speciner. *Network Security, Private Communication in a Public World*. Prentice Hall PTR, 1995.
6. H. Krawczyk, M. Bellare, and R. Canetti. RFC 2104: HMAC: Keyed-hashing for message authentication, February 1997. Status: INFORMATIONAL.
7. Tsutomu Matsumoto. Human identification through insecure channel. In *Theory and Application of Cryptographic Techniques*, pages 409–421, 1991.
8. Tsutomu Matsumoto. Human-computer cryptography: An attempt. In *ACM Conference on Computer and Communications Security*, pages 68–75, 1996.
9. Moni Naor and Benny Pinkas. Visual authentication and identification. In *CRYPTO*, pages 322–336, 1997.

SoapBox: A Platform for Ubiquitous Computing Research and Applications

Esa Tuulari and Arto Ylisaukko-oja

VTT Electronics
Kaitoväylä 1
90571 Oulu, Finland
{Esa.Tuulari, Arto.Ylisaukko-oja}@vtt.fi

Abstract. Designing, implementing and evaluating prototypes is a normal way of doing technical research. In recent years we have seen lots of research prototypes specifically designed for context awareness, future user interfaces and intelligent environment research. The problem with this type of specialised prototypes is that their lifetime is rather short and the valuable work done for them is not easily reusable. Our approach has been different as we have deliberately aimed towards a multipurpose platform that would be suitable for various ubiquitous computing related research themes. In this article we present the design and implementation of the platform that is named as SoapBox (Sensing, Operating and Activating Peripheral Box). Its main features are wired and wireless communications, in-built sensors, small size and low power consumption. We also introduce some results of research projects that have already used the platform successfully. Finally we conclude the paper with application scenarios for further work.

1   Introduction

The term Ubiquitous computing was coined in late 80's at Xerox PARC. In the beginning the technical challenges were in creating low power computers, connecting them to each other with a network and implementing software for ubiquitous applications [21]. The major goal was to put computers in the background and let people interact naturally with the real world.

To study ubiquitous computing, researchers at PARC built several prototype devices. Most famous of these was the ParcTab, which was a handsize computer with touch sensitive display, a couple of buttons and continuous wireless connection to a network [19]. It gave excellent opportunities for doing research on novel user interfaces and new type of mobile applications.

One of the many shorthand explanations for ubiquitous computing is to say that it means "computers everywhere". However, ParcTab and the other prototypes built at PARC did not address this goal directly. More close to it came ActiveBadge which was developed in Olivetti&Oracle research lab mainly by Roy Want who had later joined PARC [22]. Together ActiveBadge, ParcTab and the other research prototypes formed a working infrastructure that was used for many ubiquitous computing research projects.

The development of complete hand-held devices for research purposes has largely ceased because off-the-shelf PDAs offer good enough hardware platform as well as an

F. Mattern and M. Naghshineh (Eds.): Pervasive 2002, LNCS 2414, pp. 125-138, 2002.

open software interface for writing own applications. However, for context awareness research, several research prototypes that extend the capabilities of existing devices have been built in recent years [6, 11, 18]. Extensions have often included sensors but location tags have been used as well.

Our goal has been a bit different as we have seen it important to build a flexible and reusable platform that is applicable in several ubiquitous computing research topics including context awareness, future user interfaces and intelligent environments. This has led to the design and implementation of the SoapBox (Sensing, Operating and Activating Peripheral Box) that is a matchbox-size module with a processor, sensors, and both wired and wireless communications capabilities.

The basic building block of our system is a cell consisting of one *central SoapBox* attached to a user device (PC or PDA), together with several *remote SoapBoxes* installed in the surroundings (Fig. 1.). Remote SoapBoxes communicate wirelessly with the central SoapBox, providing a network with unobtrusive information flow. By locating remote SoapBoxes intelligently, we can collect important environmental data to the user device easily and automatically. The communication is bidirectional which makes it also possible to control the environment with the user device. Moreover, even the central SoapBox attached to the user device has in-built sensors, making also the measurement of the user easy. The sensors can also be applied to improve the usability of the user device.

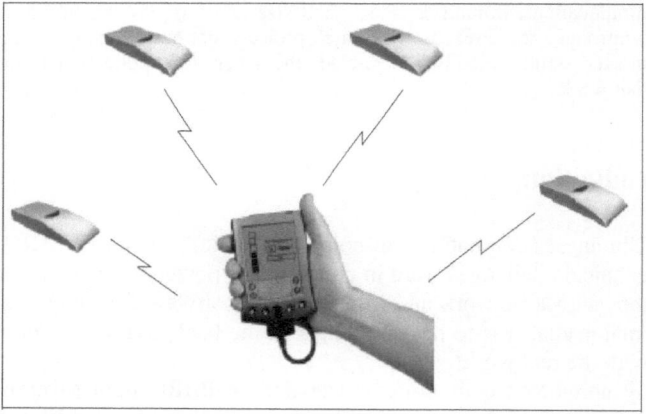

Fig. 1. SoapBox architecture consisting of one central SoapBox attached to the user device (in the back of the Palm, not shown) and several remote SoapBoxes located in the environment

2   Related Work

EasyLiving [5] and Everyday computing [1] projects study intelligent environment, which is supposed to make user - environment interaction richer. Devices that are generally included in the intelligent environment are electrical devices like light switches, refrigerators and cameras. There is very little work done where also the non-electrical everyday objects are included in the intelligent environment. Closest to this comes work done by MIT in their Things That Think consortium [10]. With the

SoapBox we enable also interaction with nonelectric objects. Especially the input direction is well supported as the only thing needed is to attach a SoapBox (with appropriate software) to the object we want to interact with. For example attaching a Soap-Box to a door gives instant access to door-related information. Depending on our interest this could be the general usage of the door (number of openings per day) or even the usage pattern (recognise the user by analysing the opening style).

Intelligent environment research is closely related to user interface research. Especially research on new user interface paradigms associated with handheld or personal technology devices has provided new ways to interact with the environment with the handheld device [1]. One common approach is to add sensors to the device in order to a) make one hand operation easier [3] or b) to make the use of the device more naturally resemble interaction with real-world objects [9,14]. At the implementation level these research prototypes have been single devices designed and built for one research project only.

Joe Paradiso first instrumented a balloon [16] and then has concentrated on instrumenting shoes [17]. The current version of the shoe electronics including sensors and wireless communication has close resemblance to our SoapBox. It uses unlicensed radio, acceleration sensors and compass as well as low cost microprocessor to mention some of the similarities. However, the starting point in Paradiso's work has been to create expressive footwear, i.e. to make research on a very specialised topic. Whereas our approach has been to make a versatile module that is suitable for many different purposes.

The digital baton is another example of a project where the user device has been equipped with sensors in order to enhance user - real world interaction [8]. For this project the user device is the baton and the real world is the orchestra. Compared to our approach where the user device is the PDA or some other personal technology device and the real world is home, office or a similar environment.

Researchers at TecO have exploited environment-sensing technologies for automated context recognition [18]. They have implemented a sensor box for measuring environmental variables. In their work, strong emphasis is in using neural network methods for context recognition purposes. Their sensor box does not include any wireless communication capabilities.

At KTH students have used SmartBadges for experimenting with ubiquitous computing [2]. As the name implies, SmartBadge is a badge having more intelligence than a normal badge. The size of the SmartBadge is somewhat bigger than the SoapBox as it has a PC-CARD slot for attaching a WLAN-card. The SmartBadge architecture is based on servers where all information is gathered for further analysis. Example applications are for example controlling access to rooms and profiling the usage of some specific door.

MicroAmps project at the University of California aimes at greating dust-size computing elements that could be used for example in military manoeuvres [20]. Their current implementation is cubic-inch sized but according to the researchers, MEMS technology will make it possible to squeeze the same functionality into the size of a dust-particle. They are not planning to use RF communication in dust-size devices because there is no room for antenna and the components for IR communication are also cheaper.

The commercial PDA manufacturers have also noticed the need and possibility to extend the devices with add-on modules. At the moment the best support for this is offered to Visor with its Springboard expansion slot [4]. Extra memories, digital cameras and wireless communication are already commercially available.

3 SoapBox Version 1.0

The design of the SoapBox started with the following general requirements:
- short-range wireless communications
- wired communications to PC, PDAs and other devices
- a versatile set of sensors
- possibility of micropower operating modes for battery-powered operation
- versatile powering options
- easy to use and program by researchers
- small size
- flexible and extensible
- cost-effective design to enable actual SoapBox-based products

Short range wireless communication is needed to enable communication between the user device and the environment, both instrumented with a SoapBox. For our purposes, the radio offers the best solution since it propagates through walls and objects and requires no pointing. For cost and bureaucracy reasons, use of an unlicensed radio band is also crucial.

Low power consumption is naturally needed when SoapBox is attached to a portable user device such as a PDA, but it is especially required when SoapBoxes are placed in the environment. Benefits of wireless communication such as ease of installation and mobility are fully obtained only if the power supply solution is also wireless, meaning typically battery-powered operation. Micropower operating modes are useful in many applications where battery life of several months or years is required. Flexibility in utilising different kinds of power sources such as internal battery or varying external supply voltage levels was also required.

As we wanted to keep the SoapBox small and simple, we wanted to have only limited number of relatively small sensors in it. However, in order to be flexible enough, a versatile set of sensors was needed. This led to the decision of including 3D acceleration measurement, magnetic sensors for enabling compass readings and a couple of different type of visible and infrared light measurements.

Modular approach was needed for flexibility: the possibility of equipping SoapBox with different sets of sensors depending on the application was considered an important feature (Fig. 2.). Finally, sufficient I/O capabilities were needed to control or read external devices, to connect to user devices, and to provide means for programming and debugging the software.

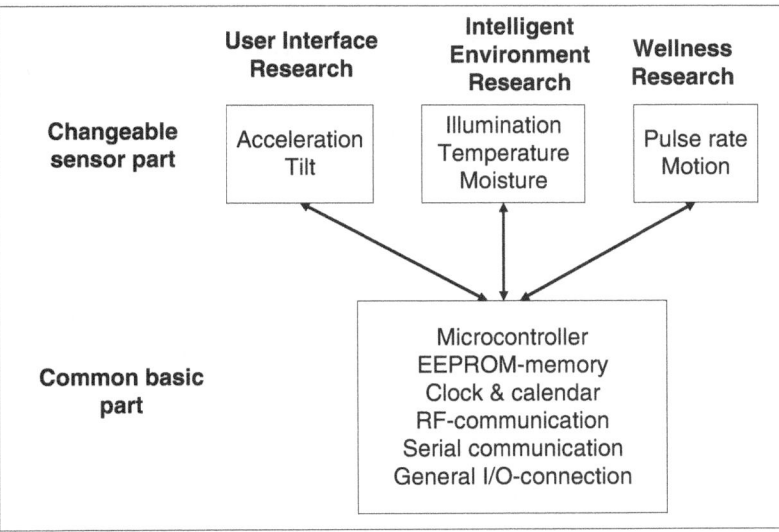

Fig. 2. Modularity of the electronics enables modifications to satisfy the needs of different research topics

3.1 Implementation

Electronics in General.
SoapBox electronics is shown in Figure 3. The electronics is contained on two printed circuit boards, one containing the sensors and the other containing all the other electronics. Basic structure of the electronics is shown in Figure 4. The requirements set for the electronics have been taken into account for example by using a high integration degree, low power microcontroller and radio module, and micropower analog and digital integrated circuits.

Fig. 3. SoapBox boards, illustrating the size and shape of the electronics

SoapBox has a single 8-bit microcontroller (PIC16LF877) which is equipped with flash program memory, enabling easy reprogramming. The microcontroller supports in-circuit debugging and also has some EEPROM. The main part of the software has been implemented in C language, but assembly language has been used in time critical parts.

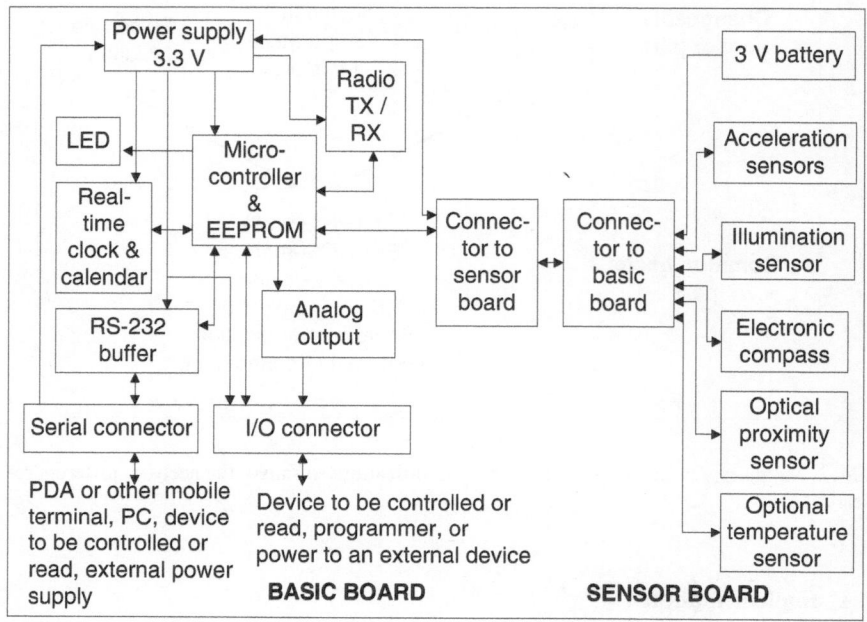

Fig. 4. SoapBox electronics (ver. 1.0) block diagram

RS-232 was selected as the form of serial communication instead of USB (Universal Serial Bus), since it is still more widespread. Analog output is provided in order to be able to control devices without digital interface. A software-controlled visible light LED is mainly used for debugging purposes.

A real-time clock and calendar circuit (Philips PCF8563) is utilised, for example, to time control functions or to record timestamps for events detected by the sensors. This circuit is also used for waking up the microcontroller from sleep mode.

Wireless Communications.
The lower protocol layers and the radio parts of Bluetooth were recently standardised by IEEE [12]. However, we do not see Bluetooth as the technology of future ubiquitous computing applications where also very cheap, simple, ultra low power items are involved. The ongoing low-rate WPAN standardisation by IEEE 802.15 Task Group 4 along with ZigBee Working Group [23] has a better approach in this sense, but the standard is not completed yet. For these reasons, we have implemented our own proprietary solution for low-rate ubiquitous radio communications.

A single channel 868.35 MHz radio (RF Monolithics TR 1001) is used for 10 kbps bidirectional, half duplex wireless communication. This radio band is license free in most European countries. The low data rate allows the microcontroller to take care of channel coding and decoding, while simultaneously running application program and wired communications in the background when necessary. The low data rate also enables slightly higher range. As the antenna we use a small helix that fits inside the SoapBox encapsulation.

The radio has two software selectable transmit power levels. The lower transmit power is intended for very short distance communication (a couple of meters). Using the maximum transmit power of 1 mW, the radio range varies from 15 m (through 4 board-constructed solid walls in very dense cubical office space) to 80 m (in corridors with line-of-sight).

The current network topology is centralised; it consists of a central SoapBox that communicates with one or more remote SoapBoxes (Fig. 5). Both types of SoapBoxes utilise the same kind of hardware, and the different communication roles are achieved by different software. Typically, the central SoapBox is connected to a user device, but it can also serve as a link to another, possibly wired, network. Several cells, consisting of one central SoapBox and at least one remote SoapBox, can coexist within the same radio range.

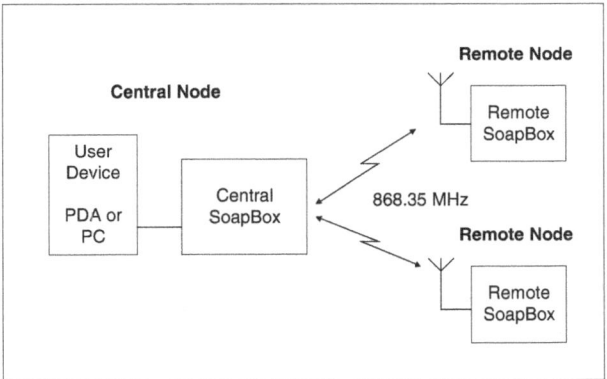

Fig. 5. A simple SoapBox network cell

The low power consumption requirements affect the MAC protocol selection since the radio is a dominant power consumer. Thus it is important to minimise the time that the radio must be kept on. The current MAC protocol is Aloha-based, with acknowledgement messages, CRC error detection and retransmissions. Communication is initiated by remote SoapBoxes. This way there is no need for the remote SoapBoxes to listen to the radio channel at other times, which is an essential pre-requisite for conserving power. The protocol is designed to support dynamic addressing in the network: nodes can be added to or removed from the network without any manual configuration.

Sensors.

The sensor board of SoapBox V1.0 includes the following sensors:

1. Acceleration sensors (2 x ADXL202E). These sensors can be used to measure 3 dimensional acceleration or tilt of the device.
2. Illumination sensor (based on Infineon SFH2400 detector). The intensity of visible light can be measured. The dynamic range of the sensor enables both indoor and outdoor measurements.

3. Magnetic sensor (Honeywell HMC1022). This sensor can be used to sense direction or magnetic objects such as steel. Enables also electronic compass.
4. Optical proximity sensor (based on Agilent Technologies HSDL-4420 IR-emitter and Infineon SFH213FA detector). This sensor is based on the sending of IR pulses and measuring of the level of reflection. Relative distances can be measured.
5. Optional temperature sensor. This can be installed whenever needed by the application, at the cost of sacrificing some external I/O pins from the I/O connector.

In a basic wireless sensor system the remote SoapBox is attached to something that we want to monitor, i.e. a door, a chair etc. The central SoapBox is attached to a PC running a Windows-program that displays all the sensor data as curves (like an oscilloscope). The PC-program is also used for calibrating the SoapBoxes. The calibration information can be stored either to a file in the PC or directly to the EEPROM of the remote SoapBox. If the SoapBox is already calibrated, it is possible to download the calibration values from the SoapBox to the PC. During calibration, the sensor signals are processed on the PC instead of the SoapBox, because this way we can send the measurement data in its basic, most universal form, and can minimise the processing needed in the SoapBox. Signals from magnetic sensors are typically processed to provide compass readings. This is presented as a needle-display. Figure 6 illustrates the sensor signals and the compass needle on the PC display.

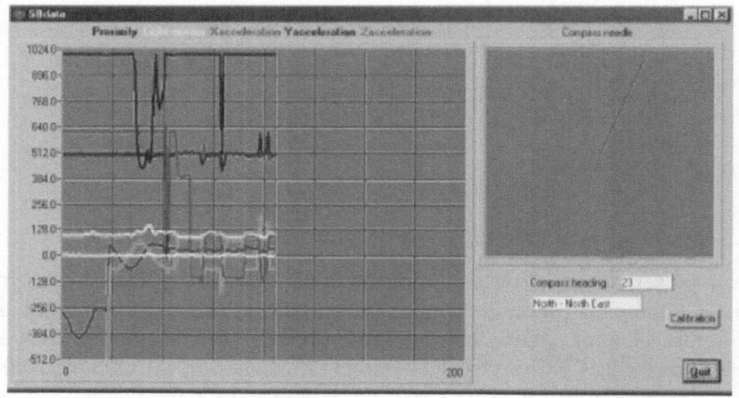

Fig. 6. PC display showing the sensor signals and the compass needle

The PC-programs have been implemented with LabWindows/CVI programming environment provided by National Instruments. This offers easy-to-use curve displays and wide range of buttons and controls, and is a good choice for everyone working with sensor data. Even if measurement cards from NI are not used.

We have also implemented similar program for Palm Vx, in order to have a more easily portable system available. In this program, only one signal is displayed at a time in order to increase readability (Fig. 7). The user can select the sensor he wants to monitor by touching the corresponding button on the touch screen. For example, "ax" for acceleration in x-direction, "ay" for acceleration in y-direction or "px" for proximity.

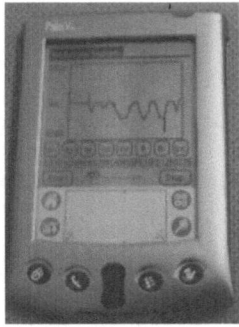

Fig. 7. Signal from a single SoapBox sensor displayed with a Palm PDA

Power Management.
SoapBox uses a 3.3V regulated supply voltage. An internal 220 mAh 3V coin-cell lithium battery is included; nevertheless, external supply voltages over a wide range of 0.9V and 28V can be utilised as well, including user device batteries as well as the RS-232 serial port. If required, SoapBox can provide a regulated 3.3V supply voltage to an external low power device.

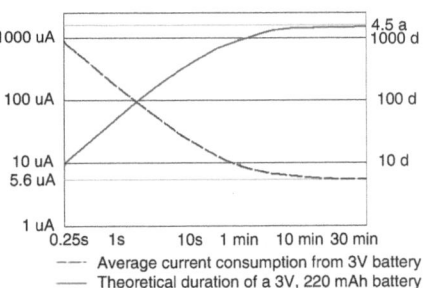

Fig. 8. Average current consumption and battery life of Remote SoapBox as a function of measurement/send interval when 3-axis acceleration is measured and sent at full RF power

The average power consumption of a remote SoapBox depends on the communication and processing activity. Figure 8 shows an example of the average current consumption and battery life as a function of measurement and send interval, when 3-axis acceleration is measured and sent at maximum power by radio, including acknowledgement reception. Major part of time is spent in a very low power standby mode. While still relatively low power, the central SoapBoxes consume more power (around 20 mW) as they have to keep the radio on continuously.

4   SoapBox Applications

SoapBox was designed as a versatile platform that is applicable to various research themes and projects. The basic idea is to speed up research projects as there is no need for every project to build their own electronics from scratch. In this chapter we describe a couple of research projects that have used SoapBoxes in this manner.

4.1   Two Player Maze

SoapBox was used as a game controller when demonstrating a multiuser game with gestural input and unobtrusive playing situation (Fig. 9.). Because the accelerometer can measure the acceleration force of the earth's gravity, it can be used as a tilt or inclination sensor [7].

Fig. 9. Two-player maze game playing in action

SoapBox offers two-dimensional control over a maze game where both the rapidity of controlling the cursor and cleverness to find the route are needed. A player navigates through the maze by tilting the SoapBox. The cursor moves when SoapBox is tilted away from neutral orientation in either or both axes. The PC program calculates tilt angles for a particular SoapBox and decides the direction and the speed of the corresponding cursor. Wireless communication is needed to make the playing situation less obtrusive. Players can change place and position freely without any limiting cables.

4.2   Personal Area Network for Recording User Activity

In context awareness research it is important to be able to collect data from the user's environment easily and unobtrusively. This is challenging as the user is mobile and does not want to carry heavy computers with him. The most obvious and most often used solution is to have a set of sensors connected to a laptop with a measurement card.

Our solution is more elegant. We have wireless sensors and the laptop is substituted with a Palm PDA device. With this system we can easily place the sensors wherever we want without the restriction that wired sensors would have (Fig. 10). The PDA fits easily in a pocket which makes it a pleasure to carry compared to a backpack with a laptop inside it. The Palm Vx has 8 Mb of memory from which about 1 Mb is used for the programs. The 7 Mb is usable for our recording purposes giving tens of minutes to hours of recording time depending on measurement rate and amount of sensors used.

Fig. 10. SoapBox attached to an ankle

4.3 Gesture Recognition

One application where wireless acceleration sensors are very handy is gesture recognition. We have done some preliminary work in implementing real-time gesture commands based on 3D acceleration signals provided by the SoapBox. Gesture recognition is done with HMM (Hidden Markov Models) and the teaching of the gestures done in off-line with a Matlab program. To enable acceleration measurements at higher frequencies, we are currently working on a higher data rate SoapBox modification.

4.4 Short Range Wireless Link

A couple of organisations outside our own laboratory have used or evaluated SoapBox as a short-range wireless connection to network devices together. In these applications, the internal sensors have not been used at all, but SoapBox has been connected to external sensor devices by the serial interface. In these applications, short messages are sent infrequently and high data rate is not needed. More important features are small size, low production costs, and low power consumption.

One example of such an approach is the demonstrator in Wireless Wellness Monitor II project. It is a joint research project by VTT Electronics, VTT Information Technology, Nokia, IST Oy and Celotron Oy. It combines different ambient intelligence and networking technologies at home environment, aiming at developing technologies for personal wellness management and home automation applications. The demonstrator system architecture is based on OSGi home server. SoapBoxes are used to wirelessly connect certain sensors to the home server. A device proxy PC is used between the home server and the central SoapBox, enabling the home server to see the SoapBox nodes as IP adressed devices. [13].

5 Conclusion and Further Work

In this paper, we have presented the SoapBox and a couple of research applications that utilise its capabilities. Its size, low power consumption and wireless range combined with its sensors and seamless connectivity to both PDAs, PCs and environmental objects makes it versatile and useful in wide range of research projects.

Our ver 1.0 SoapBox has no built-in actuators. However, we have plans for adding some sort of tactile user interface feedback device to some of the SoapBoxes in order to experiment with the subject. In our gesture recognition research [15] we noticed that it could be very beneficial for the user to have feedback from the computer about the state of the interaction. With traditional WIMP interfaces this feedback is so natural (e.g. touch and sound from the keyboard) that we do not even think about its importance.

5.1 Scenarios for New Applications

Although the SoapBox has already been used in many research projects and applications, it seems that we never have enough time to implement everything we would like to. Therefore we want to end this article by discussing about some application scenarios that are "seeking their time". In the first example the user (with the central SoapBox) is mobile while the remote SoapBox is permanently installed in the office building. In the second example both "ends" are stationary and in the last example the remote SoapBox is mobile and the user uses the information provided by the remote SoapBox mainly with his desktop. Also the use of the sensors is different in each scenario.

Mail-Watch.
One SoapBox is installed in the ceiling of a mailbox. It sends an infrared pulse with the proximity sensor and measures the light level that reflects from the bottom of the mailbox back to the SoapBox. When a letter (or anything similar) is delivered to the mailbox this light level changes. The change is detected by the software running in the SoapBox, which then transmits an RF-message indicating "new mail available" to the central SoapBox. If the owner of the mailbox is nearby with his PDA and central SoapBox, the message is received and acknowledged. It is then up to the co-operation of the central SoapBox and the PDA to inform the user about the new mail.

Occupation of a Room.
In Finland we are forced to use lights in our buildings for the bigger part of the year. This makes it possible to use the illumination sensor of the SoapBox in detecting occupation of rooms (of course it does not work if someone is sitting in a dark room

lights off). One specific modification for this general idea is to detect the occupation of a restroom, which makes it possible to implement Free/Occupied indication running on personal workstations. Making the user interface on top of an office map would give readily understandable information about the nearest free restroom, which could be convenient if you want to avoid walking around in search of one. Actually, why do we have this type of service in trains and airplanes but not in offices or buildings in general?

Location of the Coffee-Trolley.
In our office building we have centralised coffee breaks, both in time and location. The coffee is made in one of the two coffee rooms and then transported with a coffee-trolley to the other one. Usually in time and fresh. However, there have been some complaints about the trolley being late or being too long on the way. The arrival of the coffee varies from day to day, making it at least theoretically possible that the coffee is not perfectly fresh when the coffee break starts. By attaching one SoapBox to the trolley it is possible to measure the movement of the trolley by using the acceleration sensors. Surely a moving trolley would give a different signal than a trolley standing still. Ascending one floor with the elevator would give yet a different signal. Based on these signals, some coffee addict could make a program to inform him about fresh coffee before the coffee has even arrived.

Acknowledgement. This work has been partially conducted in European joint research projects Beyond the GUI and Ambience under ITEA cluster project of EUREKA network, and financially supported by TEKES (Technology Development Centre of Finland), to which organisations the authors wish to express their gratitude.

References

1. Abowd, G.D., Mynatt, E.D.: Charting Past, Present, and Future Research in Ubiquitous Computing. ACM Transactions on Computer-Human Interaction, Vol. 7, No. 1, March (2000) 29 - 58
2. Beadle, H.W.P., Maguire, G.Q. Jr., Smith, M.T.: Using location and environment awareness in mobile communications. In: Proceedings of the International Conference on Information, Communications and Signal Processing, ICICS, Part 3 (of 3), Singapore, (September 1997)
3. Björk, S., Redström, J., Ljungstrand, P., Holmquist, L.E.: PowerView; Using Information Links and Information Views to Navigate and Visualize Information on Small Displays. H.-W. Gellersen and P. Thomas (Eds.): Handheld and Ubiqtuitous Computing Symposium 2000, Lecture Notes in Computer Science 1927. (2000) 46-62
4. Brown, M.: Design a module that adds functionality to PDA's. Portable Design. (February 2001) 63 - 68
5. Brumitt, B., Myers, B., Krumm, J., Kern, A., Shafer, S.: Easy Living: Technologies for Intelligent Environments. H.-W. Gellersen and P. Thomas (Eds.): Handheld and Ubiqtuitous Computing Symposium 2000, Lecture Notes in Computer Science 1927. (2000) 12-29

6.   Dey, A.K.: Understanding and Using Context. Personal and Ubiquitous Computing No. 5 (2001) 4-7
7.   Docher J.: Using Micromachined Motion Sensors to Improve the Human/Computer Interface. Multimedia Systems Design, January Issue – Vol. 3, Nro. 1, 1999
8.   Gershenfeld, N.: When Thinks Start To Think. Henry Holt and Company Inc. New York (1999)
9.   Harrison, B.L., Fishkin, K.P., Gujar, A., Mochon, C., Want, R.: Squeeze Me, Hold Me, Tilt! An Exploration of Manipulative User Interfaces. In Procedings of ACM Conference of Computer-Human Interaction. Los Angeles (April 1998)
10.   Hawley, M.R., Dunbar, P., Tuteja, M.: Things That Think. Personal Technologies No. 1 (1997) 13 – 20
11.   Hinckley K., Pierce F., Sinclair M. Horvitz E.: Sensing Techniques for Mobile Interaction. UIST (User Interface Software and Technology): Proceedings of the ACM Symposium, 2000, 13th Annual ACM Symposium on User Interface Software and Technology, Nov 5-8 2000, San Diego, CA, pp. 91-100.
12.   The Institute of Electrical and Electronics Engineers (IEEE): http://www.ieee802.org/15/
13.   Korhonen, I.: Ambient Intelligence and Home Networking for Wellness Management and Home Automation. ERCIM News No. 47 (2001) 16
14.   Masui, T., Siio, I.: Real-World Graphical User Interfaces. H.-W. Gellersen and P. Thomas (Eds.): Handheld and Ubiqtuitous Computing Symposium 2000, Lecture Notes in Computer Science 1927 (2000) 72-84
15.   Mäntylä, V.-M., Mäntyjärvi, J., Seppänen, T., Tuulari, E..: Hand gesture recognition of a mobile device user. IEEE International Conference on Multimedia and Expo, (ICME) 2000. New York (July - August 2000)
16.   Paradiso J.: Interactive balloon: sensing, actuation, and behavior in a common object. IBM Systems Journal , Vol. 35, No.. 3-4, (1996) 473-487
17.   Paradiso J. A., Hsiao K., Benbasat A.Y., Teegarden Z.: Design and implementation of expressive footwear. IBM Systems Journal, Vol. 39, No. 3-4, (2000) 511-529
18.   Schmidt, A., Beigl, M., Gellersen, H.W.: There is more to Context than Location. In: Proceedings of the International Workshop on Interactive Applications of Mobile Computing (IMC98), Rostock, Germany (November 1998)
19.   Want, R., Schilit, B.N., Adams, N.I., Gold, R., Petersen, K., Goldberg, D., Ellis, J.R., Weiser, M.: The ParcTab Ubiquitous Computing Experiment. In: Mobile Computing by Imilienski and Korth (1996) 45 - 101
20.   Warneke B., Last M., Liebowitz B., Pister K.S.J.: Smart Dust: Communicating with a Cubic-Millimeter Computer. IEEE Computer, (January 2001) 44 - 51
21.   Weiser M.: The Computer for the 21st Century. Scientific American. Vol. 256, No. 3 (September 1991) 94-104
22.   Weiser M., Gold R., Brown J. S.: The origins of ubiquitous computing research at PARC in the late 1980's. IBM Systems Journal, Vol. 38, No. 4, (1999) 693-696
23.   The ZigBee Working Group: http://www.zigbee.com/

Pushpin Computing System Overview: A Platform for Distributed, Embedded, Ubiquitous Sensor Networks

Joshua Lifton[1], Deva Seetharam[2], Michael Broxton[1], and Joseph Paradiso[1]

[1] MIT Media Lab, Responsive Environments Group, 1 Cambridge Center 5FL,
Cambridge, MA 02142 USA
{lifton, mbroxton, joep}@media.mit.edu
http://www.media.mit.edu/resenv/
[2] MIT Media Lab, Physics & Media Group, 20 Ames Street,
Cambridge, MA 02139 USA
deva@media.mit.edu
http://www.media.mit.edu/physics/

Abstract. A hardware and software platform has been designed and implemented for modeling, testing, and deploying distributed peer-to-peer sensor networks comprised of many identical nodes. Each node possesses the tangible affordances of a commonplace pushpin to meet ease-of-use and power considerations. The sensing, computational, and communication abilities of a "Pushpin", as well as a "Pushpin" operating system supporting mobile computational processes are treated in detail. Example applications and future work are discussed.

1 Introduction

> *"A cockroach has 30,000 hairs, each of which is a sensor. The most complex robot we've built has 150 sensors and it's just about killed us. We can't expect to do as well as animals in the world until we get past that sensing barrier."*

> *Rodney Brooks in* Fast, Cheap & Out of Control [1]

Sensors to transduce physical quantities from the real world into a machine-readable digital representation are advancing to the point where size, quality of measurement, manufacturability, and cost are no longer the major stumbling blocks holding us back from creating machines equipped with as much sensory bandwidth as some animals, if not people. Rather, we are faced with a problem of our own devising – how do we communicate, coordinate, process, and react to the copious amount of sensory data now available to the machines we build? Certainly, some success in harvesting and responding to multiple data streams originating from a quantity of sensors has been demonstrated (e.g. [2]), but such examples do not scale; using traditional sensing methods, even adding one more sensor to an array of a couple dozen sensors presents a formidable challenge on

F. Mattern and M. Naghshineh (Eds.): Pervasive 2002, LNCS 2414, pp. 139–151, 2002.

both the hardware and software fronts. As the number of sensors increases to the thousands, hundreds of thousands and beyond, any tractable solution will have to rely on principles of self-organization at the level of the sensors themselves in order to guarantee the proper scaling properties. In this sense, it behooves us to begin treating sensor systems as distributed networks wherein each node is a self-sufficient sensing unit and coordination among nodes takes place locally, automatically, and without centralized supervision.

Distributed sensor networks are immediately relevant to many real world applications; robot skins, smart floors, battlefield reconnaissance, environmental monitoring, HVAC (heating, ventilation, air-conditioning) control, high-energy particle detectors, and space exploration are among the many areas that could benefit from distributed sensor networks. Perhaps the greatest use of distributed sensor networks, however, lies not in the preexisting applications they augment, but rather in the future applications they enable. Obviously, it is impossible to fully enumerate these future applications, but it is not hard to speculate that advances in any number of fields will only make that list longer.

In this paper we introduce the Pushpin Computing platform as a general purpose hardware and software toolkit for studying, designing, prototyping, and deploying dense sensor networks. Details of the hardware and programming model are given, as well as the design considerations that lead up to the current implementation. A simple example is illustrated step by step.

2   Related Work

Depending on the particular circumstances, the term *distributed sensor network* can meaningfully be attached to a large number of systems varying widely across many distinct parameters, such as physical layout, network topology, memory resources, computational throughput, sensing capabilities, communication bandwidth, and usability. Accordingly, what qualifies as research into distributed sensor networks is just as general. In such a general context, everything from tracing TCP/IP packet flow through the Internet to quantifying collective ant behavior can be considered as examples of research into distributed sensor networks. Nonetheless, there are very specific bodies of research that are either tangential or very closely related to the work presented here.

The direct inspiration for this work is Butera's Paintable Computing simulation work [3]. Paintable Computing begins with the premise that, from an engineering standpoint, we are not very far away from being able to mix thousands or millions of sand grain-sized computers into a bucket of paint, coat the walls with the resulting computationally enhanced paint, and expect a good portion of the processors to actually function and communicate with their neighbors. The main problem with this scenario, according to Butera, is that we don't yet have a compelling programming model suitable for such a system. Paintable Computing attempts to put forth just such a model, as well as a suite of example applications. To this end, Paintable Computing is a simulation of many (tens of thousands) independent computing nodes pseudo-randomly strewn across a surface. Each node is capable of communicating omnidirectionally with other nodes located within a limited radius, although no node knows *a priori* anything about its physical location on the surface. From these simple postulates, Paintable Computing demonstrates the utility of *algorithmic self-assembly* to build up complex global behavior across the system as a whole from simple local interactions among process fragments that migrate among the processing nodes. Pushpin Computing started out as an attempt to instantiate in hardware as closely as possible the Paintable simulations, each Pushpin corresponding to a single processing node. This will be discussed further in the coming sections.

Resnick's StarLogo programming language [4] provides an accessible but rich simulation environment for exploring decentralized emergent systems. The Pushpin programming model is influenced by StarLogo's intuitive approach.

Although there are surely many more examples of computer simulation research that have some bearing on distributed sensor networks, Berkeley's (now Intel Research Lab at Berkeley) SmartDust and its associated TinyOS software environment is the only known hardware platform developed in a spirit at all similar to that of the Pushpins. The SmartDust/TinyOS platform was developed from the bottom up, shaped by the real-world energy limitations placed upon nodes in a distributed sensor network [5,6]. As such, each node is relatively resource poor in terms of bandwidth and peripherals. Furthermore, the assumption is made that almost all communication within a distributed sensor network is for the purpose of communicating with a centralized base station [7]. In contrast, the Pushpin platform was built more from the top down, provides each node with a rich set of onboard peripherals, bandwidth, and software, and consumes correspondingly more energy per node.

3 Design Points

The primary motivator for the Pushpin Computing project is to achieve the one goal inaccessible to computer simulations of distributed sensor networks – to sense and react to the physical world. The goal is to devise sensor networks that self-organize in such a way so as to preprocess and condense sensory data at the local sensor level before (optionally) sending it on to more centralized systems. This idea is somewhat analogous to the way the cells making up the various layers of a retina interact locally within and across layers to preprocess some aspects of contrast and movement before passing the information on to the optic nerve and then on to the visual cortex [8].

The compelling architecture articulated and demonstrated in simulation by the Paintable Computing project provides a base set of design points for the hardware, operating system, and programming environment from which it is possible to build a distributed sensor network that achieves the goal of self-organization. Where practical, the Pushpin platform follows this blueprint closely. To paraphrase [3]:

- Each Pushpin (node) has the ability to communicate locally with its spatially proximal neighbors, the neighborhood being defined by the range of the mode of communication employed.
- Each Pushpin must reliably handle the fact that the number of addressable neighbors in the communication neighborhood can vary unpredictably.
- Each Pushpin must reliably handle the fact that messages sent to its neighbors may exhibit probabilistic transit times and are not explicitly acknowledged.
- Each Pushpin must provide for a mechanism for installing, executing, and passing on to its neighbors code and data received over the communication channel.

In addition, the Pushpin platform is designed specifically for ease of prototyping a wide range of digital and analog applications, so it can readily serve as a testbed for practitioners coming from many perspectives.

4   Hardware

The Pushpin project embeds a 20 MIPS mixed-signal microcomputer system into the form factor of a bottle cap with the tangible affordances of a thumb tack or pushpin. The Pushpin hardware platform is designed around a balanced optimization of small physical footprint, functional modularity, expandability, generality, and computational power. To this end, each Pushpin consists of four modules that separately handle power, communication, processing, and application-specific functions. Each module is contained on a printed circuit board (PCB) measuring roughly 18mm x 18mm and stacks together with other modules vertically from bottom to top in the order listed. See Fig. 1. The total stacked height of a Pushpin varies depending on the modules used, but is typically on the order of 18mm as well. A description of each module and the connections between them follow.

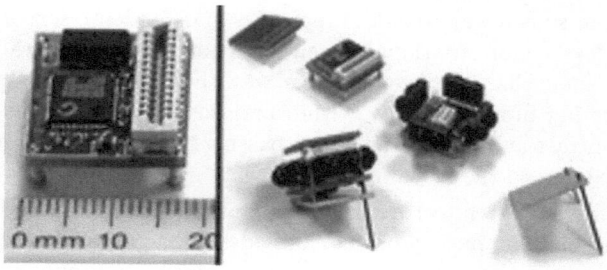

Fig. 1. Modules of a Pushpin

4.1   Power Module

The Pushpin moniker derives from the original power scheme implementation in which protruding from the underside of each Pushpin device are a pair of pins of unequal length that can be easily pushed into a laminate power plane made from two layers of aluminum foil sandwiched between insulating layers of stiff polyurethane foam [9]. One of the foil layers provides power and the other ground. This novel setup satisfies power and usability requirements (no changing of batteries or rewiring of power connections, simply push the Pushpin into the substrate) and hints at the idea of both physically and functionally merging sensing and computing networks with their surroundings. While this solution blatantly sidesteps the important issue of power consumption (the powered substrate is plugged into a power supply), it allows for very quick prototyping and minimal maintenance overhead.

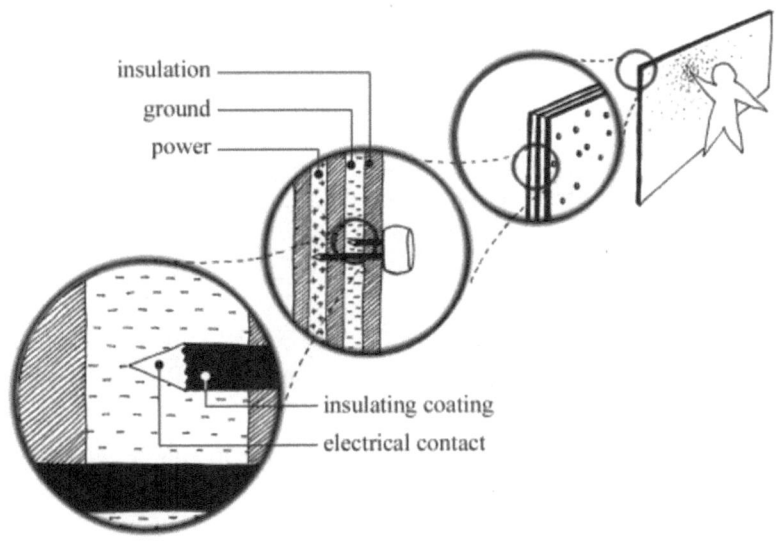

insulation

ground

power

insulating coating

electrical contact

Fig. 2. Pushpin power scheme

Other power sources can easily take the place of the pins and substrate as long as they provide 2.7VDC to 3.3VDC. Two AAA batteries in series is a simple, if bulky alternative. The total power consumed depends strongly on the particular expansion, processing, and communication modules employed and how they are used. For example, the processing module has several different modes of operation, each requiring a different amount of power. Typical current consumption of the processing module running at 22MHz with all necessary peripherals enabled is roughly 10mA, whereas the processing module running in a low-power mode off of an internal 32kHz clock requires roughly $10\mu A$. With the clock shutdown, this falls to about $5\mu A$. Accordingly, the lifespan of a power source can vary from hours to years depending on the particular circumstances.

4.2   Communication Module

Anything containing all the necessary hardware for effectively transmitting from and receiving to a typical hardware UART qualifies as a communication module. That is, the communication board consists of all communication hardware except the UART itself, which is built into the processor on the processing module. Currently, several communication modules are available for Pushpins, including a capacitive coupling module and an infrared module, both of which run at up to 166kbps. See Fig. 3. A radio module is under development. There is also an interface for RS232 communication with a PC over a serial port.

Fig. 3. Pushpins equipped with IR communication modules (and white diffuser rings) drawing power from the laminate substrate

4.3   Processing Module

The Pushpins are designed around the Cygnal C8051F016 – an 8-bit, mixed-signal, up to 25 MIPS, 8051-core microprocessor [10]. The Cygnal chip is equipped with 2.25-Kbytes of RAM and 32-Kbytes of in-system programable (ISP) flash memory. All hardware supporting the operation of the microprocessor as well as the microprocessor itself is contained on the Pushpin processing module. The microprocessor runs off of a 22.1184MHz external crystal but also has its own adjustable internal clock for lower power modes. A simple LED indicates the status of the microprocessor. Connectors providing access to the microprocessor's analog and digital peripherals comprise the remainder of the processing module. See Fig. 4.

4.4   Expansion Module

The expansion module is where most of the user hardware customization takes place for any given Pushpin. The expansion module has access to all the processing module's analog and digital peripherals not devoted to the communication module. This includes general purpose digital I/O, comparators, analog-to-digital converters, capture/compare counters, and IEEE standard JTAG programming and debugging pins, among others. The expansion module contains application-specific sensors, actuators, and external interrupt sources. Examples include sonar transducers, LED displays, microphones, light sensors, and supplementary microcontrollers.

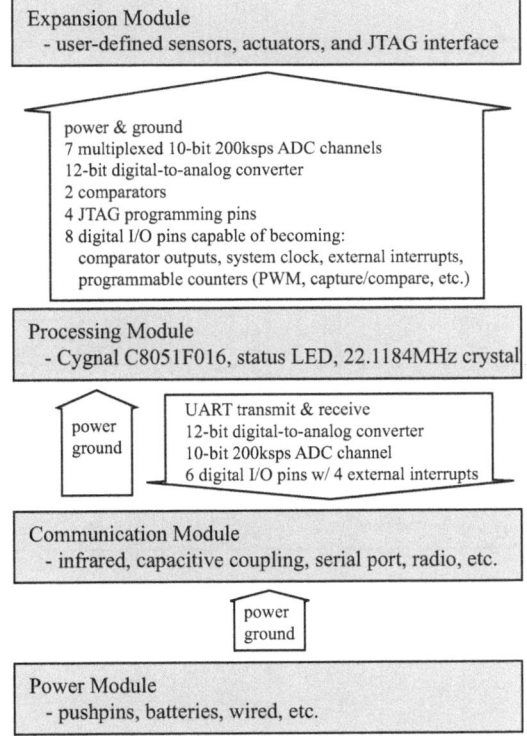

Fig. 4. The Pushpin hardware specification. The shaded boxes represent different hardware modules. The arrows represent resources that the module at the tail of the arrow provides to the module at the head of the arrow

5   Programming Model

The Pushpin programming model is heavily informed by the Paintable Computing programming model [3] and attempts to follow it as closely as possible. The occasional deviations from that model are due to somewhat limited computational resources and reasons of practicality. In essence, the programming model is based on algorithmic self-assembly; the idea that small algorithmic process fragments with simple local interactions with other process fragments can result in complex global algorithmic behavior. In a sense, algorithmic self-assembly treats algorithms in the same way thermodynamics treats gas particles [11]; when the number of particles is large, $pV = nRT$ becomes more useful than knowing the position and momentum of each particle.

The Paintable Computing project successfully demonstrated algorithmic self-assembly in simulation. The goal of the Pushpin programming model is to create a suitable tool for exploring algorithmic self-assembly as it relates to sensory data extracted from the real world. To this end, an operating system, networking protocol, and process fragment integrated development environment (IDE) have been implemented.

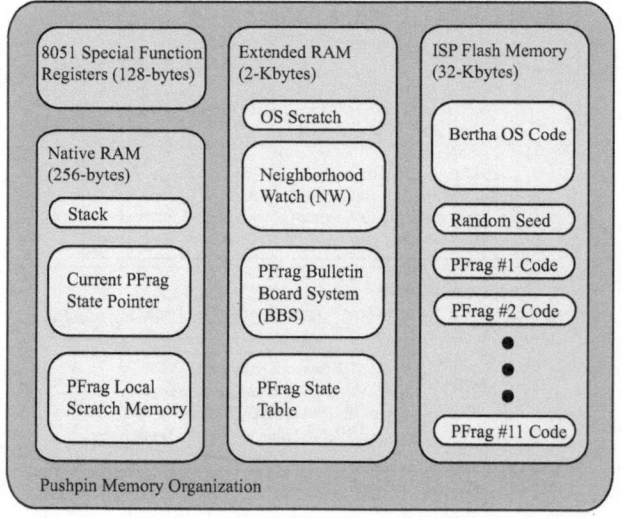

Fig. 5. A Pushpin's memory, carefully divided between process fragments and the operating system

5.1   Process Fragments

A process fragment is the atomic algorithmic unit in algorithmic self-assembly. Carrying the thermodynamics analogy further, a process fragment corresponds to a single gas particle. A process fragment ('pfrag') is defined as the coupling of state information ('state') and executable code ('code'). A pfrag's code acts on or according to the pfrag's state and has the ability to modify it. A process fragment is entirely contained and executed within a single Pushpin, but may transfer or copy itself to neighboring Pushpins and begin execution there. In order to ensure interoperation between process fragments and the Pushpin operating system (Bertha), process fragments must conform to the following constraints:

- Implement an *install* function to be called by Bertha when the process fragment is first executed in a given Pushpin.
- Implement a *deinstall* function to be called by Bertha when the process fragment is to be removed from a given Pushpin.
- Implement an *update* method to be repeatedly called by Bertha as long as the process fragment resides within a Pushpin. There is no guarantee how often the update function will be called, only that it will be called. This is where most of the functionality of a process fragment resides.
- Total process fragment code size limit of 2-Kbytes.
- Total process fragment state size limit of 445-bytes.

Aside from the required functions, process fragments may also contain as much private code as the 2-Kbyte limit allows.

5.2   Bertha: The Pushpin OS

Underlying system operation is handled by Bertha – a small, lightweight operating system developed especially for the Pushpins. Each Pushpin has its own instance of Bertha to manage processor startup, memory, access to hardware peripherals and system services, communication with neighboring Pushpins, and, its primary charge, resident process fragments.

Bertha can accommodate up to 11 process fragments at any given time. Process fragments enter a Pushpin through the communication port either wirelessly via a neighboring Pushpin or from a device pretending to be a Pushpin. The process fragment is written to memory (code to flash memory and state to RAM), checked for errors by means of a simple checksum, added to the list of resident process fragments (assuming the checksum passes), and initialized by calling its *install* function. Bertha executes the *update* function of resident process fragments using a simple round-robin scheme. Each process fragment is allowed to run its *update* function to completion each time it is called. Bertha provides various utility system functions to process fragments, such as those that return the current system time or a pseudo-random number.

Bertha also negotiates all communication on the behalf of process fragments. Specifically, it provides for communication between process fragments in the same Pushpin by means of a bulletin board system (BBS). By making system calls to Bertha, process fragments can post arbitrary messages of limited size to the BBS and read messages posted by other process fragments. A Pushpin's BBS can be posted to and read from only by process fragments within that

Pushpin. Bertha does, however, maintain a Neighborhood Watch – a list of neighboring Pushpins (those within communication range) and brief synopses of their BBSs. The information contained in each neighbor synopsis is culled from that neighbor's own BBS. Due to memory constraints, it is not possible to mirror the entirety of all neighboring Pushpins' BBSs. Instead, whenever a process fragment posts to the local BBS, it has the option of marking that post to be included in the synopsis sent out to neighboring Pushpins. Bertha is responsible for arbitrating which of these posts get included in the synopsis in the case of the synopsis filling up. Currently, Bertha gives priority to newer posts, although this does not have to be the case and process fragments should not assume any particular method for choosing what is included in the synopsis.

Process fragments can make a request of Bertha to transfer them to one of the Pushpins listed in the Neighborhood Watch. When such a request is made, Bertha adds the request to the queue, waits until all resident process fragments have been updated, and then negotiates each transfer request with the appropriate neighbor. No guarantee is made that the transfer will be granted.

At a low level, Bertha manages the Pushpin's half-duplex communication channel with its neighbors using a simple exponential back-off protocol for collision avoidance. Bertha attempts to detect collisions with a simple checksum. To help alleviate the hidden node problem, Bertha is able (when using the capacitive coupling module, but not the IR module) to listen for transmissions from neighbors at a variable threshold. When enabled, Bertha listens at a very low threshold before transmitting and a very high threshold when receiving.

An analog-to-digital converter (ADC) channel in conjunction with a simple voltage divider allows the Pushpin operating system to detect which communication and expansion modules make up the Pushpin (as each type of module produces a characteristic voltage read by the ADC), making for plug-and-play functionality. Once Bertha knows what kind of hardware it is dealing with, it provides mediated access of those resources to resident process fragments. Thus, a process fragment can request to be informed during its next update cycle of a given interrupt being triggered or of a certain condition occurring. Process fragments can also take control of certain hardware peripherals, such as general purpose I/O pins, comparators, and analog-to-digital converter channels.

Since even some of the simplest algorithms already mentioned (e.g. exponential back-off) require randomness, Bertha maintains a 1024-bit seed for use in a pseudo-random number generator. (The size of this seed is unnecessarily large due to an artifact of the hardware organization of the flash memory). This seed can be changed during runtime.

See Fig. 5 for a schematic view of the memory layout of a Pushpin and its operating system.

5.3 Pushpin IDE

Users can create custom process fragments using the Pushpin integrated development environment (IDE). The Pushpin IDE is a Java program that runs on a desktop PC. Process fragment source code is authored within the IDE using a subset of ANSI C supplemented by the system functions provided to process fragments by Bertha, preprocessor macro substitutions, and IDE pre-formatting. The IDE coordinates the formatting of source code, compilation of source code into object files, linking of object files, and transmission of complete process fragments over a serial port to an expectant Pushpin with Bertha installed and running. The IDE also enforces the process fragment structure requirements outlined in §5.1.

Currently, the Pushpin IDE calls upon a free evaluation version of the Keil C51 compiler and Keil BL51 linker [12] to compile and link process fragments. Bertha is initially installed on a Pushpin by way of an IEEE standard JTAG interface. Note that Bertha need not be compiled with any specific knowledge of the process fragments to be used; arbitrary process fragments can be introduced to Pushpins during runtime.

Of course, Pushpins can be programmed directly as a regular 8051-core microprocessor without using either Bertha or the Pushpin IDE. One of the many advantages of Bertha and the Pushpin IDE, however, is that the details of the antiquated Intel 8051 architecture are hidden from the user.

5.4   Security

One of the first observations that can be made about the Pushpin programming model is that it is incredibly insecure by almost any definition of insecure – Bertha runs any well-formed process fragment as raw machine code without any supervision. The only attempt at security is locking the flash memory containing the Bertha code so that it can't be overwritten by a process fragment. Everything else is fair game. Furthermore, there is no built-in protection against rogue process fragments with malicious intent. While security is certainly a valid concern for any system deployed in the world outside of a testbed running in a research lab, it is assumed for now that everyone authoring process fragments received the "plays well with others" stamp of approval. Although security for sensor networks is essentially ignored here, some work has been done on the subject [13]. That said, the Pushpin platform could be used in its own right to explore security issues.

6   Example: Network Gradient

To clarify the idea of process fragments and Pushpin platform operation, we present here a very simple example. The following code fragment simply copies itself to all its neighbors, keeping track of how many hops away it is from its Pushpin of origin. Its *install* routine does almost all the work. Its *update* routine copies the process fragment to neighboring Pushpins. All other required routines are implemented with default routines provided by the Pushpin IDE. The Pushpin IDE also registers this process fragment as GRADIENT with a local process fragment registry it keeps. What follows is the process fragment source code as it would appear in the IDE:

```
state {unsigned char hopsFromOrigin; unsigned char origin;}

globalID {GRADIENT;}

// Continually attempt to migrate to neighboring Pushpins.
//
unsigned int update(unsigned int eventCode, unsigned int eventValue) {
  return requestTransfer(TO_ALL_NEIGHBORS);
}

// Upon waking up in a Pushpin, check to see if there are any
// copies of this PFrag.  If so, compare hops from origin,
// keep lowest hop count, and delete yourself.  If not, check
// if you are the seed of the gradient and set hop count
// accordingly.
//
unsigned int install() {
  BBSPost post;
  getBBSPost(GRADIENT, &post);
  if (isValidBBSPost(&post)) {
    if (post.localID != getLocalID()) {
      if ((post.message[0] > state.hopsFromOrigin + 1)
          && (post.message[1] == state.origin)) {
        post.message[0] = state.hopsFromOrigin + 1;
        updateBBSPost(&post);
      }
      die();
    }
  }
  else {
    post.message[0] = state.hopsFromOrigin + 1;
    if (!isValidMessage(getNeighborMessagePostedBy(GRADIENT))) {
      post.message[1] = getPushpinID();
    }
    else {
      post.message[1] = state.origin;
    }
    postToBBS(&post, 2);
  }
  return 1;
}
```

Note that, for the sake of brevity, this process fragment is implemented in quite an inefficient manner in terms of bandwidth usage and could be improved upon with some effort.

7 Conclusions & Future Work

This paper describes the basic elements of the Pushpin Computing platform, the first hardware instantiation of an environment specifically designed to support algorithmic self-assembly for use in dense sensor networks. In particular, we have introduced the underlying Pushpin hardware and Bertha, a fully functional embedded operating system that supports mobile process fragments.

The work presented is more of a look at things to come than a culmination or conclusion of things that were. In the immediate future, there are plans to implement a Logo virtual machine on the Pushpins, improve error correction and detection, and build several complete networking and sensing applications using on the order of 100 Pushpin nodes. Longer term goals include exploring the potential of Pushpins as a tangible interface, characterizing basic algorithmic elements vital to algorithmic self-assembly in the context of dense sensor networks, and providing a theoretical foundation to describe self-assembly as a general phenomenon.

Detailed information about the Pushpin Computing project can be found at http://www.media.mit.edu/~lifton/Pushpin/.

Acknowledgments. The authors would like to give thanks to the sponsors of the MIT Media Lab for their generous support, and in particular to Steelcase, Inc. for donating the power substrate used in the Pushpin project. Special thanks goes to Dr. Bill Butera for his unflagging encouragement, openness, patience, and effort, without which this project would not be where it is today.

References

1. Morris, E.: *Fast, Cheap & Out of Control*, Sony Pictures Classics, 1997.
2. Paradiso, J.; Hsiao K.;, Strickon J.; Lifton, J.; Adler A.: *Sensor Systems for Interactive Surfaces*, IBM Systems Journal, Volume 39, Nos. 3 & 4, pp. 892-914, October 2000.
3. Butera, W.: *Programming a Paintable Computer*, MIT Media Laboratory, doctoral dissertation, 2002.
4. Resnick, M.: *Turtles, Termites, and Traffic Jams: Explorations in Massively Parallel Microworlds*, The MIT Press, 1994.
5. Culler, D; Hill, J; Buonadonna, P.; Szewczyk, R.; Woo, A.: *A Network-Centric Approach to Embedded Software for Tiny Devices*, to appear in DARPA workshop on Embedded Software.
6. Hill, J; Szewczyk, R; Woo, A.; Hollar, S.; Culler, D. Pister, K: *System Architecture Directions for Networked Sensors*, 27 April 2000.
7. Woo, A.; Culler, D.: *A Transmission Control Scheme for Media Access in Sensor Networks*, Mobicom 2001.
8. Dowling, J.: *Neurons and Networks: An Introduction to Neuroscience*, Chapter 14, Harvard University Press, 1992.
9. Dipline power panel. Donated by Steelcase, Inc.
   http://www.lightandmotion.vienna.at/eng-dipline.html
10. Cygnal C8051016 microprocessor.
    http://www.cygnal.com/products/C8051F016.htm.
11. Reif, F.: *Fundamentals of Statistical and Thermal Physics*, McGraw-Hill, 1965.
12. Keil Software, Inc. http://www.keil.com/demo/
13. Perrig, A.; Szewczyk, R.; Wen, V.; Culler, D.; Tygar, J.: *SPINS: Security Protocols for Sensor Networks*, Mobicom 2001.

Making Sensor Networks Practical with Robots

Anthony LaMarca[1], Waylon Brunette[2], David Koizumi[1], Matthew Lease[1],
Stefan B. Sigurdsson[1], Kevin Sikorski[2], Dieter Fox[2], and Gaetano Borriello[1, 2]

[1]Intel Research Laboratory @ Seattle
[2]Department of Computer Science & Engineering, University of Washington

Abstract. While wireless sensor networks offer new capabilities, there are a
number of issues that hinder their deployment in practice. We argue that
robotics can solve or greatly reduce the impact of many of these issues. Our
hypothesis has been tested in the context of an autonomous system to care for
houseplants that we have deployed in our office environment. This paper
describes what we believe is needed to make sensor networks practical, the role
robots can play in accomplishing this, and the results we have obtained in
developing our application.

1 Introduction

Wireless sensor networks offer new ways to monitor our environment and do so
continuously and invisibly. These networks have wide applicability including
medical, industrial, scientific, military, and consumer applications. Estrin et al. have
described applications to automotive telematics, precision agriculture, and defense
systems [9]. Rabaey et al. have considered management of environmental control
systems in large office buildings [28]. Schwiebert et al. have been working to
develop wireless biomedical sensors [31]. Byers and Nasser suggest using wireless
sensor networks to monitor toxicity levels in hazardous areas [7]. Krishnamurthy and
Conner have used sensor networks to implement basic office information services
such as monitoring the use of highly-coveted conference rooms [8]. From this list, it
is clear that wireless sensor networks can provide important data and context
information for a very wide range of ubiquitous computing applications.

While their potential benefits are clear, a number of open problems must be solved
in order for wireless sensor networks to become viable in practice. These problems
include issues related to deployment, security, calibration, failure detection and power
management. In the last decade, significant advances have been made in the field of
service robotics [11], and robots have become increasingly more feasible in practical
system design. Therefore, we suggest that a number of the problems with wireless
sensor networks can be solved or diminished by including a mobile robot as an
integral part of the system. Specifically, the robot can be used to deploy and calibrate
sensors, detect and react to sensor failure, deliver power to sensors, and otherwise
maintain the overall health of the wireless sensor network. The ideal is for the robot
to do all this while only engaging the user as a last resort (e.g. when new sensors are
needed to replace non-functioning ones).

F. Mattern and M. Naghshineh (Eds.): Pervasive 2002, LNCS 2414, pp. 152-166, 2002.

While this application may be new, robots have been performing these types of services for some time. Service robots are the class of mobile, autonomous robots that operate in human environments to assist and serve. The applications that employ these service robots vary greatly with respect to the level of human interaction. The scale ranges from robots that perform their tasks independently, such as janitorial robots [10] through hospital aides [21], to robots whose chief function is to interact with people, such as entertainment robots [32], museum tour guides [6, 34, 25], and robots that aid the blind [22] and the elderly [29]. Recently Howard et al. used sensors placed in the environment to help mobile robots build navigation maps [18]. However, robots have not been applied to the problem of maintaining a distributed, ubiquitous computing system, such as a sensor network, and keeping such a system running and well-calibrated over an extended period of time.

We have tested our hypothesis in the context of the PlantCare project at the Intel Research laboratory in Seattle. PlantCare is part of an experimental proactive computing platform that serves as an autonomous system to care for houseplants. While caring for houseplants may not be the most promising use of wireless sensor networks, it is a well-defined, practical problem that both encompasses many of the open problems associated with wireless sensor networks and is at the same time conducive to the use of mobile robots.

The PlantCare system consists of a wireless sensor network to measure and report environmental conditions impacting the plants, application logic to monitor these conditions and determine appropriate responsive behavior, and a mobile robot to provide system actuation. Initial results from our system suggest that this combination of robots and wireless sensors has the potential to achieve unprecedented levels of system independence, including the ability to (1) sustain the energy resources of both the robot and the sensor network indefinitely, (2) automate sensor calibration, including both configuration and dynamic response to changing hardware behavior and/or environmental conditions, and (3) detect sensor failure and inappropriate deployment.

The organization of this paper is as follows: In Section 2, we discuss what we believe are three major issues hindering the adoption of wireless sensor networks in practice. We describe the problems, as well as how robots can be used to address them. In Section 3, we discuss the PlantCare system including the sensors, robotics and overall system architecture. In Section 4, we describe our results from a simple experiment with robotic sensor calibration. Finally in Section 5, we discuss future work and conclude.

2 Practical Issues in Wireless Sensor Networks

In the context of the PlantCare project, we have had to deal with a number of practical issues relating to our wireless sensor network: How should new sensors be deployed? How do we know that we are not placing a sensor in an anomalous location that will return unrepresentative readings? How should sensors be calibrated? Should they be calibrated in a controlled environment and then deployed, running the risk of inaccuracy? Or instead, should sensors be calibrated in their deployed location, which would be more accurate but also more difficult? Since we decided to deploy sensors without a constant supply of power, how will we keep the sensors running indefinitely? How will we know when a sensor has failed? While any one of these problems can potentially be ignored or circumvented, collectively they represent an important set of issues that affect the accuracy and robustness of our overall system. We have condensed these issues into three areas: context-aware deployment, continuous calibration, and power delivery.

2.1 Context Aware Deployment of Sensors

Before a sensor can provide useful data to the system it must be deployed in a location that is contextually appropriate. While the physical placement of a sensor is often challenging, as in the case of a reactor or a water main, the issue of proper choice of location based on the application requirements is also difficult and time consuming. For example, consider the placement of a thermostat in a home. Since in most cases only one thermostat is installed, this single sensor will be used to guide all heating and cooling decisions. Thus the sensor should be placed where its readings closely correlate with the temperature of the home in general. Consequently, the placement of the sensor under a cooling vent, or on a section of wall that gets intense sunlight in the morning, will generate contextually inappropriate readings. Even if the sensor is properly calibrated, it will return data that is misleading to the system, and the residents of the home will be uncomfortable. In this example, perfect placement requires an isotherm map of the space that shows the temperature variation by location as well as time. Since this is rarely available, most sensor placement is done using simple heuristics. In the case of home thermostats, for example, the rule of thumb is to place it near the center of the house in a hallway. While this example is fairly simple, and an isotherm map is unlikely to be worth the effort, it illustrates the tradeoffs: The accuracy and relevance of the data collected are proportional to the level of understanding of the environment. Unfortunately, considerable human time is usually required to obtain this understanding.

In many cases, this problem can be solved using robots, as has already been demonstrated in the limited context of active vision [23]. Placing an accurately calibrated sensor on the robot turns it into a mobile wireless sensor. This "robo-sensor" can then move through the environment to take measurements. If an accurate map of the environment is available to the robot for navigation, then environment maps such as the temperature isotherm map mentioned above are straightforward to build. The robot simply needs to visit a series of locations determined by the required measurement density, taking a reading at each after waiting long enough for the sensor to acclimate. In many systems this is a natural extension of mapping tasks already performed by robots. For example, in PlantCare our physical environment is mapped using probabilistic techniques [12] in order to guide robotic navigation (see Figure 1). Taking readings multiple times at varying times of day will create the type of map that can strongly suggest good sensor placement to an application administrator. If robotic placement of the sensor were desired, this same data could guide an algorithm that captures the traditional heuristic wisdom. This is an example of how a robot can be used to accomplish a time-consuming, repetitive task that is likely to be unappealing to humans.

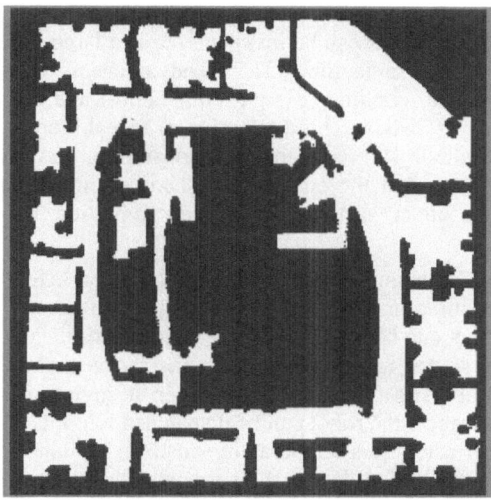

Fig. 1. This map of our office was generated by the robot's mapping system

2.2 Continuous Calibration of Sensors

Calibration is the process of deriving the function that translates the raw readings provided by the sensor into data in the correct units, while taking into account the characteristics of the particular sensor and the environment in which it has been placed. Calibration of sensors is critical to obtaining accurate measurements. Without calibration, readings produced by a sensor could range anywhere from having subtle inaccuracies to being completely meaningless. Our temperature sensors, for example, produce a voltage varying from 0 to 3 volts. The only way to know that a reading of 1.5v corresponds to 32 degrees Celsius is to obtain a number of readings from the sensor when the true temperature is known and use these calibration points to translate the voltage readings to the Celsius scale. Unfortunately, this mapping varies from sensor to sensor (especially as manufacturers endeavor to make sensors as low-cost as possible) and may even change over the lifetime of a deployed sensor (due to changes in environmental conditions or wear on the sensor). The result is that for most applications, sensors require both initial calibration and periodic recalibration to ensure accurate readings.

Sensors can be calibrated either before or after deployment. Pre-deployment calibration is performed in a controlled space in which the environmental factor measured by the sensor can be carefully regulated. The factor is then cycled through a set of values intended to represent the range the sensor will encounter in practice, and readings are taken from the sensor during this process. These actual vs. sensed values form the set of calibration data that can be used to derive a function that maps one to the other. For example, linear regression is a simple yet accurate technique for performing this mapping. In the event that the sensor data cannot be mapped into linear space, more complex algorithms can be employed. The big advantage of pre-deployment calibration is that it can be inexpensive, as a large number of sensors can often be calibrated at the same time. The disadvantage is that sensors are often sensitive to environmental conditions (e.g., temperature) and the act of deployment can change their characteristics. The alternative is to calibrate the sensor in-place, after it has been installed. This is more time consuming and inconvenient. It also suffers from the problem that the environmental factors may change slowly and it may be difficult to collect sufficiently varied calibration data to get accurate measurements.

A robot equipped with an accurate, calibrated sensor offers a novel solution: adaptive in-place calibration. Given robotic support, a sensor can be deployed uncalibrated and a robot can be used to visit the sensor periodically to take calibration readings. As long as sensor readings remain in the range of these initial calibration readings, no additional visits are required. However, if environmental factors deviate outside the calibrated range, the robot can be dispatched to collect additional readings. For example, consider a wireless temperature sensor in an unheated warehouse. The sensor is deployed uncalibrated during the summer. Over a period of five days, the robot collects a set of calibration readings. These readings combined with simple regression provide sufficient accuracy for the next three months. When fall sets in, the temperature drops, and the system determines that the summer calibration data is not sufficient to predict the characteristics of the sensor. The robot is then re-deployed to collect new calibration samples to extend the range of the regression model. Again we see the robot filling a role that a human could but would never want to: namely, as an agent on call ready to visit any sensor whenever needed.

Recalibration is another area in which a robotic solution offers advantages over the standard technique. Currently, the most common method used to ensure sensors are operating correctly is periodic manual calibration [17]. This approach has a number of drawbacks. For instance, manual labor is expensive, so sensor recalibration is performed infrequently resulting in overall inaccuracy. Complementing our robotic solution, we assign a limited lifetime to calibration data as it is collected. Rather than a task performed once or twice a year, recalibration becomes an ever-present background task of the system. Once the lifetime of a piece of calibration data has expired, it is considered inaccurate and is replaced by fresh data. The result is an adaptive mechanism in which the tradeoff of sensor accuracy vs. robotic work can be adjusted at will.

The issue of failure detection is tied closely to calibration. Detecting a failed sensor is the pathological case of calibration when it is decided that the sensor readings no longer correlate with the environmental factor in question. This detection is fairly straightforward in the case of a gross fault, in which a sensor begins reporting drastically different conditions. Regular periodic calibration by our robotic solution also removes the traditional need to detect drift errors, in which a sensor's accuracy slowly decays over time [37]. Without sophisticated models, however, it is very difficult to detect when a faulty sensor returns plausible, possibly varying, yet uncorrelated data. Because sensors are only checked periodically, faulty sensors can remain in service for periods up to the calibration frequency. Depending on the scenario in which the sensors are used, this could pose severe consequences. To mitigate this, the calibration cycle can be made more frequent by using a smaller calibration data lifetime.

Inferential sensing is an emerging technique that continuously verifies sensor accuracy via on-line, real-time modeling [17]. This technique uses historical data rather than sensor redundancy [19] to construct predictive models of sensor behavior and can significantly enhance system fault detection and handling. Whenever a sensor provides a reading, the data is compared to estimated values produced by the model, generating differences known as residuals. A decision logic module then statistically evaluates each residual to generate a health metric assigned to the corresponding sensor. Instead of performing periodic recalibration, a human operator monitors these health metrics in order to schedule maintenance proactively or provide it when necessary. Unlike simpler methods, this approach can detect sensors that fail in non-trivial ways. In addition, these predictive models can be used to generate estimated sensor readings while a sensor is offline or awaiting recalibration or replacement. These predictive modeling techniques complement our robotic calibration approach. In the case of sensor failure, the predictive model can serve as warning system and the robot can verify the failure. When the predictive model suspects a sensor of drifting, it can prematurely age the sensor's calibration data, forcing a more timely robotic recalibration.

While robotic calibration solves many problems, it has a set of issues and considerations of its own. For accurate calibration data to be collected by the robot, the system must know precisely where the sensors are located. In the case of statically positioned sensors, this requires that a detailed sensor map be provided to the system. In the case of mobile sensors, a sophisticated localization system is required. Current localization systems use ultrasound [36], RF signal strength [2], radio time-of-flight [15], or some combination of technologies to pinpoint the physical location of objects in space. Second, the sensor needs to be accessible to the robot, and not all deployment environments allow this. In some situations the robot cannot reach the sensor (e.g. a sensor on the ceiling of a room), while other environments are physically hostile to robots. Finally, some sensors require physical integration with the environment in order to obtain accurate measurement, and as a result, corroboration cannot be achieved by placing a similar sensor nearby. For example, a stress sensor installed within a concrete wall cannot be verified by a mobile sensor no matter how close it gets.

2.3 Renewable Energy Provided by Mobile Service Robots

An obvious limitation of wireless sensor networks is the lack of a continuous energy supply. To make matters worse, the additional freedom of wireless networks allows one to envision deployment environments lacking any nearby infrastructure for supplying power. Given these constraints, it is generally accepted that the majority of wireless sensor networks will face a world of severe energy restrictions [27]. We review current approaches to dealing with the power problem and describe how robots both complement and extend these techniques.

The most popular approach assumes a sensor network is deployed with batteries that will not be recharged or replaced. In this scenario, power conservation is paramount and the network is considered disposable with the rate of energy consumption determining its operational lifetime. Many alternatives for conserving power have been considered [3, 7]. For example, Heidemann et al. have proposed application-specific network topologies in order to reduce costly communications in ad-hoc wireless sensor networks. [14]. Bhadwaj et al. have even built an abstract model of such wireless sensor networks in order to derive an upper bound on their lifetimes. [4]

For some applications, however, this model of wireless sensor networks is not appropriate: The sensor network cannot be treated as disposable and it is possible to sustain the sensors by recharging or replacing batteries when needed. For example, Schwiebert et al. are considering biomedical monitoring via wireless sensors implanted in the body. Because sensors are intended for long-term use, they have proposed using radio frequencies (RF) or infrared (IR) signals to inductively charge the implanted sensors from an external power source [31]. While recharging is well-suited to this application because the patient can wear a compact, portable power supply, not all applications can depend on this level of user interaction.

Some have taken this a step further to propose extracting energy directly from the deployment environment. These "scavenging" techniques propose powering sensors via solar power [35, 20], kinetic energy [26], floor vibration, and acoustic noise [28]. However, scavenging techniques are only now becoming capable of generating the level of power required to sustain current wireless sensor applications, and not all deployment environments are conducive to such techniques.

An alternative approach is to use passive sensors. Unlike the active sensors considered thus far, passive sensors require no local power source. For example, a surface-acoustic-wave (SAW) passive sensor is powered entirely by the RF field used to read it. A broad range of SAW sensors are available for measuring temperature, pressure, magnetic field strength, torque, etc. [13, 30]. In general, passive sensors offer the advantage of lower cost and higher robustness than active sensors but tend to require more expensive infrastructure. Also, because passive sensors depend on external power, they measure their environment only when polled; an event-driven application would be required to perform such polling regularly and generate appropriate events on behalf of the sensors.

As an alternative to these approaches, we propose using mobile service robots to sustain a deployed, active sensor network. Because we require our PlantCare system to run unassisted for long periods of time, we can neither treat the sensor network as disposable nor have it depend on an administrator who can perform recharging or battery replacement. Further, passive sensors do not yet exist to measure all of the environmental conditions pertinent to plant care, such as soil moisture content. On the other hand, Michaud et al. have already demonstrated that mobile service robots are capable of recharging themselves [24], thus it appears an appropriate task to have a robot of this kind deliver power to deployed active sensors as the need arises.

The power could be delivered in a variety of forms. A sufficiently agile robot could replace weak batteries in a sensor with fresh ones. An easier approach is to outfit the robot with equipment to recharge sensors using inductance or direct electrical connection. While this idea is simple to understand and straightforward to engineer, it has the potential to greatly increase the flexibility of wireless sensor networks. With a robot integrated in the system, wireless sensor nodes can be placed in locations in which no power is available at all. Infrequent visits by the robot enable the wireless node to perpetually participate in the application without any human intervention.

In addition to delivering power to active sensors, robots can potentially improve the efficacy of passive sensor networks as well. A mobile service robot equipped to perform inductive charging via RF could read a SAW sensor using the same equipment. This makes it easier to support a mixed environment of both active and passive sensors as well as reducing the infrastructure required to read a physically disparate collection of passive sensors.

3 PlantCare

We have been exploring the relationship between wireless sensor networks and robots in the context of the PlantCare project. The goal of the project is to understand and develop solutions to the challenges facing proactive computing. Proactive computing differs from pervasive computing in the sense that proactive applications have a component that anticipates needs and provides for dealing with them without the user's attention being called to the problem unless absolutely necessary, and then only at an appropriate level of abstraction. In general, the goal of proactive computing is to develop solutions to real problems that may involve hundred or thousands of devices, but present little distraction or cognitive load on their users. Specifically, PlantCare is trying to build a zero-configuration and distraction-free system for the automatic care of houseplants. Our plan was to instrument each plant with a wireless sensor placed in its pot and employ a robot to deliver water to the plants. Soon after conceptualizing the project we realized that due to power constraints, the sensor network and robot needed as much care as the plants themselves. This realization led to the idea of robots as a worthy caretaker for wireless sensor networks, in general. In order to provide background for the work in this paper, we present a brief description of the sensors, robots, and software employed by the PlantCare project.

3.1 The Sensors

In the PlantCare system, wireless sensor nodes (see Figure 2) are placed both on the robot and in the plants being cared for. The sensors in the plants provide a continuous stream of data reflecting their state while the sensor node on the robot is used to calibrate the sensors. While the sensors in the plants and on the robot vary slightly, the wireless nodes are identical. PlantCare's sensors are built using the UC Berkeley "mote" sensor platform running TinyOS [16]. Motes operate at 3V and are assembled from off-the-shelf components that include an 8-bit microcontroller, a two-way 916MHz radio for communication, and an expansion connector that facilitates connection of environmental sensors. TinyOS is a small, real-time, modular operating system that supports ad-hoc networking to allow motes to communicate both with each other and with a base station. Our environmental sensing hardware consists of a photo-resistor for measuring light levels, a thermistor for measuring temperature, an irrometer for measuring soil moisture content, and a sensor that monitors the current charge of the power source. In addition, the sensor nodes in our plants have been augmented with a custom power system in which capacitors replace traditional batteries and can be recharged using an inductive coil to support power delivery.

Our wireless network contains a single base station mote, which by virtue of being attached to the serial port of an Internet-connected PC serves as the physical link between the wireless sensor network and the PlantCare services. The base station listens to the sensor network for messages containing sensor readings and forwards these messages to the serial port. Additional software infrastructure described in Section 3.3 handles the processing of these messages on the PC.

3.2 The Robot

The robot hardware platform (see Figure 3) consists of a Pioneer 2-DX mobile robot [1] augmented with custom hardware for watering plants, recharging the robot, recharging remote sensors, and sensing environmental conditions for calibration purposes. To deliver water to the plants, the robot has been fitted with a small water tank, dispensing spout, and pump. To deliver power to wireless sensors an inductive charging coil has been positioned near the watering spout. Similarly, another paddle-shaped inductive charge coil has been added to the robot to allow it to recharge itself at its "maintenance bay". In order to support calibration, the robot includes a sensor node that was human-calibrated. Finally, a small microcontroller board allows software on the robot to both control and read the state of this collection of custom hardware. Both this microcontroller and the laser scanner the robot uses for navigation are connected to a laptop that runs the robot's control and navigation algorithms and is in turn connected to the network via an IEEE 802.11b wireless card.

Lastly, the robot has a maintenance bay it uses to automatically charge its own batteries and refill its water reservoir. The bay has a water supply with a spout for dispensing water to the robot, and a charging system matched to the robot's induction coil. We envision that the bay would take the form of a kitchen cabinet in a more consumer-realistic deployment.

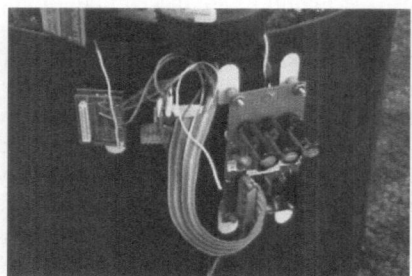
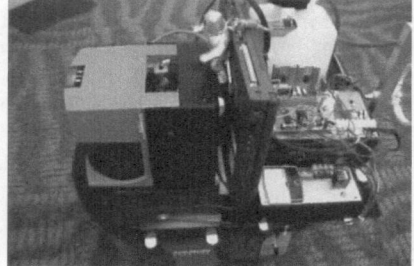

Fig. 2. A deployed plant sensor   **Fig. 3.** The robot hardware platform

We chose inductive charging for both the sensors and the robot in order to reduce the danger to people and equipment due to splashing water. The measured efficiency of inductively charging the sensors is around 70% of the baseline efficiency achieved with a shielded cable. This inefficiency reduces the amount of time the robot can function without recharging, thereby resulting in more frequent visits to the maintenance bay. This was deemed acceptable given the potential risk of the accidental meeting of water and electricity.

The main components of the robot navigation system consist of a reactive collision avoidance module, a module for map building and path planning, and a localization module. All components use probabilistic methods to deal with uncertain sensor information. The reliability of this approach has been demonstrated during the deployment of the robots Rhino and Minerva as autonomous museum tour guide robots [6, 34]. The high-level task ordering and dispatching software was custom-built for the PlantCare project.

3.3 The Software

As part of the PlantCare project we have also developed a software infrastructure called Rain to support proactive applications. The objective of Rain is to provide a framework in which to experiment with how to structure applications as a collection of cooperating services that communicate via asynchronous events. In the Rain environment, services register with a central discovery service and use this service to find other services they wish to interact with. This structure gives the application the opportunity to transparently support the highly dynamic environments envisioned for proactive systems. The asynchronous communication model allows applications to be highly responsive even in the face of widely distributed services running on hardware platforms with widely varying performance. Finally, to support heterogeneous computing environments, Rain messages are encoded in XML. Services in Rain are very similar to SOAP services that communicate by passing XML documents asynchronously [5]; The main difference between Rain and SOAP is that SOAP is as almost exclusively used as a synchronous RPC system and Rain is geared towards support of asynchronous event-based software architectures.

Our PlantCare application is composed of fifteen services that collectively provide both the high-level application logic as well as the low-level driver-like code that communicates with hardware and external software. Specific to our sensor network, there are services that independently receive data from the sensor base stations, unpack the data from its proprietary form, calibrate the data reading based on previously collected calibration data, and store the readings for future use by applications. The services pertaining to the robot consist of a low-level service that knows how to activate the robot's sensors and actuators, and a high-level service that encapsulates the understanding of our application-specific robotic tasks such as watering plants and delivering power to the motes.

4 A Simple Experiment

To demonstrate the use of a service robot for the calibration of a wireless sensor network we performed a simple experiment. The experiment emulated the lifetime of a plant sensor from initial deployment through continuous, adaptive calibration with the help of the robot, and was controlled to eliminate the effects of power constraints and navigation errors.

An uncalibrated sensor was deployed and calibrated in-place. After a period of operation we changed the environmental characteristics the sensor was measuring. The subsequently reported readings then no longer fell within the range of the previously collected calibration data, which forced the system to gather additional calibration data. This mimics the previously mentioned example of a temperature sensor that is calibrated during summertime conditions but needs further calibration at the onset of winter. Finally, we simulated a gross change in sensor behavior by physically obscuring the sensor. Once the resulting measurement error was detected by the system during a simulated maintenance check performed by the robot, the old calibration data was discarded and new calibration data was collected.

4.1 Experiment Setup

The experiment setup consists of a darkroom with a single constant intensity light source, and a single wireless sensor node equipped with a light sensor. The light source is passed through two polarizing filters, one of which is rotated at set intervals to make the intensity of the emitted light approximate a sine wave. For the first half of the experiment the rotation of the filter is limited between 0 and 45 degrees. This creates a wave that is cut off at half the possible amplitude. After two periods of this smaller wave the rotation of the filter is extended to the full range of 0 to 90 degrees to create a change in environmental conditions. This full range of light is projected onto the sensor for two full periods. The sensor is then covered by a semi-transparent filter to emulate degradation of the sensor. Finally, the full range of light is again projected onto the partially obscured sensor for two periods.

During the entire experiment, the wireless node is reporting its light readings to a base station via its radio. This data is stored in a database and post-processed using linear regression to convert these raw sensor readings into luminance values. When the algorithm needed a new calibration point, the raw sensor voltages were paired with the actual light intensity. This simulates a robot with a perfect sensor. While sensors are never perfect, in general we expect that the sensors on the robot will be significantly more accurate than those in an inexpensive wireless sensor node.

4.2 Experiment Results

Figure 4 presents three time-series graphs representing different aspects of the data collected during the experiment. In all of these graphs the vertical axis shows either volts or luminance, while the horizontal axis always shows time. In the first graph, the solid line shows the voltage values measured by the sensor during the experiment, compared with the broken line showing the actual luminance of the light source. Note that until the sensor is obscured, the voltage readings reach higher on the graph than the luminance values. Once the sensor is obscured, the voltages measured are lower though the luminance has not changed. It is clear that without recalibration, the sensor readings cannot be translated accurately across the entire time line.

The second graph in this series compares the calibrated sensor readings to the actual luminance. The markers on the broken line represent times at which the system collected calibration samples. Moving from left to right, we see that the system initially collected eight calibration points. Once the measured luminance starts to fall, no additional calibration points are collected, as the system is comfortable with the range of data it had already collected. Note also that the measured line accurately tracks the actual luminance except in the peaks and troughs of the graphs. We attribute these inaccuracies to limitations of our low-cost photo-resistor, pointing out one of the realities of using inexpensive sensors.

Midway through the third period, two things happen. The difference between the calibrated readings and the actual luminance increases. This is because the regression

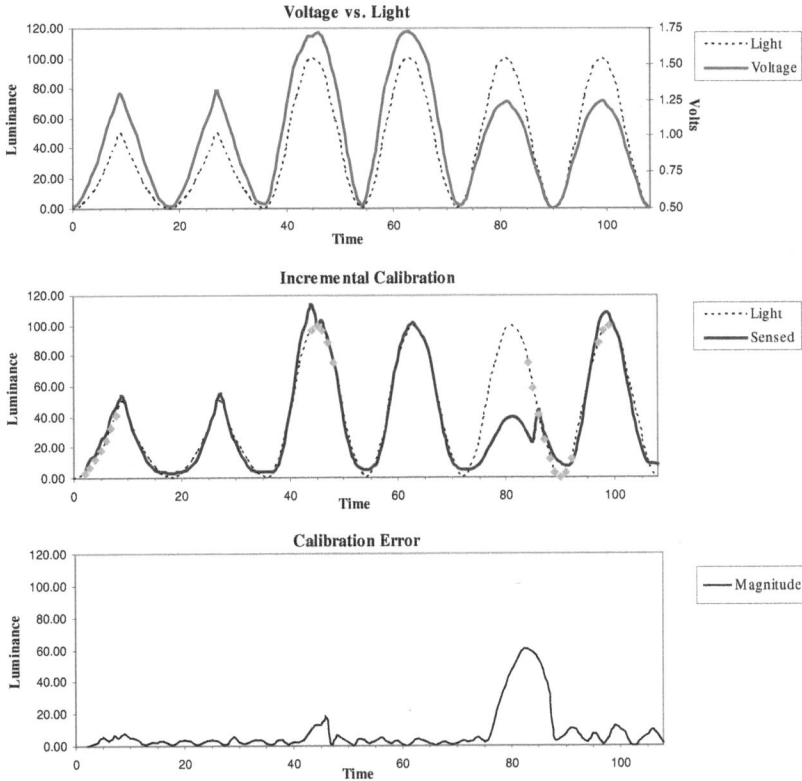

Fig. 4. The top graph shows raw voltage vs. luminance over six periods of sinusoidal variation. The second graph compares the calibrated readings vs. the actual luminance for the same time span. The marks on the graph represent times at which calibration data was collected. The third graph shows the error in the calibrated readings

algorithm was being forced to make predictions outside its range of collected data. When the system realized this, it collected an additional four calibration points after a delay approximating the navigation latency of the robot. After incorporating the additional points, the measured values again track the actual luminance well.

Two thirds of the way through the experiment the sensor was covered with a filter to change its characteristics. Since the system was unaware of this, the converted measurements are in fact quite inaccurate. This could not be detected by the system and would have continued until a routine robotic check of the sensor revealed the inaccuracy. The delay of a few readings approximates the wait until the next check of the sensor. At this point, the system realized a significant change in the sensor had occurred, discarded all of its calibration data, and began collecting new data. After the recalibration period, the readings again track the actual luminance well.

Lastly, the third graph in this series shows the error between the calibrated sensor readings and the actual luminance. During the initial calibration period there is higher than average error. Another period of increased error occurs when the environmental conditions change and only settles down once additional calibration data is gathered. At the end of the measurement there is a very large spike of error while the system continues to use calibration data that has become inaccurate. Once the error is detected and new calibration data is gathered, the error again decreases. It is interesting to note that even after recalibration the error does not diminish to the extent it did following initial calibration. This is due to the fact that obscuring the sensor rendered it less sensitive to light. While our goal was to change the scale of the voltages returned, we also inadvertently compromised its accuracy.

5 Conclusions and Future Work

In this paper, we have introduced the idea that robots have the potential to greatly increase the feasibility of practical wireless sensor networks. While sensor networks and robotics are both quickly evolving fields, the union of the two fields seems inherently symbiotic. Sensor networks have data but lack actuation, while robots have actuation but limited sensing. We have explored this concept in the context of the PlantCare system, an autonomous system for managing the health of houseplants. We have presented data from a simple initial experiment showing how robots can be used to continuously calibrate deployed sensors. In the future, we intend to more deeply explore the relationship between robots and sensor networks. We plan to leverage techniques developed in the robotics community to build spatial models from noisy sensor information and to keep track of complex dynamic systems [33]. We also plan to explore the idea of treating localization data as just another aspect of the sensed environment, enabling localization to benefit from all of the advantages of continuous calibration.

References

1.   ActivMedia Robotics, http://www.activrobots.com, visited Feb. 2002.
2.   P. Bahl and V. Padmanabhan. *RADAR: An in-building RF-based user location and tracking system. In Proceedings of IEEE INFOCOM*, volume 2, pages 775-784, March 2000.

3. P. Bhagwat et al. System Design Issues for Low-Power, Low-Cost Short Range Wireless Networking. In IEEE International Conference on Personal Wireless Communications, 1999, pp. 264-268.
4. M. Bhardwaj, A. Chandrakasan, and T. Garnett. Upper Bounds on the Lifetime of Sensor Networks. IEEE International Conference on Communications, 2001, vol.3 pp. 785-790.
5. D. Box et al. Simple Object Access Protocol (SOAP) 1.1, World Wide Web Consortium (W3C), May 2000. http://www.w3.org/TR/2000/NOTE-SOAP-20000508, visited Feb. 2002.
6. Burgard, W., A. Cremers, D. Fox, D. Haehnel, G. Lakemeyer, D. Schulz, W. Steiner and S. Thrun. 1999. Experiences with an interactive museum tour-guide robot. Artificial Intelligence.
7. J. Byers and G. Nasser. Utility-Based Decision-Making in Wireless Sensor Networks. In IEEE MobiHoc, 2000.
8. W. S. Conner, L. Krishnamurthy, and R. Want, Making Everyday Life Easier Using Dense Sensor Networks. In Proceedings of the Third International Conference on Ubiquitous Computing (Ubicomp 2001), 49-55. 2001.
9. Embedded, Everywhere: A Research Agenda for Networked Systems of Embedded Computers. Computer Science and Telecommunications Board (CSTB) Report.
10. Endres, H., W. Feiten, and G. Lawitzky. 1998. Field test of a navigation system: Autonomous cleaning in supermarkets. In Proc. of the 1998 IEEE International Conference on Robotics & Automation (ICRA 98).
11. Engelberger, G. 1999. Services. In Nof, S. Y., ed., Handbook of Industrial Robotics. John Wiley and Sons, 2nd edition. Chapter 64, 1201-1212.
12. D. Fox, S. Thrun, F. Dellaert, and W. Burgard. *Particle filters for mobile robot localization*. Springer-Verlag, New York, 2001.
13. M. Hauser, L. Kraus, and P. Ripka, *Giant magnetoimpedance sensors*. IEEE Instrumentation & Measurement Magazine, 2001, Volume 4 Issue 2, pp. 28-32.
14. J. Hiedemann et al. Building Efficient Wireless Sensor Networks with Low-Level Naming, SOSP, 2001.
15. J. Hightower, R Want, and G Borriello. SpotON: An indoor 3d location sensing technology based on RF signal strength. UW-CSE Tech Report 00-02-02, University of Washington, Department of Computer Science and Engineering, Seattle, WA, February 2000.
16. J. Hill, R. Szewcyk, A. Woo, D. Culler, S. Hollar, K. Pister. 2000. System Architecture Directions for Networked Sensors. Architectural Support for Programming Languages and Operating Systems 2000.
17. J. W. Hines, A. Gribok, and B. Rasmussen. On-Line Sensor Calibration Verification: "A Survey". 14th International Congress and Exhibition on Condition Monitoring and Diagnostic Engineering Management, September 2001.
18. A. Howard, M.J. Mataric, and G.S. Sukhatme. Relaxation on a mesh: A formalism for generalized localization. In *Proc. of the IEEE/RSJ International Conference on Intelligent Robots and Systems (IROS 2001)*, 2001.
19. C. Jaikaeo, C. Srisathapornphat, and C.-C. Shen. Communications, 2001. ICC 2001. IEEE International Conference on , Volume: 5 , 2001, pp. 1627 -1632.
20. J. M. Kahn, R. H. Katz and K. S. J. Pister. Next Century Challenges: Mobile Networking for "Smart Dust". MobiCom, 1999, pp. 271-278.
21. S. King, and C. Weiman. 1990. Helpmate autonomous mobile robot navigation system. In Proceedings of the SPIE Conference on Mobile Robots. 190-198. Volume 2352.
22. G. Lacey, and K. Dawson-Howe. 1998. The application of robotics to a mobility aid for the elderly blind. Robotics and Autonomous Systems 23:245-252.
23. P. Lehel, E. Hemayed, and A. Farag. Sensor planning for a trinocular active vision system. IEEE Computer Society Conference on Computer Vision and Pattern Recognition , 1999 -312 Vol. 2.

166   A. LaMarca et al.

24. F. Michaud et al. Experiences with with an autonomous robot attending AAAI. IEEE Intelligent Systems, 2001, Volume: 16 Issue: 5, pp. 23-29.
25. I. Nourbakhsh, J. Bobenage, S. Grange, R. Lutz, R. Meyer, and A. Soto. An affective mobile educator with a full-time job. *Artificial Intelligence* , 114(1-2):95--124, 1999.
26. J. Paradiso and M. Feldmeier, A Compact, Wireless, Self-Powered Pushbutton Controller. In Proceedings of the Third International Conference on Ubiquitous Computing (Ubicomp 2001), 299-304. 2001.
27. G. Pottie. Wireless SensorNetworks In Information Theory Workshop, 1998, pp. 139-140.
28. J. Rabaey and et al. PicoRadio Supports Ad Hoc Ultra-Low Power Wireless Networking. IEEE Computer, July 2000, Vol. 33, No. 7, pp. 42-48.
29. N. Roy, G. Baltus, D. Fox, F. Gemperle, J. Goetz, T. Hirsch, D. Margaritis, M. Montemerlo, J. Pineau, J. Schulte and S. Thrun. 2000. Towards personal service robots for the elderly. In Workshop on Interactive Robots and Entertainment (WIRE 2000).
30. G. Schimetta, F. Dollinger, and R. Weigel. A wireless pressure-measurement system using a SAW hybrid sensor. IEEE Transactions on Microwave Theory and Techniques, 2000, Volume 48 Issue 12 , pp. 2730-2735.
31. L. Schwiebert et al. Research Challenges in wireless networks of biomedical sensors. In MobiCom, 2001.
32. Sony AIBO homepage. http://www.world.sony.com/Electronics/aibo. Visited in February 2002.
33. S. Thrun, W. Burgard, and D. Fox. A real-time algorithm for mobile robot mapping with applications to multi-robot and 3D mapping. In *Proc. of the IEEE International Conference on Robotics & Automation (ICRA 2000)* , 2000.
34. S. Thrun, M. Bennewitz, W. Burgard, A. Cremers, F. Dellaert, D. Fox, D. Haehnel, C. Rosenberg, N. Roy, J. Schulte and D. Schulz. 1999. MINERVA: A second generation mobile tour-guide robot. In Proceedings of the IEEE International Conference on Robotics and Automation (ICRA).
35. E. Tunstel, R. Welch, and B. Wilcox. Embedded Control of A Miniature Science Rover For Planetary Exploration. 7th International. Symposium. on Robotics with Applications, 1998.
36. Ward A. Ward, A. Jones, A. Hopper. A New Location Technique for the Active Office. IEEE Personal Communications, Vol. 4, No. 5, October 1997, pp. 42-47.
37. Xu *et al.* On-Line Sensor Calibration Monitoring And Fault Detection For Chemical Processes. *Maintenance and Reliability Conference (MARCON 98)*, Knoxville, TN, May 12-14, 1998.

Modeling Context Information in Pervasive Computing Systems*

Karen Henricksen, Jadwiga Indulska, and Andry Rakotonirainy

School of Information Technology and Electrical Engineering
The University of Queensland
St Lucia QLD 4072 Australia
{karen, jaga, andry}@itee.uq.edu.au

Abstract. As computing becomes more pervasive, the nature of applications must change accordingly. In particular, applications must become more flexible in order to respond to highly dynamic computing environments, and more autonomous, to reflect the growing ratio of applications to users and the corresponding decline in the attention a user can devote to each. That is, applications must become more context-aware. To facilitate the programming of such applications, infrastructure is required to gather, manage, and disseminate context information to applications. This paper is concerned with the development of appropriate context modeling concepts for pervasive computing, which can form the basis for such a context management infrastructure. This model overcomes problems associated with previous context models, including their lack of formality and generality, and also tackles issues such as wide variations in information quality, the existence of complex relationships amongst context information and temporal aspects of context.

1 Motivation

The emergence of new types of mobile and embedded computing devices and developments in wireless networking are driving a spread in the domain of computing from the workplace and home office to other facets of everyday life. This trend will lead to the scenario, often termed pervasive computing, in which cheap, interconnected computing devices are ubiquitous and capable of supporting users in a range of tasks. It is now widely acknowledged that the success of pervasive computing technologies will require a radical design shift, and that it is not sufficient to simply extrapolate from existing desktop computing technologies [1, 2]. In particular, pervasive computing demands applications that are capable of operating in highly dynamic environments and of placing fewer demands on user attention. In order to meet these requirements, pervasive computing applications will need to be sensitive to context. By context, we refer to the circumstances or situation in which a computing task takes place.

* The work reported in this paper has been funded in part by the Co-operative Research Centre Program through the Department of Industry, Science and Tourism of the Commonwealth Government of Australia.

F. Mattern and M. Naghshineh (Eds.): Pervasive 2002, LNCS 2414, pp. 167–180, 2002.
© Springer-Verlag Berlin Heidelberg 2002

Currently, the programming of context-aware applications is complex and laborious. This situation could be remedied by the creation of an appropriate infrastructure that facilitates a variety of common tasks related to context-awareness, such as modeling and management of context information. This paper addresses the former issue by presenting a model of context for pervasive computing that is able to capture features such as diversity, quality and complex relationships amongst context information. The structure of the paper is as follows. Section 2 examines the nature of context information in pervasive computing environments in order to determine the requirements that a model of context must satisfy. Section 3 characterizes related work in the field of context-awareness, and evaluates its ability to support the requirements outlined in Section 2. Next, Section 4 describes our context modeling approach, and Section 5 presents some concluding remarks and outlines topics for future research.

2   Defining Context

The term context is poorly understood and overloaded with a wide variety of meanings. Various definitions of context have been put forward in the literature, but even these offer few clues of the properties that are of interest when modeling context. In this section we explore some of the characteristics of context information, using a case study as the basis for our discussion.

2.1   Case Study: Context-Aware Communication

One of the most compelling uses of context is in communications applications. Context-aware communication has been widely researched [3,4,5], and therefore is reasonably well-understood. We discuss a variation of this application here in order to illustrate the nature of context information required by pervasive computing applications, and then return to this case study throughout the paper in order to illustrate our context modeling concepts.

> Bob has finished reviewing a paper for Alice, and wishes to share his comments with her. He instructs his communication agent to initiate a discussion with Alice. Alice is in a meeting with a student, so her agent determines on her behalf that she should not be interrupted. The agent recommends that Bob either contact Alice by email or meet with her in half an hour. Bob's agent consults his schedule, and, realizing that he is not available at the time suggested by Alice's agent, prompts Bob to compose an email on the workstation he is currently using, and then dispatches it according to the instructions of Alice's agent.
> A few minutes later, Alice's supervisor, Charles, wants to know whether the report he has requested is ready. Alice's agent decides that the query needs to be answered immediately, and suggests that Charles telephone her on her office number. Charles' agent establishes the call using the mobile phone that Charles is carrying with him.

The agents in this scenario rely upon information about the participants and their communication devices and channels. For each participant, they require knowledge about the participant's activities (both current and planned), the devices he/she owns, and those he/she is currently able to use. They must also know the relationships that exist between people, such as who supervises whom and who works with whom. Finally, the agents require information about the communication channels that a participant can use and the devices that are required by each channel.

Such information can be collected from a range of sources. Some information must be explicitly supplied by users, such as that concerned with the relationships between people and the ownership of devices and communication channels. Other information may be obtained by hardware or software sensors, such as the proximity of users to their computing devices. Still other information may be derived from multiple sources; for example, user activity may be partly determined by the information stored in the user's diary and partly derived from other related context, such as the user's location. As a result, context information exhibits a diverse array of characteristics, which we now discuss.

2.2   Characteristics of Context Information

In this section, we make a number of observations about the nature of context information in pervasive computing systems. These determine the design requirements for our model of context.

Context Information Exhibits a Range of Temporal Characteristics.
Context information can be characterized as static or dynamic. Static context information describes those aspects of a pervasive system that are invariant, such as a person's date of birth. As pervasive systems are typically characterized by frequent change, the majority of information is dynamic. The persistence of dynamic context information can be highly variable; for example, relationships between colleagues typically endure for months or years, while a person's location and activity often change from one minute to the next. The persistence characteristics influence the means by which context information must be gathered. While it is reasonable to obtain largely static context directly from users, frequently changing context must be obtained by indirect means, such as through sensors.

Often, pervasive computing applications are interested in more than the current state of the context; for example, in our case study agents rely not only on information about current activity, but also activities planned for the future. Accordingly, context histories (past and future) will frequently form part of the context description.

Context Information is Imperfect. A second feature of context information in pervasive systems is imperfection. Information may be incorrect if it fails to reflect the true state of the world it models, inconsistent if it contains contradictory information, or incomplete if some aspects of the context are not known. These problems may have their roots in a number of causes. First, pervasive computing environments are highly dynamic, which means that information describing them can quickly become out of date. This problem is compounded by the fact that frequently the sources, repositories and consumers of context are distributed and information supplied by producers requires processing in order to transform it into the form required by the consumer; these factors can lead to large delays between the production and use of context information. Second, context producers, such as sensors, derivation algorithms and users, may provide faulty information. This is particularly a problem when context information must be inferred from crude sensor inputs; for example, when a person's activity must be inferred indirectly from other context information, such as location and sound level. Finally, disconnections or failures can mean that the path between the context producer and the consumer is cut, meaning that part or all of the context is unknown.

Context Has Many Alternative Representations. Much of the context information involved in pervasive systems is derived from sensors. There is usually a significant gap between sensor output and the level of information that is useful to applications, and this gap must be bridged by various kinds of processing of context information. For example, a location sensor may supply raw coordinates, whereas an application might be interested in the identity of the building or room a user is in. Moreover, requirements can vary between applications. Therefore, a context model must support multiple representations of the same context in different forms and at different levels of abstraction, and must also be able to capture the relationships that exist between the alternative representations.

Context Information is Highly Interrelated. In our case study, several relationships are evident between people, their devices and their communication channels (for example, ownership of devices and channels and proximity between users and their devices). Other less obvious types of relationships also exist amongst context information. Context information may be related by derivation rules which describe how information is obtained from one or more other pieces of information. In our case study, a person's current activity may be partially derived from other context information, such as the person's location and history of past activities. We refer to this type of relationship, where the characteristics of the derived information (its persistence, quality and so on) are intimately linked to the properties of the information it is derived from, as a dependency.

2.3 Context Modeling and Management for Pervasive Systems: Requirements

Having identified some of the features of context information in pervasive systems, we now address the issue of how to represent and manage this information.

One approach is to model context using existing data modeling techniques from the field of information systems, and to store and manage the information using a database management system. Alternatively, one of the object-modeling techniques commonly used by software engineers, such as UML, could be employed to construct a model of context information and to support the mapping of this model to an implementation in an object-oriented programming language. However, having explored these approaches, we suggest that they are neither natural nor appropriate for describing context.

We attempted to model the scenario of Section 2.1 using both the Entity-Relationship model and the class diagrams of UML, and experienced particular difficulties in distinguishing between different classes of context information (for example, static versus dynamic information, sensed information versus information supplied by users), representing the temporal and error characteristics of context and expressing relationships such as dependencies. We found the UML constructs to be more expressive than those provided by ER, but also correspondingly more cumbersome. As a result of our experiences, we suggest that the most appropriate approach to modeling context information is using special constructs designed with the characteristics of context in mind. In Section 4, we present such a modeling approach.

3   Related Work

Much of the work in the relatively new field of context-awareness is concerned with providing either frameworks that support the abstraction of context information from sensors or high-level models of context information that can be queried by context-aware applications. In this section, we review both of these areas of research and examine some of the shortcomings of the surveyed approaches.

Both the context toolkit [6] and the sensor architecture of Schmidt et al. [7] support the acquisition of context data from sensors, and the processing of this raw data to obtain high-level context information. The former is a programming toolkit that assists the developers of context-aware applications by providing abstract components (context widgets, interpreters and aggregators) that can be connected together to gather and process context information from sensors. The latter provides a layered model of context processing in which sensor output is transformed into one or more cues, which undergo processing to form an abstract context description comprising a set of values, each associated with a certainty measure that estimates the certainty that the value is correct.

Other work in the field of context-awareness largely ignores the issues of how context is derived from sensors, and focuses more upon modeling context information and delivering this information to applications. The goals of this work are closer to our own. The pioneering work in this area was carried out by Schilit et al. [8], who proposed the use of dynamic environment servers to manage and disseminate context information for an environment (where an environment might represent a person, place or community). The model of context used in this work was extremely simple, with context information being maintained by a set of environment variables. The Cooltown project [9] proposed a Web-based model of context in which each entity (person, place or thing) has a corresponding description the can be retrieved via a URL. This model is relatively informal; entity descriptions take the form of Web pages, which may be unstructured and intended for human (rather than application) consumption. The context modeling approach proposed by the Sentient Computing project [10] is more formal and is based upon an object-modeling paradigm. A conceptual model of context is constructed using a language based on the Entity-Relationship model, and context information is stored at run-time in a relational database. Gray and Salber present a model of context that aims to support design activities associated with context-awareness [11]. Their model is mainly concerned with capturing meta-information about context that describes features such as the format or representation of the context information, its quality attributes, source, transformation processes and actuation (the means by which it can be controlled). The model is informal, being concerned more with supporting the processes associated with the development of context-aware software, including requirements analysis and exploration of design issues, than with capturing context information in a format that can be queried by applications. The Owl context service, currently under development by Ebling et al. [12], aims to gather, maintain and supply context information to clients. It tackles various advanced issues, including access rights, historical context, quality, extensibility and scalability. Currently, only early research results have been published, and the underlying modeling concepts are not yet clear.

These context models exhibit a number of limitations. First, most lack the formal basis that is required in order to capture context in an unambiguous way and support reasoning about its various properties. The most formal models are those that underpin the Sentient Computing approach and the context processing framework of Schmidt et al.; however, these do not address all of the characteristics of context that we identified in Section 2.2. Additionally, many of the models are restricted to narrow classes of context; in particular, several support only sensed context information and its derivatives [6,12,11,7]. Most also ignore temporal aspects of context, including the need to represent histories of context information, and do not address context quality [6,10,9,8]. In the remainder of this paper we present a model of context that addresses these shortcomings.

4   Modeling Context Information

This section presents a collection of modeling concepts, together with an accompanying graphical notation, designed to capture many of the features of context information that are relevant to the design and construction of pervasive systems and applications. These modeling concepts provide a formal basis for representing and reasoning about some of the properties of context information that we identified in Section 2, such as its persistence and other temporal characteristics, its quality and its interdependencies.

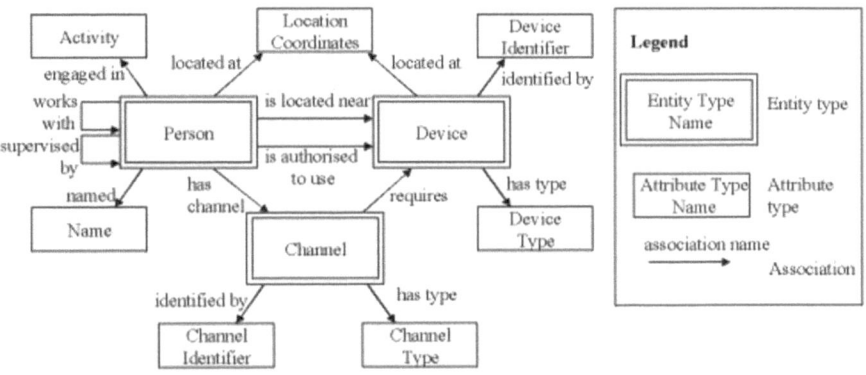

Fig. 1. Modeling the scenario of Section 2.1

The following sections present the modeling concepts incrementally, starting with the fundamental modeling concepts, and then building upon these to express more complex aspects of context. We return to the case study of Section 2.1 throughout our discussion in order to illustrate our modeling concepts by example.

4.1   Core Modeling Concepts

Our modeling concepts are founded on an object-based approach in which context information is structured around a set of entities, each describing a physical or conceptual object such as a person or communication channel. Properties of entities, such as the name of a person or the identifier of a communication channel, are represented by attributes. An entity is linked to its attributes and other entities by uni-directional relationships known as associations. Each association originates at a single entity, which we refer to as the owner of the association, and has one or more other participants. Associations can be viewed as assertions about their owning entity, and a context description can correspondingly be viewed as a set of such assertions. In the remainder of the paper, we use the terms assertion and association interchangeably.

We provide a graphical notation for our modeling concepts in order to allow context models to be specified diagrammatically. This notation takes the form of a directed graph, in which entity and attribute types form the nodes, and associations are modeled as arcs connecting these nodes. We present an example in Figure 1, based on the case study of Section 2.1, to illustrate the notation.

Our example model is constructed around three entity types: people, communication devices and communication channels. Each entity type is associated with a number of attributes: people are associated with names and activities, and channels and devices are associated with identifiers and types.

In addition to the associations between the entities and their attributes, several associations exist between the entities. These capture relationships between

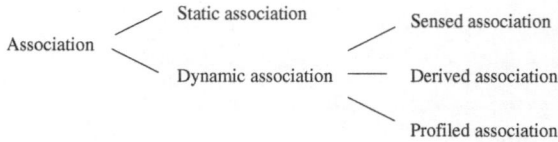

Fig. 2. A classification scheme for context associations

people (who works with whom, and who is supervised by whom), between people and devices (which devices each person is authorized to use and which devices are currently located with each person), between people and communication channels (which channels belong to each user), and between devices and channels (which devices the user requires in order to use each communication channel).

The model shown in Figure 1 captures the types of context information that are involved in the scenario, but does not describe many of the characteristics of this information that should ideally be known to context-aware applications and their developers. The following sections address this problem. Sections 4.2 and 4.3 present schemes for classifying associations according to type and structure. Section 4.4 describes our approach to capturing dependencies between associations, and Section 4.5 is concerned with characterizing the imperfection of context information.

4.2   Classifying Associations

In Section 2, we recognized the existence of several classes of context information that exhibit markedly different properties in accordance with their persistence and their source. We made the distinction between static and dynamic context, and showed that dynamic context can exhibit a wide range of persistence characteristics, which are linked to the means by which context information is obtained. In this section, we formalize these observations in a scheme for categorizing assertions about context, illustrated in Figure 2.

Static associations are relationships that remain fixed over the lifetime of the entity that owns them. The context captured by this type of association is typically known with a high degree of confidence, and in our example, includes the associations involving device and channel types.

Dynamic associations are all of those associations that are not static. We classify these according to source. Sensed associations are obtained from hardware or software sensors. Frequently, this information is not inserted directly into the model straight from the sensor, but is transformed in some way to bring it closer to the level of abstraction required by applications. Sensed context typically changes frequently, and consequently, can suffer from problems of staleness if there is a long lag between the time readings are taken at the sensor and the time that the corresponding context information is delivered to the client. Moreover, it can be subject to sensing errors arising from limitations inherent in the sensing technology. Two examples of sensed context in our case study are

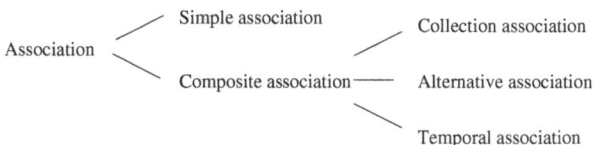

Fig. 3. Structural constraints on context associations

user and device location coordinates, which we assume are derived from location sensing mechanisms such as GPS receivers outdoors or Bats indoors [10].

Derived associations are obtained from one or more other associations using a derivation function that may range in complexity from a simple mathematical calculation to a complex AI algorithm. This type of context often assumes some of the properties of the class(es) of information it is derived from; for example, derived context information that is obtained from sensed information often has similar or magnified persistence and error characteristics. In addition, derived context as a class typically suffers from its own inherent limitations. In particular, derivation functions are often liable to draw incorrect or imprecise conclusions as a result of their reliance on crude inputs or overly simplistic classification models.

An example of derived context from our case study is the *is located near* relationship that, for each person, describes the set of devices located nearby. Relationships of this type need not be modeled explicitly, but can be derived for a given person by examining the *Location Coordinates* attribute of every device, and comparing it with the *Location Coordinates* attribute of the person.

The third class of dynamic association captures profiled information; that is, information that has been supplied by users. This class of information is typically more reliable than sensed and derived context and longer-lived, but can still suffer from staleness, as users may neglect to update information as it becomes out of date. Examples of profiled context include user names, and the *works with* and *supervised by* associations that exist between people.

The main benefit of classifying context information as we have described is that reasoning about information persistence and quality becomes possible. For example, conflicts can be resolved by favoring the classes of context that are most reliable (static followed by profiled) over those that are more often subject to error (sensed and derived).

4.3   Structural Constraints on Associations

Context information can vary from simple, atomic facts to complex histories. We support these different types of context by further categorizing associations according to structure, as shown in Figure 3.

An association is simple if each entity participating as owner of the association participates no more than once in this role. An example of this type of association is the *named* association of *Person*.

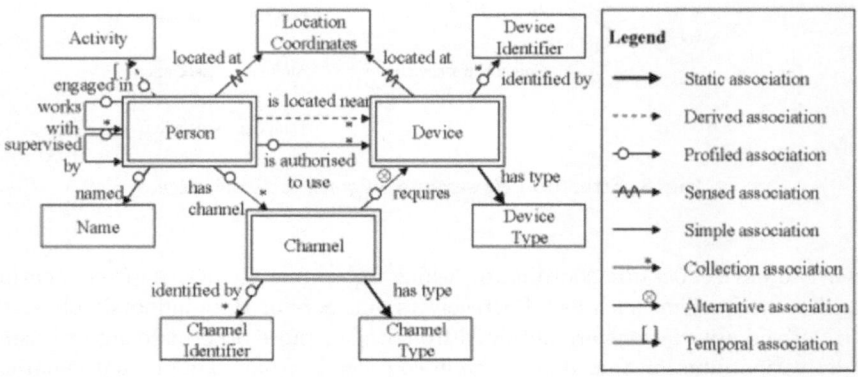

Fig. 4. Modeling the different association types for our case study

An association is composite if it is not simple. We refine composite associations into collection, alternative and temporal associations. Collections are used to represent the fact that the owning entity can simultaneously be associated with several attribute values and/or other entities, for example, people may work with many other people, and may have several communication channels. Alternatives differ from collections in that they describe alternative possibilities that can be considered to be logically linked by the 'or' operator rather than the 'and' operator. One example of an association of this type is the *requires* relationship between channels and devices. By classifying the role as an alternative rather than a collection, it acquires the semantics that a channel requires one of the devices it is associated with, rather than all. This type of association is useful when the context model must capture a number of different representations of the same information as described in Section 2.2, or when two or more sources of context information supply contradicting information and it is desirable to capture each of the different possibilities. Finally, a temporal association is also associated with a set of alternative values, but each of these is attached to a given time interval. This type of association can be viewed as a function mapping each point in time to a unique value. In our example, user activity is captured by a temporal association.

We distinguish the various different types of associations we have discussed in this and the preceding section diagrammatically by annotating the association arcs as shown in Figure 4.

4.4 Modeling Dependencies

A dependency is a special type of relationship, common amongst context information, that exists not between entities and attributes, as in the case of associations, but between associations themselves. A dependency captures the existence of a reliance of one association upon another. We say that an associa-

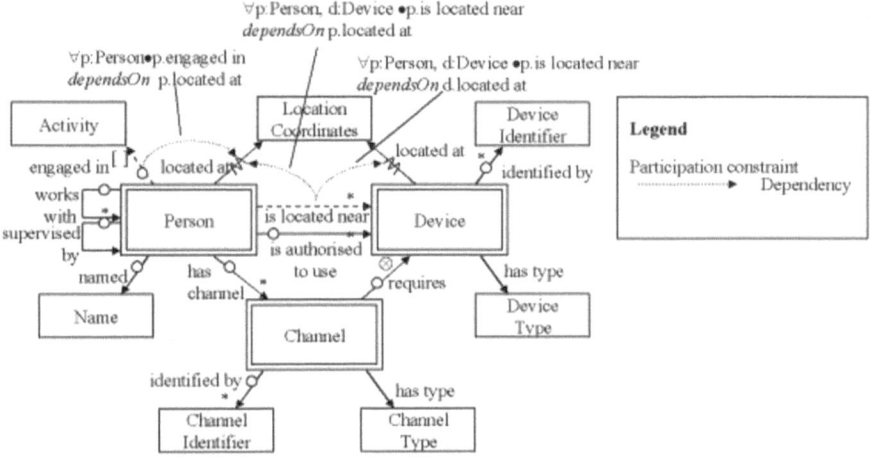

Fig. 5. Context model for our case study, showing the derivation dependencies

tion, a_1, *dependsOn* another association, a_2, iff a change to a_2 has the potential to cause a change in a_1. Each derived association is accompanied by at least one dependency; however, dependencies can also exist independently of derived associations. For example, on a mobile device a change in usage of network bandwidth can influence the battery life; that is, battery life *dependsOn* bandwidth.

The importance of capturing dependencies is pointed out by Efstratiou et al. [13]. Without knowledge of such dependencies, inappropriate decisions can be made by context-aware applications that lead to instability. Moreover, knowledge of dependencies is important from a context management perspective, as it can assist in the detection of context information that has become out-of-date.

We model a dependency, a_1 *dependsOn* a_2, as a directed arc leading from a_1 to a_2 , as shown in Figure 5. A dependency can be qualified by a participation constraint, which limits the pairs of associations to which the dependency applies. We capture three derivation dependencies in the figure. First, we show that the *engaged in* association that links people with activities is dependent upon the *located at* association. We qualify this association to indicate that associations of these two types are linked only if they describe the same person (that is, a person's activity is only dependent on that same person's location, and not on any other person's location). Similarly, we show that the set of devices located near a person is dependent on that person's location as well as the location of all devices.

4.5   Modeling Context Quality

In Section 2.2, we identified imperfection as one of the characteristics of context information in pervasive systems. Errors in context information may arise as a result of sensing and classification errors, changes in the environment leading to staleness, and so on. As context information is relied upon by applications to make decisions on the user's behalf, it is essential that applications have some means by which to judge the reliability of the information. For this reason, we incorporate measures of information quality into our model of context.

The need to address the varying quality of context information has been widely recognized [14,15,12,11,7], yet none of the existing work addresses the problem in an adequate or general way. Dey et al. suggest that ambiguous information can be resolved by a mediation process involving the user [15]. However, considering the potentially large quantities of context information involved in pervasive computing environments and the rapid rate at which context can change, this approach places an unreasonable burden on the user. Ebling et al. describe a context service that allows context information to be associated with quality metrics, such as freshness and confidence [12], but their model of context is incomplete and lacks formality. Castro et al. have a well-defined notion of quality based on measures of accuracy and confidence [14], but their work considers only location information. Schmidt et al. associate each of their context values with a certainty measure that captures the likelihood that the value accurately reflects reality [7]. They are concerned only with sensed context information, and moreover take a rather narrow view of context quality. Finally, Gray and Salber include information quality as a type of meta-information in their context model, and describe six quality attributes: coverage, resolution, accuracy, repeatability, frequency and timeliness [11]. Neither their information model nor their quality model are formally defined, as they are intended to support requirements analysis and the exploration of design issues, rather than to support the development of a context model that can be populated with data and queried by applications.

Quality modeling has been more extensively researched by the information systems community. Our modeling concepts borrow ideas from the work of Wang et al., who describe a quality model in which attributes are tagged with various quality indicators [16]. In our model, we support quality by allowing associations to be annotated with a number of quality parameters, which capture the dimensions of quality considered relevant to that association. Each parameter is described by one or more appropriate quality metrics, which represent precise ways of measuring context quality with respect to the parameter.

The types of quality parameters and metrics that are relevant are dependent on the nature of the association. For example, the quality of information about a user's location can be characterized by its accuracy, measured by the standard error of the location system, and freshness, determined by the time the location information was produced and the average lifetime of information related to user location. On the other hand, the quality of an assertion about user activity can be described by the certainty the information source has about the supplied information, measured as a probability estimate, and the overall accuracy of that information source, also described by a probability value. We illustrate the tagging of associations with quality parameters and metrics in Figure 6.

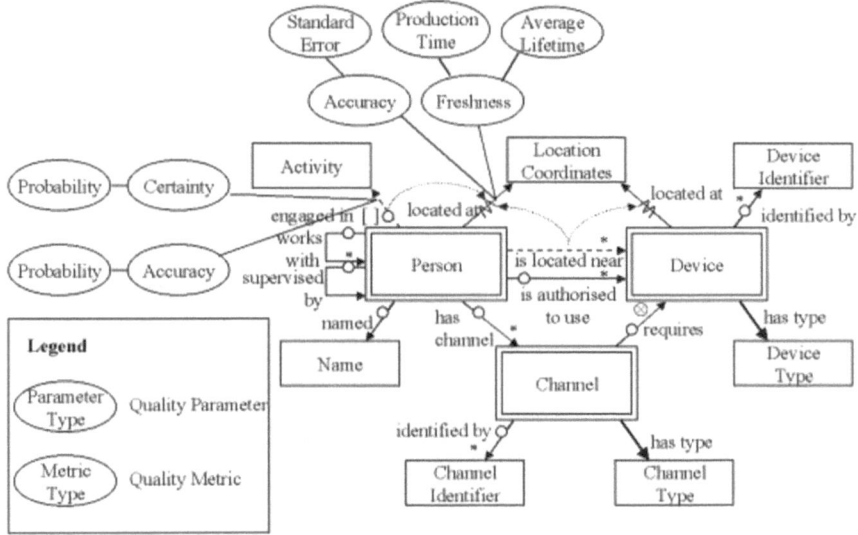

Fig. 6. Context model for our case study, annotated with quality parameters and metrics for two of the associations

5   Concluding Remarks

In this paper, we explored the characteristics of context information in pervasive systems and described a set of context modeling concepts designed to accommodate these. Our concepts were presented using the case study of Section 2.1, but are sufficiently generic to capture arbitrary types of context information, and thus to support a diverse range of context-aware applications.

We are currently in the process of developing a context management system founded upon the modeling constructs that we have presented. This system will allow abstract models described in our notation to be mapped with little effort to corresponding implementation models that can be populated with context information and queried by applications. It will be responsible for a range of management tasks, such as integration of context information from a variety of sources, management of sensors and derived context, detection of conflicting information, and so on. Concurrently, we are implementing the context-aware communication application that we have described. We have already used this case study to validate the context modeling concepts we presented in this paper, and, next, we hope to use the implementation of the case study to validate our context management infrastructure.

Aside from our implementation efforts, we envisage several areas for future work. These involve the extension of our context modeling concepts in order to address key issues for pervasive computing systems, such as privacy and distribution of context information. A privacy model is required in order to prevent abuses of context information, particularly personal information, by limiting its dissemination. Similarly, a distribution model is needed to support the appropriate partitioning and replication of context information across pervasive systems. This model must balance the requirement for a globally consistent view of context with the need for timely retrieval and continued access to information during periods of network disconnection.

References

1. Norman, D.: The Invisible Computer. MIT Press, Cambridge, Massachusetts (1998)
2. Henricksen, K., Indulska, J., Rakotonirainy, A.: Infrastructure for pervasive computing: Challenges. In: Informatik 2001: Workshop on Pervasive Computing, Vienna (2001)
3. Hong, J., Landay, J.: A context/communication information agent. Personal and Ubiquitous Computing: Special Issue on Situated Interaction and Context-Aware Computing 5 (2001)
4. Schmandt, C.: Everywhere messaging. In: 1st International Symposium on Handheld and Ubiquitous Computing (HUC'99). (1999)
5. A, S., Takaluoma, A., Mäntyjärvi, J.: Context-aware telephony over wap. Personal Technologies 4 (2000)
6. Dey, A., Salber, D., Abowd, G.: A context-based infrastructure for smart environments. In: 1st International Workshop on Managing Interactions in Smart Environments (MANSE'99). (1999)
7. Schmidt, A., et al.: Advanced interaction in context. In: 1st International Symposium on Handheld and Ubiquitous Computing (HUC'99), Karlsruhe (1999)
8. Schilit, B., Theimer, M., Welch, B.: Customising mobile applications. In: USENIX Symposium on Mobile and Location-Independent Computing. (1993)
9. Kindberg, T., et al.: People, places, things: Web presence for the real world. Technical Report HPL-2000-16, Hewlett-Packard Labs (2000)
10. Harter, A., Hopper, A., Steggles, P., Ward, A., Webster, P.: The anatomy of a context-aware application. In: Mobile Computing and Networking. (1999) 59–68
11. Gray, P., Salber, D.: Modelling and using sensed context in the design of interactive applications. In: 8th IFIP Conference on Engineering for Human-Computer Interaction, Toronto (2001)
12. Ebling, M., Hunt, G.D.H., Lei, H.: Issues for context services for pervasive computing. In: Middleware 2001 Workshop on Middleware for Mobile Computing, Heidelberg (2001)
13. Efstratiou, C., Cheverst, K., Davies, N., Friday, A.: An architecture for the effective support of adaptive context-aware applications. In: Mobile Data Management (MDM), Hong Kong, China, Springer (2001) 15–26
14. Castro, P., Chiu, P., Kremenek, T., Muntz, R.: A probabilistic room location service for wireless networked environments. In: UbiComp 2001 Conference, Atlanta (2001)
15. Dey, A., Mankoff, J., Abowd, G.: Distributed mediation of imperfectly sensed context in aware environments. Technical Report GIT-GVU-00-14, Georgia Institute of Technology (2000)
16. Wang, R., Reddy, M.P., Kon, H.: Towards quality data: An attribute-based approach. Decision Support Systems 13 (1995) 349–372

A Model for Software Configuration in Ubiquitous Computing Environments

Simon Schubiger-Banz and Béat Hirsbrunner

Department of Informatics (DIUF), University of Fribourg, Chemin du Musée 3,
CH-1700 Fribourg
simon.schubiger@swisscom.com beat.hirsbrunner@unifr.ch

Abstract. Software configuration in a heterogeneous and dynamic environment such as ubiquitous computing is a challenging task. This paper presents the COCA model, which transforms heterogeneous ubiquitous computing resources through a process called classification into a conceptualized representation, which allows high-level manipulation and configuration by ubiquitous computing applications. A multi-modal ubiquitous computing application serves as a sample implementation of the model that uses an automatic software configuration process to dynamically adapt to changes in the environment.

1 Introduction

Research in ubiquitous computing is towards the development of an environment able to deal with the mobility and interactions of both users and devices. The vision of ubiquitous computing [23,33] relies on the presence of environments enriched by computers embedded in everyday objects (blackboards, pens, clothes, etc.) and by sensors able to acquire information from the context.

Ubiquitous computing devices exhibit a tight coupling between hardware and software, on the one hand because of the resource limitations of the device itself and on the other hand due tosoftwarae-hardware dependencies such as pen-input or temperature measurement. For software running on ubiquitous computing devices, the abstraction requirements are twofold:

- The *intra*-device abstraction is similar to today's computers, solved by small-operating system kernels. Additional constraints such as screen size, memory size, CPU-performance, or power consumption may limit the application but basically intra-device abstraction has to deal with a static environment known at build time.
- The *inter*-device abstraction is a much bigger problem. Beside the decision of which services should be accessible from the outside, different hardware and software interfaces have to be considered as well as "smart" cooperation of devices. Additional difficulties arise because the cooperating devices may be unknown or untrusted, there are no standards to follow due to innovation, there is no global accessible state, and there is usually nothing like a "system administrator" keeping a birds eye view of the system.

F. Mattern and M. Naghshineh (Eds.): Pervasive 2002, LNCS 2414, pp. 181–194, 2002.

On the user level, using an application and doing useful work should not require to be an expert in the organization of the software infrastructure. As much as possible, the software should be self-managing and self-repairing in the face of simple transient faults.

For developers, hardware and protocol heterogeneity must be hidden from ubiquitous computing application as good as possible. Thanks to a common design methodology the application programmer should not have to understand the entire software and hardware infrastructure, which may not be known in advance.

Ubiquitous computing applications go beyond the services offered by a single device. They depend on resources provided by a context, which is composed of numerous devices implementing various services. An application running in such an environment is a collection of services linked at runtime. These links may change during execution especially if the application is composed of services running on mobile devices, which may leave or join the application context or devices that are simply turned off independently.

Finding and accessing resources and services becomes a major problem because the application context may contain devices and features not known at the time the application was developed. A common approach is standardizing interfaces for resource access and selection. Several architectures rely on directory services combined with attribute based query services for resource selection and access protocols such as Jini [32] or Ninja [9]. Unfortunately, standardization takes a long time, usually involves several compromises and the resulting specifications are seldom open enough to allow innovation. Because of that, it may be claimed that standardization will be unable to keep pace with the highly dynamic evolution of ubiquitous computing devices.

Instead of enforcing a standardized view for the application, a ubiquitous computing middleware should decouple the high-level concepts (abstractions) from the instances implemented by a context [17]. The concept "nearest printer" for example may be used no matter how a context supplies the corresponding implementation. This means that an application expresses its resource requirements in terms of its concepts instead of addressing specific resources directly (e.g. by an URL). The application concepts are instantiated by the middleware in function of the context and may change with time and location. For example when a users walks around, the middleware will keep track of the concept "nearest printer", switching from one physical printer to another as needed and without user intervention.

One of the fundamental problems prohibiting "smart" cooperation of services is the lack of machine processable semantics in interfaces. Adding semantics to interfaces became thus one of the key goals of our research, which finally resulted in the COCA model. Additionally, the model should help the developer to structure its ubiquitous computing application and provide a high level of abstraction to reason about the problem without paying too much attention to the underlying technical details.

The rest of this paper introduces the COCA model and presents an implementation of the model. Some final remarks conclude this paper. See [26] for a more detailed discussion of the model.

2   The COCA Model

The COCA (*C*oncepts, *O*ntologies, *C*lassifers, and *A*ctions)[1] model not only emerged out of the desire of adding semantics to interfaces but also borrows aspects from cognitive psychology [29]. The motivation for looking at human cognition and perception was the observation that humans perform extremely well in heterogeneous and dynamic environments. The idea was thus to adapt existing models from cognitive psychology to computer systems, in the hope to obtain a model that helps dealing with heterogeneity and dynamisms in computer systems.

Ross Quillian [24] was the first who introduced *semantic nets* [30] as a way of talking about the organization of human semantic memory. Semantic nets are a memory of concepts, where each concept is in relation with other concepts. COCA extended this model by adding *actions* as transformations between concepts and *classifiers* for symbol grounding and replacement for human sensory input.

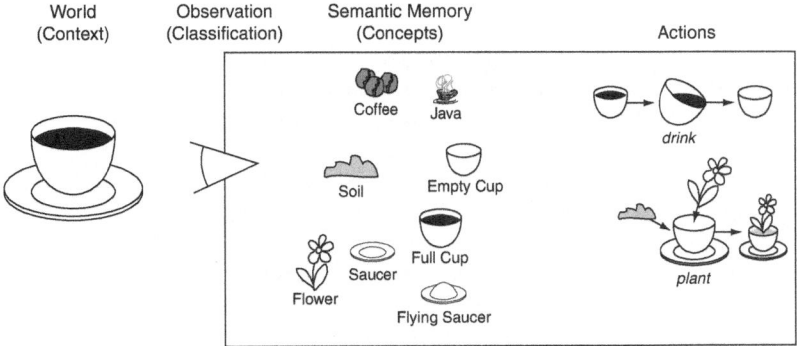

Fig. 1. The basic idea behind COCA: The world is classified into concepts and actions are derived from matching concepts

Figure 1 illustrates the COCA model in the light of human perception. The world is perceived as sensory input, which is interpreted or classified. Classification associates concepts with the input, abstracting the input and giving it a

---

[1] The COCA acronym is only one half of the story. Coordination language research at the DIUF resulted in various variants of CoLa (for *Co*ordination *La*nguage) [10, 15,28]. Because these languages usually deal with high-level abstractions, COCA is one way of providing them these abstractions through classifiers. This puts COCA in front of CoLa.

semantic. Thus classification transforms the perception of the real world into a meaningful symbolic representation.

Actions transform between concepts, like "drinking" from a "full cup" to an "empty cup". Concepts such as "full cup" can be matched thanks to the classification with the objects in the real world and define in turn the set of actions that can be applied to them such as "drinking".

The following sections introduce the key elements of the model in more detail.

2.1   Resources and Context

A basic part of COCA is its access to the environment, which is composed of *resources*. Resources have a representation in computer systems (e.g. a name) and are accessible in some form (e.g. as a byte stream). Sets of resources are collected in a *context* that defines a namespace and the resource access. Example of contexts are the world wide Web with its URL [1] namespace and HTTP [7] as access, a database with its tables and columns as namespace and SQL [6] as access, the internet domain name system (DNS [21]), a POP server [22], a file system, and that like.

In order to extend or restrict the perception of the environment, COCA allows two operations on contexts. A new, extended context can be constructed by the union of contexts and a context can be restricted by building a sub-context that contains resources of a specific type only.

Examples of restricting a context are for example a Web search engine that only returns Web pages that contain a certain keyword, a SQL "SELECT" statement, a segmentation algorithm of an optical character recognition (OCR) program, all JPEG files in a context, etc. Sub-context creation can be applied recursively, creating a hierarchy of sub-contexts where each contains resources with a specific characteristic.

The motivation behind sub-contexts is that large and dynamic resource spaces can be partitioned by an application to have only resources exhibiting certain properties in scope. An example might be a context browser that allows the user to see resources with a certain semantic only such as "Java source written by me", "five nearest color printers" or "emails from Joe".

Resources and contexts have the following properties:

- *Resources* represent the "real world" objects perceived by a computer system.
- *Access:* A resource is accessible through at least one context, the context defines and abstracts the resource access.
- *Namespace:* The context that makes a resource accessible defines also a namespace for the resource.
- *Unique name:* A resource has an unique name in every context.
- *Immutable:* A resource is immutable, resource "changes" result in a changed name.[2]

---

[2] Because COCA actions are purely functional, no state change of the manipulated objects (resources) is allowed. The immutability property also helps to provide versioning where every version of a resource can be clearly identified and is guaranteed to remain unchanged.

– *Context operations:* New contexts can be constructed from existing ones either by union or restriction.

The classifiers presented in the next section link the semantic-poor world of resources with a conceptualized high-level view used by the application for resource manipulation and configuration. Classifiers also provide the semantics for resource selection in sub-context construction.

2.2   Classifiers and Concepts

In contrast to the classical approach of adding semantics by the introduction of a new formalism[3], COCA only defines the internal representation (concepts) of semantics and the relation to the environment (context and resources) through classifiers. Classifiers and concepts are a solution of the *symbol-grounding problem* [25], a problem intrinsic to manipulation of a symbolic representation. The symbol-grounding problem initially appears when one studies the relation between the real world and a system that uses some form of model or representation of reality to make decisions about a real world problem. The key question is how do these entities relate to each other. Can the symbols manipulated by the computer systems acquire an intrinsic meaning related to the problem?

The "real world" in COCA consists of resources that are related to concepts (symbols) by classifiers, which give the intrinsic meaning to resources. More specifically, classifiers in COCA are functions of the form $R \mapsto C$, $R \mapsto <C, I\!\!R>$, or $R \mapsto <C, S>$ where R denotes the set of all resources, C the set of concepts defined by the classifier, $I\!\!R$ the set of real values, and S the set of character strings. Every classifier defines the set of concepts C it is able to classify, giving implicitly the semantic of each concept by the execution of the classifier. Figure 2 illustrates the operation of a classifier taking a resource as input and associating the concepts the resource is instance of with the resource in question.

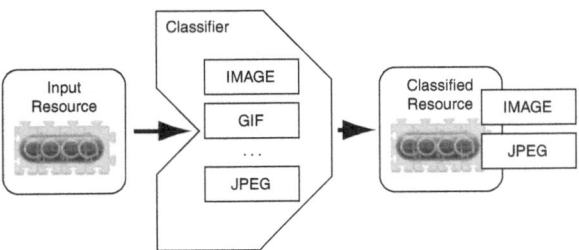

Fig. 2. Classifier operation

In the simplest case ($R \mapsto C$, figure 2) a classifier only tells of which concept a given resource is an instance of. For example a classifier that defines a set

---

[3] Most of these approaches are based on an extended first-order logic like the knowledge interchange format (KIF [8]) or the Semantic Web [2].

of MIME types may state that a given resource is an instance of the concept "image/jpeg" if the resource is a JPEG iamge.

The $<C, I\!R>$ classification can be seen as associating a weighted concept with a resource. For example a classifier able to classify colors could assign weights in the interval $[0..1]$ for the concepts "red", "green", and "blue".

Classifiers mapping resources to <concept, string> pairs $(R \mapsto <C, S>)$ are used to extract textual information stored in resources like names or document content. A classifier for paragraphs for example may associate the concept "paragraph" and its content with a resource.

Because classifiers define concepts by their execution, every concept has an exact semantic. The rationale behind restricting classifiers to three types of functions is the observation that a lot of semantic is already available in one of these forms. For example does it not make much sense to redefine the semantics of the concept "C-source" when there are C-compilers available that can easily decide if a file in question really contains C-source code or something else. Thus building a classifier for the concept "C-source" simply means running a C-compiler with the resource in question and looking at the exit code. This leads on the one hand to direct reuse of semantics and knowledge, encoded and available in computer programs today and it helps on the other hand constructing new semantic out of the existing pool of programs.

Classifiers have some interesting practical properties. For example they can be seen as test software assuring certain properties of a resource before using it, helping the construction of robust software. Because mapping from existing software to classifiers is relatively simple (for example by looking at the output of a program), a pool of classifiers is quickly constructed. If multiple applications share the same pool of classifiers, mutual understanding on a semantics (concept) level between the applications is possible; all applications share implicitly the semantics of the concepts defined by the shared classifier pool. If classifiers are written in a platform independent language like Java, applications may even "learn" new semantics by downloading appropriate classifiers.

2.3   Ontologies and Relations

COCA uses the term *ontology* for its knowledge base which consists of interrelated concepts. Concepts have a well-defined meaning given by classifiers that associate concepts with resources.

Knowledge representation may range from very domain specific to holistic knowledge. Various projects try to construct a holistic ontology such as the OntoWeb (www.ontoweb.org) or the SENSUS [14] project. A holistic ontology often resembles an encyclopedia [18], something used for centuries by humans. In practice, representation of encyclopedic information in a machine processable form turned out to be difficult, one reason why COCA focus only on applications in computer systems such as ubiquitous computing, not aiming at a "one-for-all" solution.

Although a holistic ontology is not per-se excluded by COCA, a domain-specific approach is favored. Every application based on COCA usually defines its own ontology specific for the problem domain addressed by the application, keeping the ontology small and the application focused on the problem at hand. This implies local evolution of ontologies without global or semi-local coordination, or adherence to some standard. In the case where inter-application communication is required, ontologies can be shared for example by:

- *Merging the concept name spaces:* A very simple solution which may nevertheless make sense for some domains, but it may also result in conflicting semantics for concepts having the same concept name.
- *Mutual resource classification:* Applications give mutual access to each other (context union) and each application classifies the resources of the other application according to the individual ontology.
- *Explicit declaration of concept equivalence:* Meta ontologies may be used to explicitly state equivalence of concepts in different ontologies. These meta-ontologies will require careful maintenance, likely done by humans.
- *Proving concept equivalence:* If two concepts in different ontology are defined by the same classifier they are considered equivalent allowing inference of further properties through each ontology's relations.
- *Inferring concept equivalence:* Concept equivalence may also be automatically inferred by observing classifiers by so-called meta-classifiers. That is, the automated version of building meta ontologies. Concepts may be considered equivalent as long as different classifiers consistently output the same concepts for a set of resources under observation.

Concepts in COCA are interrelated by relations that add inter-concept semantic. Relations in COCA are of the form $C \mapsto C$, where C denotes the set of concepts in an ontology. A relation may be either defined by enumeration (e.g. a database table), or by executable code (e.g. a program). Relations define by themselves their semantic through the enumerated concepts or the program execution. Examples of relations are the Java method **Class. assignableFrom(Class)** that defines the "is-a" graph of Java classes. Another example is the "salary" relation in an employee database.

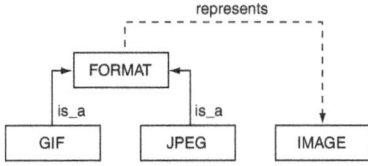

Fig. 3. An example of an ontology

Figure 3 gives yet another example of a simple ontology consisting of four concepts (FORMAT, GIF, JPEG, and IMAGE) and two relations (*represents*, dashed and *is_a*, continuous). The relation *is_a* relates the concepts GIF and FORMAT as well as the concepts JPEG and FORMAT. The relation *represents* relates the concept FORMAT with IMAGE. A classifier that recognizes file formats may define the GIF and JPEG concepts. Their relation with the other concepts defines the IMAGE and FORMAT concepts.

2.4 Actions

The previous sections addressed the static aspects captured by an ontology. In a dynamic environment, resources not only exists, they are created, deleted, and manipulated.

In COCA, a resource manipulation is seen as a transformation of a resource, yielding a new resource. Such a resource transformation is called an *action* and is a relation of the form $R \mapsto R$ where R denotes the set of resources. Additionally, an action is constraint by the concept it accepts as argument (*input concept*) and the concept the result of the action is an instances of (*output concept*).

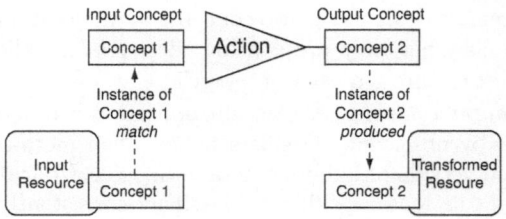

Fig. 4. A resource transformation through an action

This requires a resource to be instance of the input concept of the action which produces an instance of the output concept as depicted in figure 4. The input concept acts as a precondition and the output concept as a postcondition as also found for example in an Eiffel software contract [19]. A COCA action has the following properties:

- *Relation:* An action is a relation of the form $R \mapsto R$ where R denotes the set of resources.
- *Input constraint:* The input of an action is constraint by a the input concept. An action accepts only a resource that is an instance of the input concept as argument. The input constraint defines the *source domain* of the action.
- *Output constraint:* The output of an action is guaranteed to be an instance of the output concept. An action only produces a resource that is an instance of the output concept as result. The output constraint defines the *target domain* of the action.
- *Stateless:* An action is stateless and side-effect free. If state has to be preserved, it has to be stored as part of a resource.

Actions can be linked when the output concept of an action matches the input concept of another action. Such a chain of actions is expected to execute under transactional semantics and appears to the application as an atomic action transforming from the input concept of the first action in the chain to the output concept of the last action in the chain.

This chaining of actions is also implemented by the Path [12] prototype which is part of the Ninja project. The main difference is that Path is based on pre-existing type information whereas COCA classifiers establish this information at runtime. In that respect, COCA classifiers are similar to Microsoft Smart Tags recognizers [5,20] or the Java activation framework (JAF [4]).

3   Ubiquitous Computing Demonstrator

In order to have a verification of the model a simple demonstrator using COCA was realized. In the mean time two other project based on COCA are under the way, emphasizing the use of the model.[4]

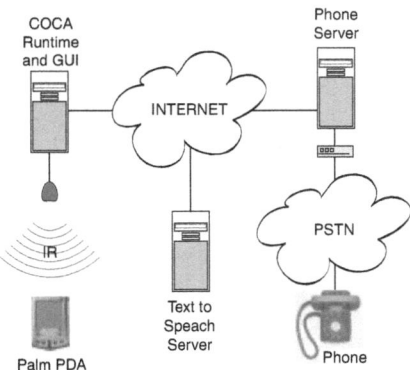

Fig. 5. An overview of the ubiquitous computing demonstrator

The multi-modal interface scenario implemented by the ubiquitous computing demonstrator allows a user to establish relations between messages and devices, with the COCA runtime automatically transforming the message to fit the user's current device. For example think of a user who associates a text file with a PDA and a phone. As soon as this file changes, the user receives a rendered version of the file on its PDA. If he is also available on the phone, the message is synthesized as voice, the user called, and the message is played over the phone.

Figure 5 gives an overview of the ubiquitous computing demonstrator. The key parts of the implementation are:

- *COCA runtime:* The COCA runtime controls and interacts with the environment over the Internet and an infrared port.

---

[4] One is a diploma thesis that explores the use of COCA in a more classic AI style application for document classification. The other is a smart wireless application protocol (WAP) proxy that classifies and enhances wireless markup language (WML) decks under development at Swisscom Corporate Technology.

- *COCA GUI:* The COCA GUI allows interaction and visualization of the operation of the COCA runtime. Figure 6 shows a screenshot of the the COCA GUI.
- *PDAs:* A set of PalmOS based PDAs run a tiny COCA runtime that allows them to be classified and exchange messages with the COCA runtime over the infrared port.
- *Text-to-speech:* A text to speech server located somewhere on the Internet is used as an example of a Web based action.
- *Phone server:* The phone server allows the COCA runtime to interact with the public switched telephony network (PSTN).

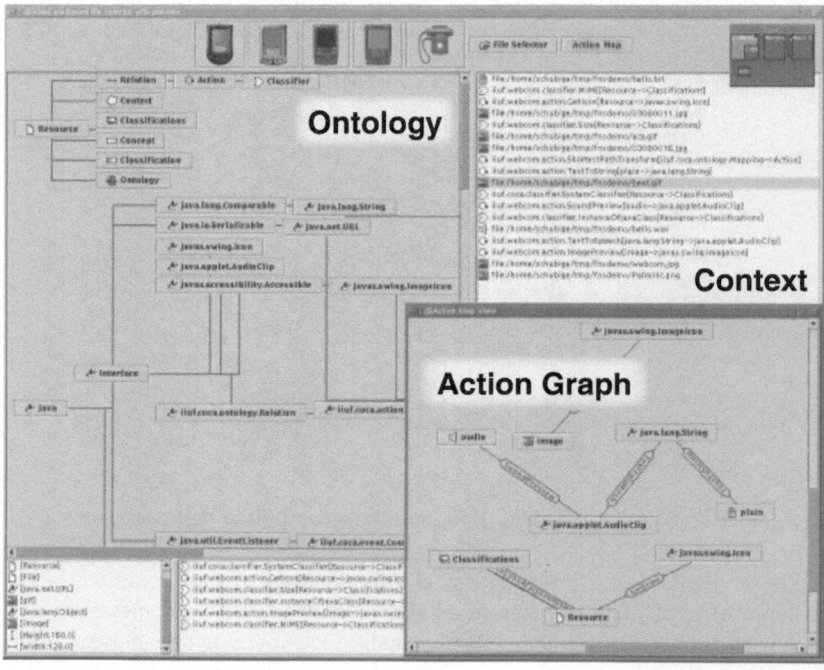

Fig. 6. A screenshot of the ubiquitous computing demonstrator

COCA is applied in the demonstrator to address the heterogeneity and dynamism of the environment as well as for dynamically adapt the software configuration to the set of available services and current user requirements. Heterogeneity appears basically on the hardware level in the ubiquitous computing demonstrator composed of PDAs, phones, and workstations.

Classifiers are used extensively to detect changes in the environment and to dynamically adjust the set of available resources in the contexts. The ubiquitous computing demonstrator implements a classifier for the various Palm models in order to figure out their hardware capabilities (such as color or monochrome displays). Another classifier is able to locate an appropriate text-to-speech server on the Internet. Yet another classifiers associates MIME-types [3] with messages created by the user. For better integration with the Java environment, all Java classes loaded by the demonstrator end up in the ontology as concepts, again through a classifier. It is important to note that only the classifiers know the details of resource lookup, configuration, and access. There is no standardized "interface" imposed on resources. The application deals with resources on a conceptual level only and is agnostic about the underlying instances.

Dynamism in the ubiquitous computing demonstrator appears mainly in three areas: PDAs can be reached by infrared if within range, the text-to-speech service on the Internet may be up or down, and messages are created. All the above items are resources in a corresponding context. In consequence, the ubiquitous computing demonstrator implements three contexts: one for what is reachable by infrared, one for the Web and one for the filesystem where messages are stored. Every context is responsible for tracking changes occurring in it and to provide context access to the COCA runtime. The COCA runtime unifies these three contexts and sees changes only in the unified context, which is hiding the peculiarities and access details involved by the specific contexts. Parts of the demonstrator only interested in particular resources such as the GUI use sub-contexts to see only PDAs for example.

Dynamic software configuration for the ubiquitous computing demonstrator means the automatic composition of software components in order to achieve some higher-level service. This higher-level service is only described by its properties such as "output on phone" but not how these properties can be provided. Automatic software configuration is therefore inherently declarative, the requested service exists only as a description and it is up to the system to provide a fitting implementation. Even the description itself is implicit, for example when a PDA is within range, it is assumed that the user wants the message rendered on its PDA.

Services in COCA are modeled by actions, so automatic software configuration means composition of actions. In fact, automatic software configuration itself is just another action that has as its input-concept a <concept, concept> pair describing the requested service and as its output an action representing the transformation between these concepts.[5]

---

[5] There are many ways to implement automatic software configuration. A Prolog-like inference engine may be one solution. Another much simpler is the action graph used by the ubiquitous computing demonstrator. Maintaining the action graph is relatively simple; it is sufficient to create a sub-context with actions only. These actions then form the edges of the action graph and the input and output concepts of the actions are the nodes of the graph.

The automatic software configuration action implemented by the ubiquitous computing demonstrator is also very simple but sufficient for that case: for a <resource,concept> pair an action is returned that allows the input resource to be transformed to the requested concept. The action is found by computing the shortest path from all the input concepts the resource is an instance of to the target concept. Then the shortest of all these paths is returned as an action. The returned action may be a composition (a path) of other actions.

4   Conclusion

The ubiquitous computing demonstrator [27] is the first application of the COCA model. The most important result of this implementation is that the model as such is realistic and indeed appropriate for ubiquitous computing settings. Another result is that the design of the application was largely simplified by thinking in COCA terms. Every problem encountered had its clear solution in COCA, helping the construction of the demonstrator in a relatively short time.[6] Another interesting property of COCA is that actions can be developed with very local scope. An action basically only has to know about its input and output concepts, the integration in the "big-picture" is done by the COCA runtime and the application. This allowed for example to develop and integrate the text-to-speech service in a few hours and all parts of the demonstrator dealing with text were thereafter speech enabled without touching them.

To conclude, COCA is a suitable model for providing service in a heterogeneous and dynamic environment such as ubiquitous computing. Integrating existing systems into COCA based applications should not be to difficult because several prominent models have mappings to COCA [26] such as object oriented, functional and logic programming as well as the entity relationship model and autonomous agents. The COCA implementation used for the ubiquitous computing demonstrator emphasizes the practical value of the model.

Because a COCA implementation will be used for the UbiDev [31] ubiquitous computing middleware under development at the DIUF as well as parts of it contribute to the Focale project [16], the model will certainly evolve towards a even more solid foundation for ubiquitous computing applications. The UbiDev as well as the XCM [11] projects focus on the high-level coordination aspects involved in highly dynamic scenarios and clearly benefit form the abstractions provided by COCA.

References

1. T. Berners-Lee. *Universal Resource Identifiers in WWW*, 1994. RFC 1630.
2. T. Berners-Lee, J. Hendler, and O. Lassila. The Semantic Web. *Scientific American*, May 2001.
3. N. Borenstein and N. Freed. *MIME - Multipurpose Internet Mail Extension*, 1993. RFC 1521.
4. B. Calder and B. Shannon. *JavaBeans Activation Framework Specification Version 1.0a*. Sun Microsystems, Inc., 2550 Garcia Avenue, Mountain View, CA 94042-1100 U.S.A, May 1999.
5. Microsoft Corporation. BRIDGE Reaches New Customers with Office XP Smart Tags and Real-Time Data. Technical report, Microsoft Corporation, Redmond, WA 98052-6399, USA, June 2001.
6. C. J. Date. *An Introduction to Database Systems*. Addison-Wesely Publishing Company, Inc., Reading, Massachusetts, 6th edition, August 1995.

---

[6] It took about 4 weeks to implement the COCA model and to build the ubiquitous computing demonstrator on top of it.

7. R. Fielding, J. Gettys, J. Mogul, H. Frystyk, L. Masinter, P. Leach, and T. Berners-Lee. *Hypertext Transfer Protocol - HTTP/1.1*, 1999. RFC 2616.
8. M. R. Genesereth and R. E. Fikes. Knowledge Interchange Format, Version 3.0. Reference Manual. Technical Report Logic-92-1, Computer Science Department, Stanford University, 1992.
9. S. D. Gribble, M. W. R. von Behren, E. A. Brewer, D. Culler, N. Borisov, S. Czerwinski, R. Gummadi, J. Hill, A. Joseph, R.H. Katz, Z. M. Mao, S. Ross, , and B. Zha. The Ninja Architecture for Robust Internet-Scale Systems and Services. *Computer Networks*, 1999. Special Issue on Pervasive Computing.
10. B. Hirsbrunner, M. Aguilar, and O. Krone. CoLa: A Coordination Language for Massive Parallelism. In *Proceedings of the ACM Symposium on Principles of Distributed Computing (PODC)*, Los Angeles, California, August 14–17 1994.
11. B. Hirsbrunner, A. Tafat-Bouzid, and M. Courant. XCM: A Unified Coordination Model. Technical Report 02-02, DIUF, March 2002.
12. E. Kiciman and A. Fox. Using Dynamic Mediator to Integrate COTS Entities in a Ubiquitous Computing Environmen. *Springer-Verlag*, 2000.
13. M. Klein, D. Fensel, F. van Harmelen, and I. Horrocks. The Relation between Ontologies and Schema-Languages: Translating OIL-specifications in XML-Schema. In *Proceedings of the ECAI'00 Workshop on Applications of Ontologies and Problem-Solving Methods*, Berlin, August 2000.
14. K. Knight and S. Luk. Building a Large Knowledge Base for Machine Translation. In *Proceedings of the American Association of Artificial Intelligence Conference AAAI-94*, Seattle, WA, 1994.
15. O. Krone. *STL and Pt-PVM: Concepts and Tools for Coordination of Multi-threaded Applications*. PhD thesis, University of Fribourg, 1997. No. 1191.
16. S. Le Peutrec and M. Courant. Instruments pour la vie artificielle. In *Proceedings of the Eleventh French-speaking Congress of Human Computer Interaction*, Montpellier, France, November 1999.
17. S. Maffioletti, S. Schubiger, and B. Hirsbrunner. Towards a Homogeneous Environment for Ubiquitous Interactive Devices. Technical report, Department of Informatics, University of Fribourg, Switzerland, 2001.
18. G. Mazzola. Humanities@EncycloSpace - Der enzyklopädische Wissensraum zur Informationstechnologie, February 1997. Empfehlungen an den Schweizerischen Wissenschaftsrat, Bern.
19. B. Meyer. *Object Oriented Software Construction*. Prentice-Hall, Inc, Upper Saddle River, New Jersey, 2nd edition, 1997.
20. Microsoft Corporation, Redmond, WA 98052-6399, USA. *Microsoft Office XP Smart Tag SDK v1.1*, 2001. http://www.microsoft.com/Office/developer/platform/smartag.asp.
21. P. Mockapetris. *Domain Names - Implementation and Specification*, 1987. RFC 1035.
22. J. Myers and M. Rose. *Post Office Protocol - Version 3*, 1996. RFC 1939.
23. D. A. Normann. *The Invisible Computer*. MIT Press, 1999.
24. M. R. Quillian. Computational Linguistics. *Communications of the ACM*, 12(8), August 1969.
25. A. Robert. *EMuds: Adaption in Text-Based Virtual Worlds*. PhD thesis, University of Fribourg, Fribourg, Switzerland, 2000. No. 1272.
26. S. Schubiger. *Automatic Software Configuration - A Model for Service Provision in a Dynamic and Heterogenous Environment*. PhD thesis, University of Fribourg, February 2002. submitted.

27. S. Schubiger, O. Hitz, L. Robadey, and D. Rossier. The IIUF Java Package. `http://diuf.unifr.ch/iiufdev/doc`.
28. M. Schumacher. *Objective Coordination in Multi-Agent System Engineering*. LNAI 2039. Springer, 2001.
29. M. Sharples. *Computer and Thought: A Practical Introduction to Artificial Intelligence*. The MIT Press, Cambridge, Massachusetts, 1989.
30. J. F. Sowa, editor. *Principles of Semantic Networks*. Morgan Kaufmann Publishers, San Mateo, California, 1991.
31. S.Schubiger, S.Maffioletti, A. Tafat-Bouzid, and B. Hirbrunner. Providing Service in a Changing Ubiquitous Computing Environment. In *Workshop on Infrastructure for Smart Devices - How to Make Ubiquity an Actuality*, September 2000.
32. J. Waldo. The Jini Architecture for Network-Centric Computing. *Communication of the ACM*, 42(7), July 1999.
33. M. Weiser. The computer for the 21st century. *Scientific American*, September 1991.

INS/Twine: A Scalable Peer-to-Peer Architecture for Intentional Resource Discovery

Magdalena Balazinska, Hari Balakrishnan, and David Karger

MIT Laboratory for Computer Science
Cambridge, MA 02139
{mbalazin,hari,karger}@lcs.mit.edu

Abstract. The decreasing cost of computing technology is speeding the deployment of abundant ubiquitous computation and communication. With increasingly large and dynamic computing environments comes the challenge of scalable resource discovery, where client applications search for resources (services, devices, etc.) on the network by describing some attributes of what they are looking for. This is normally achieved through directory services (also called resolvers), which store resource information and resolve queries. This paper describes the design, implementation, and evaluation of INS/Twine, an approach to scalable intentional resource discovery, where resolvers collaborate as peers to distribute resource information and to resolve queries. Our system maps resources to resolvers by transforming descriptions into numeric keys in a manner that preserves their expressiveness, facilitates even data distribution and enables efficient query resolution. Additionally, INS/Twine handles resource and resolver dynamism by treating all data as soft-state.

1 Introduction

An important challenge facing pervasive computing systems is the development of scalable resource discovery techniques that allow client applications to locate services and devices, in increasingly large-scale environments, using detailed *intentional* [1] descriptions. Resource discovery systems should achieve three main goals: (i) handle sophisticated resource descriptions and query patterns; (ii) handle dynamism in the operating environment, including changes in resource state and network attachment point; and (iii) scale to large numbers of resources spread throughout a wide network across administrative domains. While some systems have addressed limited combinations of these properties, we address all three in this paper. Resource discovery efforts have been largely geared toward expressive resource descriptions and complex query predicates [1,2,3,4,5,6, 7]. These approaches differ in the details of how they name resources and how these names resolve to the appropriate network location. However, they all essentially rely on semistructured resource descriptions [8], *attribute-based* naming schemes with orthogonal attribute-value bindings, in which some attributes are hierarchically dependent on others.

F. Mattern and M. Naghshineh (Eds.): Pervasive 2002, LNCS 2414, pp. 195–210, 2002.

Many resource discovery schemes have been designed primarily for small networks [1,2,3,4,7], *or* for networks where dynamic updates are relatively uncommon or infrequent (e.t., DNS [9], LDAP [10]). They do not work well when the number of resources grows, *and* updates are common.

Static resource partitioning [11] and hierarchical organization of resolvers [5, 6,12] solve scalable and dynamic resource discovery. Static partitioning relies on some application-defined attribute to divide resource information among resolvers. However, static partitioning does not guarantee good load distribution and burdens clients with selecting the relevant partitions.

Hierarchical approaches [5] organize resolvers around increasingly large domains for which they are responsible. These domains are created around particular attributes of resource descriptions, such as geographical location. Other hierarchical schemes keep data information in local resolvers and create hierarchies to filter out irrelevant queries as they travel toward the leaves [6,12]. However, even if many hierarchies coexist, or if the hierarchies are created dynamically, root nodes may become bottlenecks. If queries are not propagated through root nodes to avoid bottlenecks, results become dependent on the origin of each query. Also hierarchies may not efficiently resolve queries that involve multiple orthogonal attributes. For example, imagine a metropolitan resource discovery system, where resolvers are hierarchically organized by institutions, neighborhoods, cities, and finally the metropolitan level. A client may be interested in locating all cameras filming main points of congestion in the metropolis, independent of location. The hierarchy described would not handle this query in a scalable manner, since it is based on location.

We describe the architecture, implementation, and evaluation of *Twine*, an approach to resource discovery that achieves scalability via hash-based partitioning of resource descriptions amongst a set of symmetric peer resolvers. Twine works with arbitrary attribute sets. It handles queries based on orthogonal and hierarchical attributes, with no content or location constraints. It also handles *partial queries*, queries that contain only a subset of the attributes originally advertised by resources (considering the other attributes as wildcards). Twine evenly distributes resource information and queries among participating resolvers. Finally, our system efficiently handles both resource and resolver dynamism. Twine is integrated with INS [1], the Intentional Naming System from MIT and now forms the core of its architecture. Therefore, we refer to INS/Twine nodes as *Intentional Name Resolvers* (INRs).

INS/Twine leverages recent work in peer-to-peer document publishing and distribution ([13,14,15,16]). Peer-to-peer systems do not rely on any hierarchical organization or central authority for distributing content or searching for documents. However, current peer-to-peer applications lack adequate querying capability for complex resource queries.

INS/Twine is designed to achieve scalable resource discovery in an environment where *all resources are equally useful*. We could imagine deploying INS/Twine throughout the Internet and letting everyone announce resources of global interest to users around the world. Such resources may be file servers, cameras showing the weather in different cities, Web services [17], and so on. Similarly we could imagine deploying INS/Twine within a city and letting users access resources such as air quality sensors, water temperature/quality indicators at public beaches, business information (e.g., number of currently available cars at a car rental company), and so on. In each deployment scenario different resources are advertised, but in both cases, all resources are potentially equally useful to clients. An important goal in INS/Twine is therefore to make all resources available to all users independent of location. However, location-dependent queries are handled well by specifying "location" as an attribute.

In both examples, the number of resources could be considerable. There are around 10^5 establishments in a city the size of Los Angeles or New York [18]. Each establishment could easily offer as many as a few thousand resources. INS/Twine should therefore scale to $O(10^8)$ resources and around $O(10^5)$ resolvers (assuming each establishment could run at least one resolver). For this, each resolver should hold only a small subset of all resource information. Most importantly, only a small subset of resolvers should be contacted to resolve queries or to update resource information.

To achieve these goals, INS/Twine relies on an efficient distributed hash table process (such as Chord [15], CAN [14] or Pastry [16]), which it uses as a building block. Twine transforms each resource description into a set of numeric keys. It does so in a way that preserves the expressiveness of semistructured data, facilitates an even data distribution through the network, and enables efficient query resolution even in the case where queries contain a subset of attributes originally advertised by any resource. From a resource description, Twine extracts each unique subsequence of attributes and values. Each such subsequence is called a *strand*. Twine then computes a hash value for each strand, which constitutes the numeric keys.

A peer-to-peer approach to resource discovery creates new challenges for data freshness and consistency. Indeed, as resolvers fail or new ones join the network, the mapping from resource descriptions to resolvers changes. To maintain consistency in the face of network changes and resource mobility, resolvers treat all resource information as *soft state*. If a resource (or a proxy acting on its behalf) does not refresh its presence within a certain interval, the corresponding description is removed from the network. To achieve scalability, the refreshing frequency in the *core* of an INS/Twine network (i.e., among resolvers) is significantly lower than the refreshing frequency at the *edge* (i.e., between client applications and resolvers).

We evaluated INS/Twine by running 75 instances of the resolver and inserting various types of descriptions into the network. We find that both resource information and queries are evenly distributed among resolvers. Each resolver receives only a small subset of resource information and queries. The size of the set is proportional to the number of strands and resource descriptions but inversely proportional to the number of resolvers on the network. Resolvers associated with resource descriptions are located within $O(\log N)$ hops through the network, where N is the total number of resolvers. The query success rate is

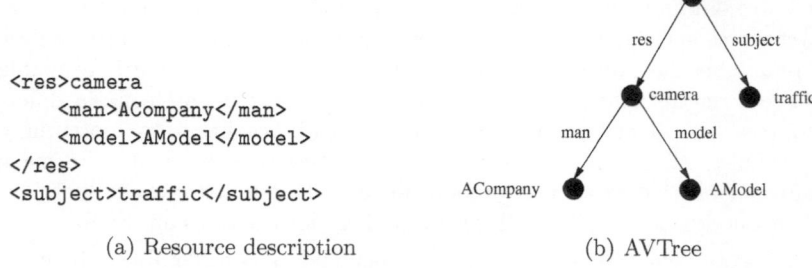

```
<res>camera
    <man>ACompany</man>
    <model>AModel</model>
</res>
<subject>traffic</subject>
```

(a) Resource description   (b) AVTree

Fig. 1. Example of a very simple resource description and its corresponding AVTree. The resource is a camera, manufactured by ACompany and filming traffic

100% when less than $k - 1$ resolvers (where k is a configurable replication level) fail or join the network within one refresh interval. Additionally, for a fraction F of failed resolvers, query failures decrease exponentially with k.

2 INS/Twine System Architecture

In this section, we first describe the details of resource descriptions. We then present the system architecture of INS/Twine and the algorithms for transforming resource descriptions into numeric keys, distributing information, and resolving queries.

2.1 Resource Descriptions

Resources in INS/Twine are described with hierarchies of attribute-value pairs in a convenient language (e.g., XML, INS name-specifiers [1], etc.). Our approach is to convert any such description into a canonical form: an attribute-value tree (*AVTree*). Figure 1 shows an example of a very simple resource description and its AVTree. All resources that can be annotated with meta-data descriptions, can be represented with an AVTree.

Each resource description points to a *name-record*, which contains information about the current network location of the advertised resource, including its IP address, transport/application protocol, and transport port number.

In INS/Twine, a resource matches a query if the AVTree formed by the query is the same as the AVTree of the original description, with zero or more truncated attribute-value pairs. For example, the device from Figure 1 would match the query: `<res>camera<man>ACompany</man></res>` or even the query: `<res>camera</res>`. This implies that an important class of queries that INS/Twine must support is *partial queries*, in addition to *complete queries* that specify the exact advertised resource descriptions.

Therefore, like resource descriptions, client queries are described using hierarchies of attribute-values and are converted to AVTrees. INS/Twine provides a way for queries to reach the resolver nodes best-equipped to handle them, based on the attributes and values being sought. The ultimate results of query matching depend on the local query processing engine attached to each resolver. Examples of this include INS's subtree-matching algorithm [1], UnQL [8] or the XSet query engine for XML [19].

Since query routing relies on exact matches of both attributes and values, it is possible to allow more flexible queries by separating string values into several attribute-value pairs. For example, `<model>AModel Camcorder 123</model>` could be divided into `<modelw>AModel</modelw>`, `<modelw>Camcorder</modelw>` and, `<modelw>123</modelw>`, allowing queries of type `<modelw>123</modelw>`.

2.2   Architecture Overview

INS/Twine uses a set of resolvers that organize themselves into an overlay network to route resource descriptions to each other for storage, and to collaboratively resolve client queries. Each resolver knows about a subset of other resolvers on the network.

Devices and users communicate with resolvers to advertise resources or submit queries. When a resource periodically advertises itself through a particular resolver, it is considered *directly connected* to that resolver. When a client issues a query to a resolver, it receives the response from that resolver. Communication between client applications and resolvers happens *at the edge* of the INS/Twine network. Communication between resolvers takes place in the network *core*.

The architecture of INS/Twine has three layers, as shown in Figure 2. The top-most layer, the *Resolver*, interfaces with a local AVTree storage and query engine, which holds resource descriptions and implements query processing, returning sets of name-records corresponding to (partial) queries.

When the Resolver receives an advertisement from its client application, it stores it locally using that engine. Local storage of information about directly connected resources serves for state management as described in Section 3. The resolver then splits the advertised resource description into strands and passes each one to the *StrandMapper* layer. The details of strand-splitting are discussed in Section 2.3. The StrandMapper maps the strand onto one or more numeric m-bit keys using a hash function. It then passes each key, together with the complete advertisement (the *value* corresponding to the key), to the *KeyRouter* layer. Finally, given a key, the KeyRouter determines which resolver in the network should receive the corresponding value and forwards the information to the selected peer. Hence, for data distribution, our approach boils down to inserting resource descriptions using each prefix-strand as a separate key.

As complete resource descriptions are transmitted during resource advertisement, any resolver specializing in a key computed from a query should be able to resolve the query without requiring any joins or extra data transfers. Hence, when a client submits a query, the resolver that first receives it randomly selects

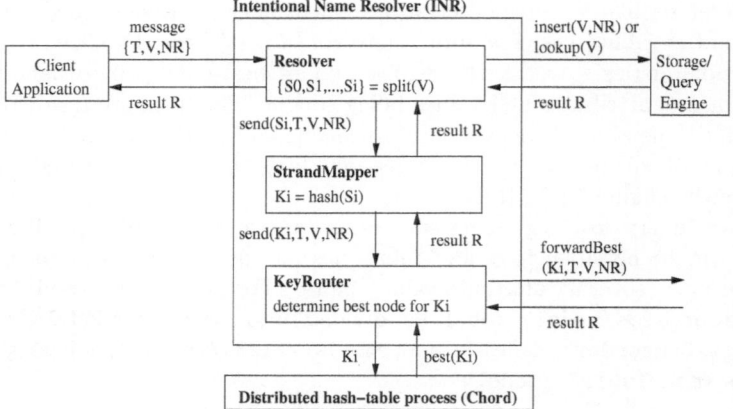

Fig. 2. An INR is composed of three layers. The Resolver receives messages from client applications. These messages include a type T (advertisement or query) an AVTree V and a name-record NR. The Resolver stores the description or performs the lookup locally. It then extracts all strands S_i from V, and for each one, the StrandMapper computes a key K_i. The KeyRouter finally maps each key onto an INR using a distributed hash table process like Chord. The message is then forwarded to that resolver. For queries, the resolver sends results back to the originating INR which in turn sends them to the application

one of the longest strands from the query AVTree. This strand serves to determine which resolver should solve the query. Query results are later returned to the originating resolver, which forwards them to the client.

A client application may also specify that it is interested in any resource matching a description. In that case, a single answer is returned. It is the resource that matches the given description and that has the lowest application-level metric. This feature comes from the original INS design [1].

Splitting resource descriptions into strands is critical to INS/Twine's ability to scale well. It enables resolvers to specialize in holding information and answering queries about a subset of resources. The choice of an adequate distributed hash table process at the KeyRouter layer is critical to achieving high query success rates while minimizing the number of resolvers contacted on each query. We discuss our choice in Section 2.5. The following sections describe each layer and its main algorithms in more detail.

2.3   Resolver Layer

At the core of an INS/Twine resolver is a strand-splitting algorithm that extracts strands from a description. The goal of the algorithm is to break descriptions into meaningful pieces so resolvers can specialize around subsets of descriptions. At the same time, the splitting must preserve the description structure and support partial queries.

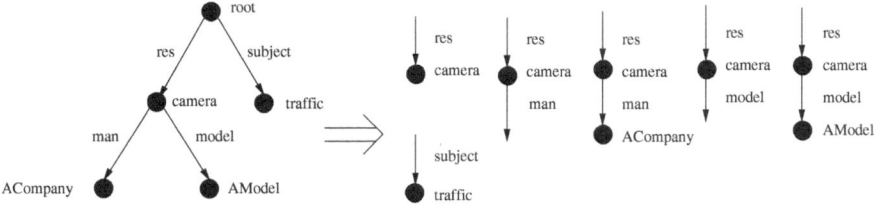

Fig. 3. Splitting a resource description into strands

A simple strand-splitting method would be to extract each attribute-value pair from the AVTree, and independently map it onto a key. However, this scheme would lose the richness of hierarchical descriptions and would not allow expressive queries to be performed. For example, *format* in the attribute sequence *printer-paper-format* would become indistinguishable from format in *video-cassette-format*.

The Twine algorithm for strand-splitting preserves the description structure and supports partial queries. It extracts each unique prefix subsequence of attributes and values from a description (advertisement or query) as illustrated in Figure 3. Each subsequence is called a *strand*. Each strand is then used to produce a separate key. For example:

```
Input strand: res-camera-man-ACompany
h1 = hash(res-camera)
h2 = hash(res-camera-man)
h3 = hash(res-camera-man-ACompany)
Output keys: h1, h2 and h3
```

We consider each top-level attribute-value pair as the minimal unit for resource descriptions and queries. We therefore omit strands composed of a single attribute and start strand-splitting after the first value. We believe that single top-level attributes would be too general to be useful in wide-area applications.

The Twine strand-splitting algorithm effectively extracts one strand for each attribute and each value in the AVTree, except for root attributes. Therefore, the number of strands depends on the number of attributes and values in the AVtree rather than its structure. More precisely, given a resource description with a attribute-value pairs, t at the root level, the total number of strands is given by:

$$s = 2a - t \tag{1}$$

With this scheme, partial queries are easily handled since each possible subsequence of attributes and values maps to a separate key, which in turn maps to a single resolver. The expectation of the storage requirement Z at each resolver is given by:

$$Z = \frac{(RSK)}{N} \tag{2}$$

where R is the number of resources in the system, S is the average number of strands per resource description, K is a configurable resource information replication level, and N is the number of resolvers in the network. This relation holds for $SK \ll N$.

Some strands such as `<resource>file server</resource>` may be extremely popular in resource descriptions. Advertisements can then overwhelm the node that is in charge of the popular strand. We tackle this problem by allowing each node to set a threshold (determined by the node's capacity) on the maximum number of resources for each key. When the threshold is exceeded for some key, no new resource is accepted under that key. The node could also start randomly replacing some entries with new advertisements. In both cases, this effectively renders that strand unusable for query purposes.

Since a query containing several strands is solved using one of the longest strands, the threshold restriction does not affect most queries. During query resolution, if a resolver returns an incomplete response (due to a threshold), a new strand is selected, and the process repeats until the response exhaustively lists all resources matching the description or no more strands are left in the query. In the latter case, the *partial* list of all accumulated matches is returned. It is flagged as being a partial answer, letting the application or user refine the query if necessary.

Additionally, in the rare case where a query containing only a very short and popular strand becomes extremely popular itself, edge resolvers may cache a few results to avoid flooding the node responsible for the strand. Caching is not yet implemented in INS/Twine.

2.4   StrandMapper Layer

Each strand extracted from the description is independently passed to the *StrandMapper* layer, together with the *complete* resource description or query. The StrandMapper is responsible for associating numeric keys with each strand. It does this by concatenating the attributes and values of the strand into a single string, and computing a 128-bit MD5 hash for the string.

2.5   KeyRouter Layer

The StrandMapper passes the key to the *KeyRouter* layer which uses it to determine which other resolvers should store information about the resource, or should participate in solving the query. The choice of an appropriate scheme for the KeyRouter is critical as it may easily become the limiting factor of INS/Twine's performance.

The KeyRouter may be thought of as a distributed hash table, where each node on the network keeps key-value bindings within a dynamically determined key range. Several efficient peer-to-peer algorithms have recently been proposed for this purpose: CAN [14], Chord [15], or Pastry [16]. Given a key, these systems find the node on the network that should store the corresponding value. Chord and Pastry are based on some variant of consistent hashing [20], where a node is responsible for all keys whose identifier falls between the node identifier and the closest preceding node identifier currently on the network. Hence only local disruptions occur when nodes join or leave the system. CAN uses a similar scheme, although not consistent hashing.

To achieve scalability, these systems require that each node only keeps in its routing table information about a subset of other nodes. This set is determined by the node unique identifier. Finding a node for a given key is then achieved by hopping from node to node in the appropriate direction, until the destination node is reached. This operation typically requires $O(\log N)$ hops, where N is the size of the overlay network of resolvers.

Our implementation is built on Chord, which efficiently rebuilds its overlay network in the presence of failures. INS/Twine uses Chord to efficiently identify which node, or set of k consecutive nodes, should store a given key.

3   State Management

The following consistency goals guided the design of the resource information management mechanisms in INS/Twine:

- When a resource joins a network, or moves or modifies its description, the update is propagated to the appropriate resolvers immediately. The new information replaces the old, ensuring that neither the old description nor the old location are ever returned as result of a query. While resource information propagates through the network, it is possible that both the old and new information be returned in response to a query.
- When a resource leaves or fails, its information is deleted at all resolvers.
- Query results are not affected by new resolvers joining the network, or by resolvers failing (up to a level of fault-tolerance determined by a configurable parameter k).

There are several ways to achieve consistency as defined above. If we use hard-state and require resources to always keep their information updated, we are not resilient to resources failing without prior de-registration.

Resolvers can also treat all resource information as *soft state*, requiring resources to refresh their information periodically throughout the network. If a resource does not refresh its presence within a certain interval, the corresponding description is removed from the network. A resource is free to leave the network at any time; if it does not de-register its description, the soft-state expiration mechanism will cause the resource description to be deleted. However, to keep information up-to-date, the refresh interval must be small, which imposes a high bandwidth overhead.

Periodically refreshing resource information also provides some degree of fault-tolerance by periodically sending each description not to one, but to $k > 1$ nodes per strand. This scheme relies on the capability of the underlying distributed hash table process (Chord [15] in our case) to rebuild the overlay network of interconnections as nodes join and leave. It also relies on the fact that

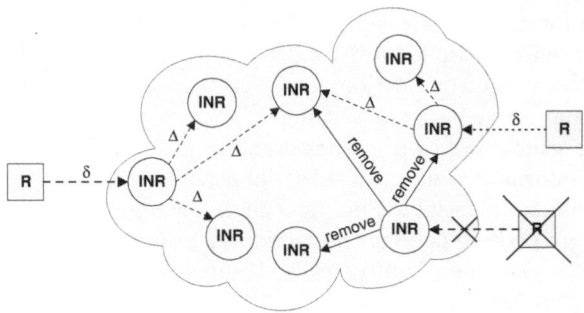

Fig. 4. State management in INS/Twine. Each resource R refreshes its information to a resolver INR at a high frequency defined by the refresh interval δ. Resolvers refresh resource information within the network at a lower rate defined by the refresh interval Δ. If a resource leaves the network without de-registering, the closest resolver detects the departure within δ and sends explicit remove messages to other nodes. Similarly when a resource updates its information. When a resolver crashes, corresponding resource information will remain in the network no longer than Δ

the set k is computed dynamically, so a failed node from the set is automatically replaced on subsequent resource advertisements.

We adopt a hybrid scheme in INS/Twine that combines the management simplicity of soft-state with the low-bandwidth requirements of hard-state. Figure 4 illustrates the approach. Each resolver at the edge of the network is responsible for resources that communicate directly with it. It acts as a proxy for these resources, keeping their states updated at the appropriate locations. Edge resolvers receive updates about the states of their resources at a fine time granularity δ (a few seconds in our implementation). They propagate any changes to appropriate resolvers. If a resource does not advertise itself within δ time units, an edge resolver assumes it has left the network and sends explicit remove messages to appropriate resolvers.

Resolvers in the network core also preserve soft-state, but they use a much longer period Δ (on the order of a few hours, for example). At every Δ time units, and for each directly connected resource, each edge resolver recomputes the set of resolvers in charge of that resource. It then transmits advertisements to every new resolver in a set. It transmits de-registration messages to resolvers no longer in a set. For resolvers that did not change, a simple ping is transmitted.

With this scheme, if a resource fails, it is de-registered from the network within δ time units. If a resolver acting on behalf of some directly connected resources fails, these resources can re-connect to another resolver while remaining available to clients. Finally, in the rare case where both the resolver and some directly connected resources fail, the resource information will be deleted from the network at most within Δ time units.

For increased fault-tolerance, each strand is mapped onto k replicas. Since we never transfer data between replicas, it is not guaranteed that each of the replicas knows about all resources matching any given strand. However, for up to $k-1$ failures within a refresh interval Δ, at least one of them does. Therefore queries are sent to all k nodes responsible for a given key, and the union of all results is returned to the client application. For larger number of failures within the interval Δ, the query failure rate becomes dependent on the *fraction* of failed resolvers. However, the query failure rate decreases exponentially with k.

Although we handle resolvers joining and leaving the network at any time, in case of network partitions and healing, information at different resolvers may become inconsistent. We currently do not handle this case, but we could use timestamps assigned by resources to their advertisements to determine which replicas hold the most recent information for each resource.

4   Evaluation

In this section, we first evaluate the strand-splitting algorithm by examining the distribution of strands from splitting real resource descriptions. We then examine how data and queries are distributed among resolvers. We finally evaluate the query success rate in the presence of failures.

4.1   Splitting Descriptions into Strands

In the first experiment, we evaluate the strand-splitting algorithm applied to real resource descriptions. Our goal is to determine how many strands are produced by such descriptions and how often the same strands come up. It is very difficult to obtain a large quantity of real resource descriptions. For our experiment, we used two sets of data. The first set contained 4318 bibliographical entries taken from latex bibliography files obtained from our own repositories as well as from Netlib [21]. The second set contained descriptions extracted from 883 mp3 files taken from our private collections. In both cases, we extracted word values from each string to enable hierarchical searches based on keywords in the title, author, and other fields. Although the data used does not describe devices, we believe it gives an intuition of the strand distribution that may appear in real data descriptions.

Figure 5(a) shows the strand distribution obtained from each data set. Bibliographical entries contain 12.9 strands on average, whereas mp3 tags produce an average of 8.7 strands. Although different resources will have descriptions of various complexity, it is interesting to note that splitting these real descriptions produced a reasonable number of strands. Figure 5(b) shows that, as expected, some strands are very popular - three strands come-up in over 10% of all resource descriptions for mp3 files. One of them is <artist>boys</artist>. Nine strands appear in over 10% of bibliographical descriptions, and three of them are in almost 30% of the entries. One example is <type>article</type>. However, this represents a very small fraction of all strands (less than 0.06%).

We evaluated INS/Twine on these two data sets as well as on workloads consisting of diverse synthetic descriptions. Our scheme mostly depends on the num-

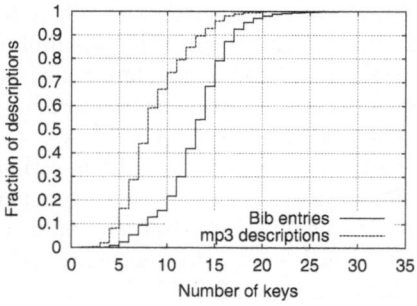
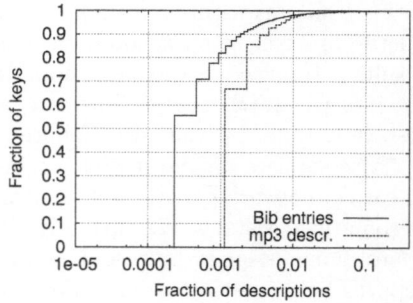

(a) Number of strands in resource descriptions

(b) Frequency of identical strands in different descriptions

Fig. 5. (a) When splitting real descriptions into strands, most descriptions tend to contain a small amount of strands compared to our scalability goals of $O(10^5)$ resolvers. 80% of the mp3 descriptions produce 12 strands or less. 80% of the bibliographical entries generate 16 strands or less. The medians are 8 and 13 respectively. (b) Most descriptions are composed of unique strands. However, a few strands may appear in as much as one third of all descriptions

ber of strands in resource descriptions and not on the structure of the AVTrees. We therefore present only the results obtained for the real data sets.

4.2   Distributing Data among Resolvers

In the second experiment, we evaluated the quality of data distribution in INS/Twine. We ran 75 independent INR instances on 15 machines. We connected 10 client applications to 10 resolvers. Each client application inserted between 60 and 160 resource descriptions, for a total of 883 resources (from the mp3 set) and 7,668 strands. For comparison purposes, we also ran experiments on 800 of the bibliographical entries, inserted by 8 clients (100 resources per client). The main difference between the data sets is the number of strands in resource descriptions which is 30% lower on average for mp3 tags. At the Key-Router layer, each resolver advertised itself as 20 virtual nodes to create a more uniform distribution of resolvers throughout the whole key space. Since each experiment is deterministic given a set of resolvers and resources, each curve in Figure 6(a) presents the results from a single run. Different sets with the same number of resolvers give almost identical graphs.

INS/Twine dynamically specializes nodes around specific resource descriptions, so each peer stores only a small fraction of the complete directory. Figure 6(a) shows that over half of resolvers hold information about less than 15% of resources for both data sets.

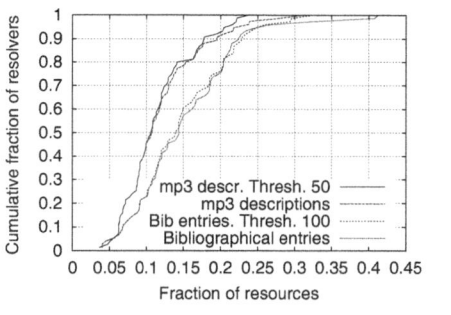   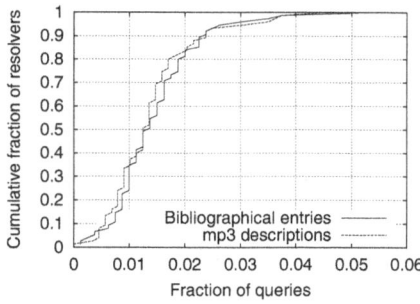

(a) Resource information distribution   (b) Query distribution

Fig. 6. Cumulative distribution of resource information and queries in a network of 75 resolvers. Resource information is evenly distributed among resolvers. Increasing the proportion of number of strands to number of resolvers increases the fraction of resources known by each node. Queries are distributed evenly among all resolvers with a median of 1.2% queries per resolver, equal to $\frac{1}{N}$ where N is the number of resolvers. No resolver solves an excessive number of queries

The expected value for the fraction of resource information stored at each resolver is $\frac{(SRK)}{N} * \frac{1}{R}$, where S is the average number of strands in resource descriptions, R is the total number of resource descriptions, K is the configurable replication level, and N is the number of resolvers in the network. We compare the value from this theoretical model to the actual values obtained from the experiments. In the experiment, $K = 1$ and $N = 75$. For mp3 files, $S = 8.7$, so resolvers should know about 11.6% of all resource descriptions. This is indeed the average obtained in the experiment. The median is a little lower at 10.4%. For bibliographical entries $S = 12.9$ on average, so resolvers should know about 17.2% of all resources. The actual average and median are just slightly lower at 15.2% and 14.0%, respectively. When the number of strands is high compared to the network size, there is a higher probability that multiple strands from the same description get assigned to the same resolver. Overall, the experimental values match the model.

The long tails of the distributions are due to popular strands. To alleviate their impact, we imposed a node-based fixed threshold value for the number of resources accepted for any particular strand. We set the threshold at 50 resources for the mp3 data set and 100 resources for the bibliographical entries, since the latter contain significantly more strands than the former. In reality, these thresholds would be determined by the capacity of each node. Figure 6(a) shows that for both data sets, the overall distribution remains similar, while the tail gets significantly cut. The maximum amount of information known by any given node drops under 24% for the mp3 files and under 33% for the bibliographical entries. No node knows about more resources than twice the average.

Table 1. Query success rates function of replication level. Each resource was requested using a single randomly chosen strand, but the complete description was used for the actual lookup. Values shown are averages of three runs

Replication	Fraction failed nodes					
	0	0.013 (1 node)	0.027 (2 nodes)	0.067 (5)	0.13 (10)	0.27 (20)
None($k = 1$)	1.0	0.98	0.92	0.88	0.83	0.70
$k = 2$	1.0	1.0	0.96	0.95	0.95	0.94

Hence data is evenly distributed in INS/Twine with each resolver holding only a small subset of resource descriptions. Thresholds are also efficient at eliminating overly popular strands without changing the overall resource information distribution. Taking the city example from the introduction, with around 10^8 resources, and 10^5 resolvers, considering an average of 13 strands per description, and an additional replication level of 3, each resolver would need to know about $\frac{(SRK)}{N} = \frac{(13*10^8*3)}{10^5}$ or as few as $40 * 10^3$ resources.

4.3 Resolving Queries

Without failures, Twine finds any resource present in the network with the same performance as the underlying KeyRouter layer (Chord in our implementation). For all queries, $O(\log N)$ resolvers are contacted at the KeyRouter layer [15] to find the set of nodes associated with a given key. When replication is used, k resolvers then resolve the query in parallel.

To evaluate the distribution of queries among resolvers, we used 800 descriptions from each data set as queries. We submitted all queries through one randomly selected resolver. Figure 6(b) shows how queries were distributed among resolvers. The average fraction of queries solved by each resolver should be $\frac{1}{N}$ where N is the number of resolvers. For $N = 75$, this gives 1.3% of all resources. The distribution obtained shows that the queries are in fact evenly distributed with 80% of resolvers receiving less than 2% of the queries. No resolver solves significantly more queries than the average, since the maximum number of queries received by any resolver was just a little over 5%. We also find that the distribution is independent of the number of strands in resource descriptions since we obtain the same graph for both data sets.

Replicating each key onto $k > 1$ nodes allows Twine to support up to $k - 1$ nodes joining or leaving the system within an in-core refresh interval Δ. Table 1 shows the query success rate function of the fraction of failed resolvers. To show the worst case scenario, only one strand was randomly selected from each resource description to serve as query. The table shows that increasing the replication level improves success rates, as all replicas for a given key have to be down for the query to fail. For example, for 20 failed resolvers out of 75, and for $k = 2$, the probability to pick a strand whose replicas both map to a failed resolver is approximately $(\frac{20}{75})^2 = 7\%$.

The latency of query resolution (as well as resource information updates) is determined by the time taken by peers to exchange information. Therefore,

INS/Twine latency and responsiveness will improve when proximity-based routing heuristics are used in the underlying key-routing system. For queries, since each description is replicated at several places in the network (at least one per prefix), there are many possible nodes that can resolve a query, and Twine may itself be able to choose a good node if it had information about network path latencies between nodes.

5 Conclusion

This paper described INS/Twine, a scalable resource discovery system using intentional naming and a peer-to-peer network of resolvers. The peer-to-peer architecture of INS/Twine facilitates a dynamic and even distribution of resource information and queries among resolvers. Central bottlenecks are avoided and results are independent of the location where queries are issued.

INS/Twine achieves scalability through a hash-based mapping of resource descriptions to resolvers. It manipulates AVTrees which are canonical resource descriptions. It does not require any a priori knowledge of attributes AVTrees may contain. Twine transforms descriptions into numeric keys in a manner that preserves their expressiveness, facilitates even data distribution and enables efficient query resolution. Resolver nodes hence dynamically specialize in storing information about subsets of all the resources in the system. Queries are resolved by contacting only a small number of nodes.

Additionally, INS/Twine handles resource and resolver dynamism responsively and scalably by using replication, by considering all data as soft-state and by applying much slower refresh rates in the core of an INS/Twine overlay network than at the edges.

INS/Twine scales to large numbers of resources and resolvers. Our experimental results show that resource information and query loads get evenly distributed among resolvers which demonstrates the ability to scale incrementally by adding more resolvers as needed.

Acknowledgments. The authors thank Godfrey Tan and Allen Miu for numerous helpful discussions. We thank Nick Feamster, Dina Katabi, Dave Andersen, and Chuck Blake for their comments on early drafts of this paper. This work was funded by NTT Inc. under the NTT-MIT research collaboration, by Acer Inc., Delta Electronics Inc., HP Corp., NTT Inc., Nokia Research Center, and Philips Research under the MIT Project Oxygen partnership, by Intel Corp., and by IBM Corp. under a university faculty award. Magdalena Balazinska is partially supported by the Natural Sciences and Engineering Research Council of Canada (NSERC).

References

1. Adjie-Winoto, W., Schwartz, E., Balakrishnan, H., Lilley, J.: The design and implementation of an intentional naming system. In: Proc. ACM Symposium on Operating Systems Principles. (1999) 186–201

2. Guttman, E., Perkins, C.: Service Location Protocol, Version2. RFC2608. http://www.ietf.org/rfc/rfc2608.txt (1999)
3. Sun Microsystems: Jini technology architectural overview. http://www.sun.com/jini/whitepapers/architecture.pdf (1999)
4. UPnP Forum: Understanding Universal Plug and Play: A white paper. http://upnp.org/download/UPNP_UnderstandingUPNP.doc (2000)
5. Czerwinski, S., Zhao, B., Hodes, T., Joseph, A., Katz, R.: An architecture for a Secure Service Discovery Service. In: Proc. of the Fifth Annual Int. Conf. on Mobile Computing and Networking (MobiCom), ACM Press (1999) 24–35
6. Castro, P., Greenstein, B., Muntz, R., Bisdikian, C., Kermani, P., Papadopouli, M.: Locating application data across service discovery domains. In: Proc. of the Seventh Annual Int. Conf. on Mobile Computing and Networking (MobiCom). (2001) 28–42
7. Hermann, R., Husemann, D., Moser, M., Nidd, M., Rohner, C., Schade, A.: DEAPspace – Transient ad-hoc networking of pervasive devices. In: Proc. of the ACM Symposium on Mobile Ad Hoc Networking & Computing (MobiHoc). (2000)
8. Abiteboul, S.: Querying semi-structured data. In: ICDT. Volume 6. (1997) 1–18
9. Mockapetris, P.V., Dunlap, K.J.: Development of the Domain Name System. In: Proc. of the ACM SIGCOMM Conference, Standford, CA (1988) 123–133
10. Yeong, W.: Lightweight Directory Access Protocol. (1995) RFC 1777.
11. Lilley, J.: Scalability in an intentional naming system. Master's thesis, Massachusetts Institute of Technology (2000)
12. Castro, P., Muntz, R.: An adaptive approach to indexing pervasive data. In: Second ACM international workshop on Data engineering for wireless and mobile access (MobiDE), Santa Barbara, CA (2001)
13. Gnutella: website. http://gnutella.com/ (2001)
14. Ratnasamy, S., Francis, P., M.Handley, Karp, R., Shenker, S.: A scalable content-addressable network. In: Proc. of the ACM SIGCOMM Conference, San Diego, CA (2001) 161–172
15. Stoica, I., Morris, R., Karger, D., Kaashoek, M.F., Balakrishnan, H.: Chord: A scalable peer-to-peer lookup service for Internet applications. In: Proc. of the ACM SIGCOMM Conference, San Diego, CA (2001) 149–160
16. Rowstron, A., Druschel, P.: Pastry: Scalable, distributed object location and routing for large-scale peer-to-peer systems. In: IFIP/ACM Middleware, Heidelberg, Germany (2001)
17. W3C: Web Services activity. http://www.w3.org/2002/ws/ (2002)
18. U.S. Census Bureau: United States census 2000. http://www.census.gov/ (2002)
19. Zhao, B., Joseph, A.: Xset: A lightweight database for internet applications. http://www.cs.berkeley.edu/~ravenben/publications/saint.pdf10 (2000)
20. Karger, D., Lehman, E., Leighton, T., Levine, M., Lewin, D., Panigrahy, R.: Consistent hashing and random trees: Distributed caching protocols for relieving hot spots on the world wide web. In: Proc. of the 29th Annual ASM Symposium on Theory of Computing. (1997) 654–663
21. Netlib: Netlib repository at UTK and ORNL. http://www.netlib.org/ (2002)

Location Estimation Indoors by Means of Small Computing Power Devices, Accelerometers, Magnetic Sensors, and Map Knowledge

Elena Vildjiounaite, Esko-Juhani Malm, Jouni Kaartinen, and Petteri Alahuhta

Technical Research Center of Finland, Kaitovayla 1, P.O.Box 1100, FIN-90571, Oulu, Finland
{Elena.Vildjiounaite, Esko-Juhani.Malm, Jouni.Kaartinen, Petteri.Alahuhta}@vtt.fi
http://www.vtt.fi/ele/indexe.htm

Abstract. A distributed real-time system, based on wearable accelerometers and magnetic sensors, is proposed for location estimation and recognition of walking behaviors. Suitable for both outdoor and indoor navigation, the system is especially adjusted for irregular movements indoors. The algorithm, which demands only small computing resources, performs step detection and classification in the time domain, allowing the estimation of the size of each separate step independently. Since the system finds the user's position relative to an initial position, it is intended to be supplemented with different types of absolute positioning information. Making use of map knowledge, as an easily available source of this information, is analyzed. The conclusion is drawn that referring to the locations of the corridors and stairways increases the positioning accuracy and reduces the effect of magnetic field distortions encountered inside buildings. The positioning error of different system configurations was 3–10 % from traveled distance.

1 Introduction

The awareness of a user's location and activity has principle significance for mobile computing. Most of the current context aware applications are based on location knowledge, such as smart workspaces [1,2], guides (e.g. tourists [3,4] and visually impaired people [5]), and memory aid tools [6-8].

Many researchers are involved in the development of different types of positioning systems [9]. The dead reckoning systems, based on inertial and magnetic sensors, are fairly specific with respect to other systems. On the one hand, location estimation errors could be huge (for example, if magnetic field disturbances effect heading determination). On the other hand these systems don't need special installation and could be easily used to complement other types: outdoors, the Global Positioning System [5,10-13]. Indoors, a dead reckoning system could be combined with infrared beacons

F. Mattern and M. Naghshineh (Eds.): Pervasive 2002, LNCS 2414, pp. 211–224, 2002.
© Springer-Verlag Berlin Heidelberg 2002

or bar code readers, allowing either to improve the precision of positioning or to decrease the density of infrared beacons. Additionally, the accelerometers and magnetic sensors are valuable for the user's activity recognition such as not moving - walking - running - going up- and downstairs [14-15], and they are fairly cheap.

In this paper, we describe a simple location estimation algorithm and a recognition method for the moving activities of the user. Unlike the outdoors style of human walking, where step length does not vary much [10], the style of human displacements indoors is fairly irregular. Thus, an indoor positioning system should not rely hard on the style of previous steps while considering the new one. In the work of Levi and Judd [16] step determination is done by analyzing accelerometer data. Step size estimation is based on step frequency only, and data processing includes Fast Fourier Transform, which requires a large number of samples to be stored and does not allow each step to be analyzed separately.

The methods of Legat et al. [12,13] and Lee and Mase [14] are very similar to that of Levi and Judd: processing of accelerometer data for step detection, step size estimation based on the step frequency. The last one is calculated using the autocorrelation function of vertical acceleration. Thus, at least two consecutive steps are needed for analysis. In the work of Lee and Mase [15] the location estimation is done by step counting only, and the method is able to recognize only certain locations if the user has come to these locations along a regular route. Step detection is done by using the peak detection algorithm, applied to vertical acceleration. In the work of Legat et al. acceleration is sampled at a 150Hz frequency. In the work of Lee and Mase [14,15] the sampling frequency is lower: 50 Hz.

In our system, the sampling frequency is only 12 Hz. We are interested in the reduction of sampling rate and simplification of calculations, because we, inspired by the notion of "invisible computing" [17], are aiming at embedded applications. We propose light-weighted step detection and an activity recognition algorithm, based on the changes in the tilt of the human leg while moving. The algorithm needs a fairly small amount of data to be stored, so that the algorithm fits in a PIC microcontroller with only 368 bytes of RAM. Analysis of tilt changes allows finding time duration for each step independently which is very important for irregular displacements. Tilt of the leg could be estimated by different means. We tested the use of accelerometers and an infrared proximity sensor, but finally we have chosen to use the magnetic sensor data, since tilt is well known to be a hindrance for compass heading determination. Processing both magnetic sensor and accelerometer data simplifies step detection and the recognition of different types of human activity. In addition, the activity recognition is very fast since all data processing is done in the time domain.

Unlike the work that has been referred to earlier, our step size estimation method takes into account not only the timing characteristics of the step, but also forward acceleration, which proved to be more robust for the indoors style of walking. For example, while opening doors, people are making both forward and backward steps.

We also tested the possibility to use the map knowledge as absolute position information. Map matching is widely used for the correction of navigation errors in positioning systems for cars [18] and mobile robots [19], but there are differences compared with our system. As for robots, they are usually equipped with sonars, cameras, lasers, etc. and thus are able to detect more features of environment than a system based on accelerometers and magnetic sensors. Another difference is that unlike robots and cars, the function of which has mainly to do with avoiding obstacles while travelling towards certain destination, humans are used to walk inside their offices in a non-regular manner. The main difference with using this method for cars is that in a normal office environment the density of intersections is higher than on the road. Due to these differences, test results have shown that referring to detailed map knowledge is not worth additional work, but using knowledge only about locations of corridors and stairways is helpful even in cases of the magnetic field distortions encountered inside buildings.

2 System Description

Unlike the previous work, we have not tried to combine all sensors and data processing in a single module. Instead, we were interested in the collective context awareness of a number of smart devices. Fig. 1 presents the current prototype together with our vision of the system after further development.

Fig. 1. Current prototype (on the left) and the future system (on the right)

Sensor boxes called SoapBoxes [20] were developed in our research institute for research purposes. This is a generic hardware, and its description is presented in the Pervasive 2002 conference in the paper of E. Tuulari et al. "SoapBox: A Platform for Ubiquitous Computing Research and Applications". SoapBoxes include a PIC 16F877 microcontroller, a short range radio, two 2-axis accelerometers ADXL 202E (together they provide 3-axis accelerometer data), 2-axis magnetic sensor HMC 1022, light sensor and infrared proximity sensor. The last one could also work as an infrared light detector, so that the positioning system would be able to correct the accumulated errors by detecting signals from infrared transmitters fixed at certain places. This feature is not implemented yet. The light sensor has not proved to be useful for us because of the dark northern winter, thus the current system uses accelerometer and magnetic sensor data. Our system consists of two or three sensor boxes and a notebook PC. All sensor boxes are small and light (approximately the size of a matchbox) and capable of wireless and wired communication and some data processing. Data is sampled at the rate of 12 Hz. One of the boxes is placed on the back of user's waist and the other one or two - on the legs. Currently, we attach the sensor box to the user's ankle, but looking forward to the time when sensor boxes of extremely small size (one cubic millimeter, as proposed in the "Smart Dust" project [21]) will be widespread and cheap, we are thinking about location-aware shoes.

Accelerometers in our sensor box have a fairly high noise level, up to $0.5 m/s^2$. Therefore, we have to use a digital "moving average" filter [22] with a sliding time window of five samples. We apply the same size of window to all data.

The sensor box on the leg is used for step detection and step size estimation. The sensor box is attached to the side of the ankle in such a way that when the user is standing, one of the magnetic sensors (along X-axis) lies parallel to the ground and points forward. It is intended to be fixed on the side surface of a shoe, near the heel. In this position, the accelerometer sensors point forward (X-axis), upward (Z-axis) and sideward (Y-axis). Currently, we assume that the sensor box will be fixed firmly. In the case of boxes on both legs each box data gives independent results. In the case of only one box the assumption is made that the one leg displacement is always followed by exactly the same step of the other leg. Such an assumption is not that wrong as might seem from the first impression. First of all, many people are used to start walking from the same leg most of the time. Second, one extra step matters only if the user stops very many times while walking or turns very often, which is not always the case. Finally, analysis of data from the box on the back can help to eliminate such extra steps.

The sensor box fixed on the back is used for heading determination. It could be also used for distinguishing between standing and sitting, which is impossible to do with boxes on the legs. While walking, magnetic sensors on the back undergo some tilt. In order to decrease the tilt influence on the heading determination, we use a low-pass filter. However, a better compensation for tilt should be added to the heading determination algorithm.

The personal computer is responsible for data synchronization. Each time data from the sensor box on the back arrives, the algorithm checks if the direction of movement has been changed (currently we use an 8-point compass in order to simplify calculations). The algorithm keeps the array of compass headings together with corresponding time stamps of the start and end of moving in the same direction. When the new step is detected, the computer checks compass heading information during the time period of the step and calculates the displacement. If the compass heading has been changed while stepping, the resulting direction and the length of the step are calculated according to simple rules, depending on the time duration of moving in each direction. In cases of neighboring directions being close to each other, the weighed sum of them is taken as the resulting direction.

After calculating the new position, the personal computer checks if it is consistent with the map knowledge, provided that such knowledge is available. The system is capable of both using the detailed map knowledge, including all door and wall coordinates, and using only partial map knowledge, namely, locations of corridors and stairways. In the case of using the entire map knowledge, any displacement inside a room is allowed as long as the estimated path does not go through a wall, but leaving the room is possible only through a door. The map-matching method used is curve-to-curve topological matching [23].

In case of referring only to the position of corridors and stairways, the first thing for the system to do is to recognize one of the two special types of user behavior: "walking along a corridor" and "walking up or down the stairs". Similarly to the map matching algorithms for cars [18], it is assumed that if the user has traveled reasonable distance in the direction of the corridor, he is in the corridor. In the cases when the system recognizes "walking along a corridor" behavior, it uses curve-to-curve topological matching [23] in order to correct the only coordinate (orthogonal to the direction of corridor). If the user's activity is recognized as "walking up or down the stairs", the user is assumed to be in a stairway. In the cases when the system recognizes an "end of stairs" reference point, it corrects both coordinates and counts how many staircases up or down the user has walked. In this case the map is much simpler and does not contain information about rooms and doors locations. Only corridors (as arcs) and corridor intersections and stairways (as nodes) are presented. At present, the system does not recalibrate step length from the map matches. Since it is a relative positioning system, it should be initialized by means of coordinates of starting point.

In our current prototype most of data processing is performed by the PC, but this is subject to change. The algorithm for step detection and classification fits in the microcontroller inside the sensor box. Currently sensor boxes communicate with the PC via serial cables and the communication protocol for wireless communication is under development.

The data analysis process is shown in Fig. 2.

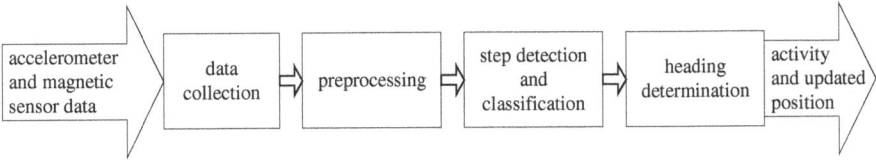

Fig. 2. The data analysis process

3   Step Detection and Activity Recognition

Unfortunately for the heading determination, the earth's magnetic field has a vertical component which influences magnetic sensor readings significantly in the case of the tilt. In the Northern Hemisphere, the angle of the magnetic field to the surface of the earth is roughly 70 degrees down toward the north [24]. This feature has appeared to be useful, however, for step count. While walking, the human leg tilt occurs naturally. On the graph of the magnetic sensor data (see Fig. 3) one step corresponds to one period of oscillations. Accordingly, the time duration of each step could be found as the time difference between two adjacent minima independently from the previous steps. If the person is standing, no pronounced peaks in magnetic sensor data appear. Fig. 3 demonstrates that step count by processing magnetic sensor data is easier than by processing accelerometer data, especially for slow steps.

In order to count steps, the algorithm looks for minima and maxima in the incoming magnetic sensor data and calculates both the difference between the peak values in the adjacent minimum and maximum and time interval between the neighboring minima. A step has occurred, if the amplitude difference is higher than a certain threshold amplitude A_{th} and the time duration of the step is shorter than time threshold t_{th}. The threshold t_{th} is chosen to be 2.5 s., since typical walking frequencies are known to be within the range of 0.5 to 3 Hz [13]. The step size is estimated based on a simple look-up table, and depends on the time duration of each step and peak value of forward acceleration in the vicinity of the magnetic sensor minimum. The method does not need a lot of data to be stored. As for acceleration data, we need only the maximum of forward acceleration during the last five samples. When the minimum of a magnetic sensor signal is detected, we look at the next five acceleration data samples and find the maximum value among these ten samples. This value serves as the acceleration peak for the look-up table. This method allows reducing the rate of erroneous step detection when the user is sitting and moving his legs. It also allows assigning the correct step size to slow, but long steps and short fast steps, which would be impossible based only on the time value. (Both cases are not typical and don't appear often in normal walking, but they happen while moving in a crowded area.) Currently our look-up table consists of five possible step lengths for walking. During our tests we have concluded that certain improvements in the resolution may be achieved with a more detailed look-up table. It is important to remember that values in the table are highly dependent on the personal style of walking. Our table was created experimentally: the user was asked to walk slowly, normally and fast along the corridor. The table includes the average step sizes of fast and slow walking together with two values for moderate walking and one for very slow walking. In the future, these values should be adapted automatically for different users by making use of absolute positioning information, for example from infrared beacons.

Most office buildings consist of more than one floor, so navigation systems should be able to notice if the user is going upstairs or downstairs. Knowing the stairs location could help to compensate for relative location errors. The proposed algorithm distinguish between level walking, ascending and descending stairs by comparison of peak values of magnetic sensor data, forward and vertical acceleration data. Smaller amplitude of forward acceleration and magnetic sensor data, but higher amplitude of vertical acceleration characterizes upstairs and downstairs walking. Running is very easily recognizable by the highest forward acceleration values and highest frequency of steps.

Fig. 3 presents the patterns of different types of activity. Unlike the accelerometer data, there is a shift in the graph of magnetic sensor data along the vertical axis. The shift is due to the change of direction of movement and has nothing to do with activity recognition.

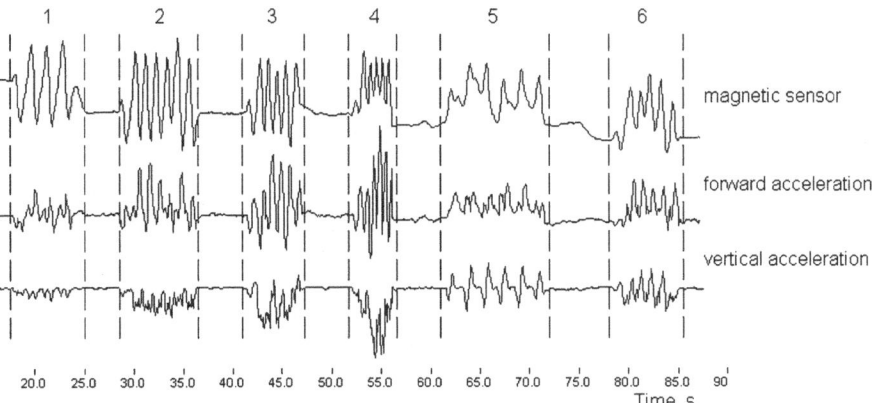

Fig. 3. Magnetic sensor, forward and vertical acceleration data during slow (1), normal (2), fast (3) walking, running (4) and going up- (5) and downstairs (6) with short periods of standing between these activities. The sensors are fixed on a leg. Forward and vertical acceleration stands for the data from two orthogonal accelerometer axes, X and Z (see Fig. 1), magnetic sensor stands for the data from the sensor pointing forward along X axis

4   Experimental Results

Here we present the results for three experimental routes shown in Fig. 4 and Fig. 5. To begin with, the first route consists of long and short curvy paths on level ground. Next, the activities of going upstairs and downstairs were added to the second one. Finally, the act of passing nearby a source of strong magnetic field distortions was added to the third route. All the routes started and finished sitting in the chair near a computer in the user's working room (position A in Fig. 4 and Fig. 5). Thus, they all consist of standing up, visiting rooms of two colleagues (the positions where the user stopped in these rooms are marked on the plan as B and C), returning back to the position A and sitting down in the same chair. Similarly to a real working situation, the door to the colleague's room had to be opened sometimes, and sometimes the door had already been opened. In the second and third routes, the heavy doors across the corridor and to the stairway needed to be always opened, two of them with a magnetic key. This means additional steps forwards and backwards. Thus, the style of displacements was really irregular. In the start of each run, the system was initiated with the coordinates of the user's chair in the working room. In the future smart office, the system should discover the initial coordinates automatically, for instance, with the help of infrared or RF beacons.

The coordinates given by the positioning system in each test position were compared with the correct coordinates. Absolute positioning error is calculated as $\sqrt{e_x^2 + e_y^2}$, where e_x and e_y are the errors in the corresponding coordinate. Relative error is the ratio of the absolute error to the full pass. The accuracy of absolute error estimation is approximately 0.3 m since with a natural style of walking it is impossible to stop exactly in the same place and since the human body occupies some space.

The first test was performed in three different modes: without using any kind of map knowledge, by referring to the positions of corridors, and by using all the map knowledge, including all door and wall coordinates. The second and third tests were performed in two modes: without using any kind of map knowledge, and by referring to the positions of corridors and stairways. One user performed all the tests 20 times in each mode.

4.1 System Performance on Level Ground

The first route (Fig. 4) was chosen to lie on the same floor. It allowed us to evaluate the accumulation of errors both on long paths and on short curvy paths. The distances between the rooms are indicated on the plan.

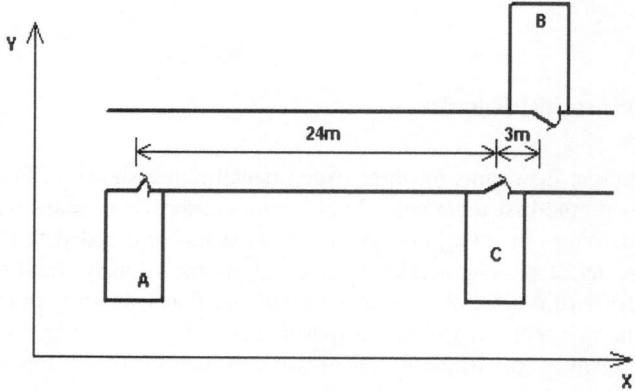

Fig. 4. Schematic plan of the first experimental route. The user traveled from position A to positions B and C and back to position A

In the case where the sensor box was attached to one leg only, it was assumed that after one leg step exactly the same step of the other leg followed. Experiments show that, on long relatively straight paths, such assumption is acceptable, but the situation changes if there are many turns on the short path, like from position B to position C. The accuracy of the location estimation was the worst in position C for that reason, and the contribution of errors in the main walking direction (along the X-axis) becomes comparable with errors in the orthogonal direction. However, it could be amended by separate treatment of steps while turning, the first steps after standing and last steps before stops. Step count accuracy could also be improved by analyzing of tilt of the sensor box on the back.

Table 1. Test results for the sensor box attached to one leg only, first route

Path	without using any map knowledge, absolute/relative error		referring to corridor coordinates only, absolute/relative error		using all the map knowledge, absolute/relative error	
	Average	worst case	average	worst case	average	worst case
A-B	1.5m/4.8%	3.0m/9.7%	1.3m/4.2%	2.5m/8.1%	1.0m/3.2%	2.8m/9.0%
A-B-C	2.1m/5.3%	5.1m/13%	1.8m/4.6%	3.6m/9.1%	2.2m/5.6%	3.7m/9.4%
A-B-C-A	3.4m/5.1%	5.5m/8.3%	3.0m/4.5%	4.0m/6.0%	2.2m/3.3%	4.3m/6.5%

From the table above we draw the conclusion that referring to the entire map knowledge is not worth the amount of data stored and additional reasoning. When compared with the case of referring to corridor coordinates only, the average error is smaller in case of entire map knowledge, but the worst case is worse.

Referring to the corridor coordinates does not allow improving accuracy in position estimation in the X direction, but it removes part of errors in the Y direction. It is especially important when the disturbances in the magnetic field briefly distort heading determination.

When sensor boxes are attached to both legs, accuracy does not increase significantly for long paths, as expected. For short paths, like that from position B to position C, the difference in results is more noticeable, but not as much as we were expecting. The reason is that finding the direction of the step is not easy when a person is turning. The sensor boxes on both legs have proved to be most useful for such irregular displacements like opening and closing doors and intermittent walks, which are fairly common in office environment.

Table 2. Test results for sensor boxes attached to both legs, first route

Path	without using any map knowledge, absolute/relative error		referring to corridor coordinates only, absolute/relative error		using all the map knowledge, absolute/relative error	
	Average	worst case	average	worst case	average	worst case
A-B	1.4m/4.5%	2.8m/9.1%	0.9m/3.0%	2.6m/8.4%	1.0m/3.2%	2.5m/8.1%
A-B-C	2.1m/5.3%	3.9m/9.8%	1.5m/3.8%	3.2m/8.1%	1.7m/4.3%	2.7m/6.8%
A-B-C-A	3.0m/4.5%	5.0m/7.5%	2.1m/3.1%	4.1m/6.1%	2.2m/3.3%	4.0m/6.0%

The results presented in Table 2 show again that referring to the entire map knowledge is not reasonable. If we count the number of occasions when position estimation error was less than 5% from the path traveled, we get the following numbers: in the case of entire map knowledge in position B there were 90% of such occasions, in position C 40%, and in case of corridor coordinates knowledge 75% and 70% correspondingly. Thus, referring to the entire map knowledge increases the accuracy of position estimation only when the positioning errors are less than half the distance between neighboring doors. For some office buildings this means that the location estimation error should be less than 1 meter, which is not easily achievable by means of dead reckoning.

4.2   System Performance with Upstairs and Downstairs Walking Included

The two other routes (the second and the third ones) were performed in two modes: without using any map knowledge and with referral to stairway and corridor coordinates. The schematic plan of the routes is presented in Fig. 5.

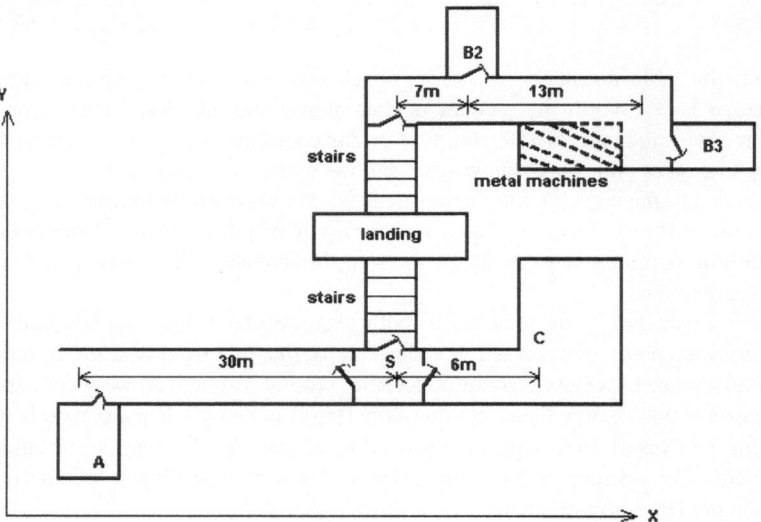

Fig. 5. Schematic plan of the second and third experimental routes. The positions A, S and C are same in both routes. The position B_2 belongs to the second route, the position B_3 to the third one. The user traveled from position A to the position S, then upstairs to the position B_2 or B_3, then downstairs to the position C and back to the position A

The second and third experimental routes were longer than the first one and included going up and down the stairs. Similar to the first route, these routes also started sitting in the chair near a computer in the user's working room (position A). The first point where positioning error was estimated (position S) was chosen to be near the door to the stairway on the same floor. In the second route the next point (position B_2)

was one floor up and seven meters away from the stairs. After visiting position B_2 the user turned back, went downstairs and checked his coordinates in position C, located six meters away from the stairs on the floor from which the user started the trip. After that the user returned to his room (position A) and sat down in the chair.

The second route allowed us to estimate the recognition of upstairs and downstairs walking. Some of the steps were not recognized correctly; this was the case especially with the first steps after standing, and with slow steps. This was due to the small tilt of the leg. For the first steps after landings it was difficult to distinguish upstairs from downstairs walking, but the overall performance was fairly good. No errors were recorded in switching from one floor to another either on the way upstairs, or downstairs. Our experiments have shown that step detection works reliably also in cases when the angle between a magnetic field and a magnetic sensor changes, but because the values in the look-up table for the classification of upstairs and downstairs walking are dependent on this angle, some additional reasoning needs to be included in the algorithm.

The results presented in Table 3 show that referring to the locations of stairways and corridors significantly improves positioning accuracy in spite of the fact that the recognition errors for "end of stairs" reference points are one-two meters. These "end of stairs" errors partly depend on the way the user opens the heavy doors from the stairs to the corridor. While opening these doors, the user has to make extra forward and backward steps, most of them short and slow. Such steps are most difficult to detect and classify correctly, mainly due to lower acceleration. As was mentioned before, the accelerometer noise level is fairly high in our sensor boxes, which results in a significantly distorted acceleration peaks of slow steps.

Table 3. Test results for sensor boxes attached to one leg only, second route

Path	without using any map knowledge, absolute/relative error		referring to stairway and corridor locations, absolute error		the kind of map knowledge used on the way from previous position
	average	worst case	average	Worst case	
A-S	2.3m/6.8%	3.4m/10.0%	1.4m	3.1m	referring to corridor
A-S-B_2	3.7m/6.9%	6.3m/11.8%	1.6m	3.0m	referring to stairway on the way up
A-S-B_2-C	5.2m/6.6%	7.5m/9.5%	2.0m	3.8m	referring to stairway on the way down
A-S-B_2-C-A	6.4m/5.5%	8.5m/7.3%	2.9m	5.5m	referring to corridor

4.3 System Performance in the Presence of Strong Magnetic Field Distortions

The third experimental route (see Fig. 5) was similar to the second one, with the only difference that position B_3 was moved further along the corridor so that the user had to walk beside a group of large copy machines and printers. The results of this test are summarized in Table 4. Since the test did not bring any new information about paths

A-S and C-A, they are not included in the table. With this route we tested the effect of strong magnetic field distortions on position estimation and got two positive results. First, the step detection algorithm was not influenced by these distortions. Second, referring to the locations of corridors and stairways decreases positioning errors even in the presence of big errors in heading determination (up to 90 degrees). The improvement in position B_3 is less noticeable because the algorithm was not able to recognize "walking along the corridor" behavior correctly when the compass gave wrong data on several consecutive steps. Location estimation error in position C in the case of referring to the stairways coordinates is little bit higher than in the previous test, probably because of additional movements while opening the doors.

Table 4. Test results for sensor boxes attached to one leg only, third route

Path	without using any map knowledge, absolute/relative error		referring to stairway and corridor locations, absolute error		the kind of map knowledge used on the way from previous position
	average	worst case	average	Worst case	
A-S-B_3	5.6m/8.6%	10.1m/15.4%	3.1m	7.8m	referring to stairway on the way up and corridor
A-S-B_3-C	8.1m/7.9%	13.0m/12.6%	2.3m	4.1m	referring to stairway on the way down

5   Conclusions and Future Work

The distributed walking behavior recognition and location estimation system, proposed in this paper, showed reliable performance for step detection, step size estimation and classification of running, level walking, ascending and descending of stairs for one user. No errors were recorded in switching from one floor to another either on the way upstairs, or downstairs. The average error was found to be around 5% from the total pass in case of level walking, and around 7% when upstairs and downstairs walking was included (the increase of average error is due to the imprecise recognition of upstairs and downstairs activities). In the presence of magnetic field distortions, which influences the heading determination, the average error was found to be around 9%. These values were obtained by averaging over 20 passes for each of the three different routes. As a result, the accuracy of location estimation without absolute positioning information does not allow neighboring rooms to be distinguished reliably in our office environment.

The rationality of using map matching in an indoor environment was also tested. Experimental results show that referring to the locations of corridors and stairways significantly increases positioning accuracy, especially in cases of strong magnetic field disturbances. In these tests the average error does not exceed 5%.

The system is intended to be combined with an absolute positioning system based either on infrared beacons or radio communication, and the comparison of these two types should be investigated. The current prototype is built with the help of generic hardware, and step detection and activity recognition algorithm can run on the micro-controller inside the sensor box or on some specific hardware embedded into shoes, for example. Also a new wireless sensor board called Smart-It could be used for context recognition [25]. Smart-Its sensor boards are developed in EC project called Smart-Its.

Our future work includes changes in data processing algorithms, improvement of synchronization between different parts of the system and analyses of data from other sensors, mainly infrared proximity sensors. Future work also includes searching for fast ways of tuning parameters for different walking styles and doing it online according to absolute positioning information. Currently we are working on modifying the prototype to utilize the wireless capabilities of the sensor boxes and to achieve stronger distribution of computations. The algorithm demands only small computing resources and is developed for use in small-scale embedded devices. It is planned to be used in a distributed context recognition system for collective context awareness in office and home environments.

Acknowledgements. The Smart-Its project is funded in part by the Commission of European Union under contract IST-2000-25428, and by the Technical Research Center of Finland. We would like to thank the sensor box team at the Technical Research Center of Finland, especially Mr. Arto Ylisaukko-oja, for their support.

References

1. Want, R., Shilit, B.N., Adams, N.I., Gold, R., Petersen, K., Goldberg, D., Ellis, J.R., Weiser, M., "An Overview of the PARCTAB Ubiquitous Computing Environment", IEEE Personal Communications, Vol. 2, No. 6, Dec 1995, pp. 28-43
2. Harter, A., Hopper, A., "A Distributed Location System for the Active Office", IEEE Network, Jan./Feb. 1994, pp. 62-70
3. Davies, N., Cheverst, K., Mitchell, K., Efrat, A., "Using and determining location in a context-sensitive tour guide", Computer , Volume: 34 Issue: 8 , Aug. 2001, pp. 35 -41
4. Oppermann, R., Specht, M., "A Context-Sensitive Nomadic Exhibition Guide", in H.-W. Gellersen and P. Thomas (Eds.): HUC 2000, LNCS 1927, pp. 172-186, 2000, Springer-Verlag Berlin Heidelberg 2000
5. Shinoda, Y.; Yakabe, Y.; Magatani, K.; Yanashima, K.; Sato, R. "Development of navigation system for the visually impaired", Engineering in Medicine and Biology Society, 1996, Bridging Disciplines for Biomedicine., 18th Annual International Conference of the IEEE , Volume: 1 , 1997, pp. 399 -400 Vol.1
6. Brown, P.J., "Triggering information by context", Personal Technologies, 2(1):1-9, September 1998
7. Dey, A.K., Abowd G.D., "CybreMinder: A Context-Aware System for Supporting Reminders", in H.-W. Gellersen and P. Thomas (Eds.): HUC 2000, LNCS 1927, pp. 172-186, 2000, Springer-Verlag Berlin Heidelberg 2000

8. Marmasse, N., Schmandt, Ch., "Location-Aware Information Delivery with ComMotion", in H.-W. Gellersen and P. Thomas (Eds.): HUC 2000, LNCS 1927, pp. 172-186, 2000, Springer-Verlag Berlin Heidelberg 2000
9. Hightower, J., Borriello, G., "Location systems for ubiquitous computing", Computer , Volume: 34 Issue: 8 , Aug. 2001, p. 57 -66
10. Gabaglio, V., Merminod, B., "Real Time Calibration of Length of Steps with GPS and Accelerometers", GNSS, Genova, Italia, 1999,.October 5-8
11. Ladetto, O., Gabaglio, V., Merminod, B., "Two different Appproaches for Augmented GPS Pedestrian Navigation", International Symposium on Location Based Services for Cellular Users, Locellus 2001
12. Legat, K., Lechner, W.,"Integrated navigation for pedestrians", GNSS 2001
13. Legat, K., Lechner, W., "Navigation systems for pedestrians - a basis for various value-added services", ION GPS 2000
14. Seon-Woo Lee; Mase, K., "Recognition of walking behaviors for pedestrian navigation", Control Applications, 2001. (CCA '01). Proceedings of the 2001 IEEE International Conference on , 2001, pp. 1152 -1155
15. Seon-Woo Lee; Mase, K., "Incremental motion-based location recognition", Wearable Computers, 2001. Proceedings. Fifth International Symposium on , 2001, pp. 123 -130
16. Levi, R.W., Judd, T., "Dead reckoning navigational system using accelerometer to measure foot impacts", US Patent 5583776
17. Borriello, G., "The challenges to Invisible Computing", Computer, Volume: 33 Issue: 11, Nov. 2000, pp. 123-125
18. Sweeney, L., "Comparative Benefits Of Various Automotive Navigation And Routing Technologies", Position Location and Navigation Symposium, IEEE 1996, pp. 415-421
19. Golfarelli, M., Maio, D., Rizzi, S., "Correction of Dead-Reckoning Errors in Map Building for Mobile Robots", IEEE Transactions on Robotics and Automation, Vol. 17, No. 1, February 2001, pp. 37-47
20. Tuulari, E., "Enabling ambient intelligence research with soapbox platform", Ercim news (2001) No. 47, pp. 18 - 19
21. http://www-bsac.eecs.berkeley.edu/~warneke/SmartDust/
22. Smith, Steven W., "The scientist and engineer's guide to digital signal processing", California Technical Publishing, San Diego, CA. 1997
23. Bernstein, D., Kornhauser, A., "An Introduction to Map Matching for Personal Navigation Assistants", New Jersey TIDE Center, 1996
24. Caruso, M., "Applications of Magnetoresistive Sensors in Navigation Systems", http://www.ssec.honeywell.com/magnetic/datasheets/sae.pdf.
25. http://www.smart-its.org/

Estimating the Benefit of Location-Awareness for Mobile Data Management Mechanisms

Uwe Kubach and Kurt Rothermel

Institute of Parallel and Distributed High-Performance Systems (IPVR),
University of Stuttgart, Breitwiesenstr. 20-22, 70565 Stuttgart, Germany
Uwe.Kubach@informatik.uni-stuttgart.de

Abstract. With the increasing popularity of mobile computing devices, the need to access information in mobile environments has also grown rapidly. In order to support such mobile information accesses, location-based services and mobile information systems often rely on location-aware data management mechanisms like location-aware caching, data dissemination or prefetching. As we explain in this paper, the location-awareness of such mechanisms is only useful, if the accessed information is location-dependent, i.e. if the probability with that a certain information object is accessed depends on the user's location.
Although the location-dependency of the accessed information is crucial for the efficiency of location-aware data management mechanisms and the benefit they can get out of their location-awareness, no metric to measure the location-dependency of information has been proposed so far. In this paper, we describe such a metric together with a second one for a further important characteristic of mobile information accesses, the so-called focus.

1 Introduction

Location-based services provide their users with local information depending on their current geographic position. For example, a user can ask for nearby restaurants or the shopping centers in his proximity. An important requirement for such location-based services to be beneficial is that the offered information is location-dependent, i.e. that the relevance of each information object for the user depends on his location.

Such a location-dependency is not only exploited for the pre-selection of information in location-based services but also in many other data management mechanisms supporting mobile information systems. Since they consider a user's location, these mechanisms are called location-aware. Examples are location-aware caching [11,5], dissemination [6], prefetching [4], and hoarding mechanisms [8].

So far, the location-dependency has been mostly considered as a binary characteristic of an information system, i.e. an information system respectively the information offered by the system was said to be location-dependent or not. However, we claim that there is a complete spectrum of mobile information systems, which differ in the degree of their location-dependency. This spectrum ranges from inherently location-dependent systems to location-independent systems with many nuances in between.

In inherently location-dependent systems each information object belongs to a fixed location and is only accessed when the user is located there. An example for such a

F. Mattern and M. Naghshineh (Eds.): Pervasive 2002, LNCS 2414, pp. 225–238, 2002.
© Springer-Verlag Berlin Heidelberg 2002

system is a map application showing the user a map of his environment [13]. A browser-based mobile tourist guide is an example of a system that is neither inherently location-dependent nor location-independent. Although the user can potentially access all information objects available in the information space from any location, he will in most cases preferably access information about his current environment. In contrast, wireless web browsers [3] will mostly be used in a location-independent manner. For example, if a user looks for his stock quotes, his location will usually have no influence on which quotes he requests.

The location-dependency of the information available in a mobile information system strongly influences the amount of data a user needs at a certain location and with it the amount of data that has to be cached, disseminated or prefetched for a certain location. Therefore, the location-dependency is crucial for the efficiency of location-aware data management mechanisms. In this paper, we propose a metric that allows to quantify the location-dependency of a single information object as well as that of a whole information space. Thereby we provide a means to better understand what location-dependency is and to decide whether using a location-aware data management mechanism is suitable or not.

In addition to the location-dependency, we discuss a further characteristic of mobile information systems, which we call the focus. It also has a strong effect on the efficiency of location-aware data management mechanisms.

Finally, we use a hoarding mechanism [8], which we developed for a platform supporting location-aware applications [7], as an example to illustrate how the discussed characteristics influence the efficiency of a location-aware mobile data management mechanism.

The remainder of this paper is structured as follows: in the following section, we discuss the related work, before we introduce two metrics to measure the disparity of frequency distributions in Section 3. Next, in Section 4 and Section 5, we describe our metrics for the location-dependency and the focus. Afterwards, we evaluate the metrics and give an example for their application in Section 6. Finally, Section 7 concludes our paper.

2   Related Work

To our knowledge, no measure for the location-dependency of an information space has been proposed so far. For the location-dependency it is important, how information requests are distributed over the area, in which the information system is accessible. Such a spatial distribution also plays an important role in wireless ad hoc sensor networks. There, the distribution of the sensors strongly influences the coverage of the sensor network.

However, most work in this area is focused on the question of how to arrange a set of sensors to observe an area, room, or building as good as possible. Only a few work has been done on measuring the coverage [9,10]. In these articles, coverage is measured in terms of how well an object that crosses the area covered by the sensor network can be observed. The inequalities in the spatial distribution of the sensors are not considered. In contrast, the inequalities in the spatial distribution of the information requests have a strong impact on the location-dependency and must therefore be considered when measuring this dependency.

Inequalities in spatial distributions are often considered in social sciences, e.g. income or wealth inequalities [1,12]. There, the Herfindahl coefficient and the Gini coefficient are widely used to measure the inequalities. These two coefficients are also the basis of our metric. In the following section, they are discussed in detail.

3   Preliminaries

Basically, our idea to measure the location-dependency of an information object is to consider how the requests for the object are distributed over the area in which the information system can be accessed. If many requests are concentrated on a few locations, the object is obviously only relevant at certain locations, i.e. it is strongly location-dependent. However, if the requests are equally distributed over the plane, the object's relevance for the users does not depend on their location, i.e. the object is location-independent.

In spatial statistics two coefficients, namely the Herfindahl and the Gini coefficient, are commonly used to describe such inequalities in frequency distributions. Since we also use these coefficients for our metrics, we describe them in this section.

A frequency distribution is a function $f : O \rightarrow \mathbb{N}$ that assigns to each unit of observation $o \in O$ the absolute frequency with that it occurred during an observation. An example for a frequency distribution is a function that assigns to each information object within an information space the number of requests that occurred for the object during a certain period of time. In the following, $N = |O|$ denotes the number of different units of observation that might occur. We assume that the units of observation are consecutively numbered from 1 to N. Hence, $f(i)$ denotes the frequency with that the unit of observation i occurred.

3.1   Herfindahl Coefficient

The Herfindahl coefficient HC is defined as the sum of the squares of the relative frequencies with which the units of observation occur:

$$HC = \sum_{i=1}^{N} \left(\frac{f(i)}{\sum_{j=1}^{N} f(j)} \right)^2 \tag{1}$$

For the Herfindahl coefficient values between $\frac{1}{N}$ and 1 are possible. It is minimal, when the frequencies are equally distributed, whereas the maximum value is reached, when only one certain unit of observation occurs at all.

3.2   Gini Coefficient

The Gini coefficient is based on the Lorenz curve, which can be used to depict inequalities in frequency distributions. To construct a Lorenz curve, the frequencies $f(i)$ first have to be sorted in an increasing order, i.e. we have to ensure that

$$f(1) \leq f(2) \leq \ldots \leq f(N-1) \leq f(N). \tag{2}$$

The Lorenz curve is then constructed as polygon through the points

$$(0,0), (u_1, v_1), (u_2, v_2), \ldots, (u_{N-1}, v_{N-1}), (u_N, v_N). \quad (3)$$

The coordinates of each point (u_j, v_j) are calculated as follows:

$$u_j = \frac{j}{N} \quad \text{and} \quad v_j = \sum_{i=1}^{j} \frac{f(i)}{\sum_{k=1}^{N} f(k)} \quad (4)$$

This means that the Lorenz curve is obtained by plotting the cumulative relative frequencies with that the units of observation occur during the observation period against the share of considered units of observation. If the considered frequencies are equally distributed, the Lorenz curve is equivalent to the 45 degree line. The bigger the area between the Lorenz curve and the 45 degree line is, the higher is the inequality in the corresponding frequency distribution.

In Table 1 an example for a frequency distribution is given. It shows how 100 information requests might be distributed over an information space containing 10 information objects, i.e. $N = 10$. Besides the absolute request frequencies $f(i)$, the share of the considered units of observation, and the cumulative relative frequencies are included in the table. Figure 1 shows the corresponding Lorenz curve together with the 45 degree line.

Table 1. A frequency distribution

Object ID	Number of Requests	Share	cumulative rel. frequencies
1	4	0.1	0.04
2	5	0.2	0.09
3	5	0.3	0.14
4	6	0.4	0.20
5	7	0.5	0.27
6	8	0.6	0.35
7	9	0.7	0.44
8	12	0.8	0.56
9	16	0.9	0.72
10	28	1.0	1.0

The Gini coefficient GC is defined as the ratio of the area enclosed by the Lorenz curve and the 45 degree line to the area enclosed by the 45 degree line and the x-axis. It can be calculated as follows:

$$GC = \frac{2 \cdot \sum_{i=1}^{N} i \cdot f(i)}{N \cdot \sum_{i=1}^{N} f(i)} - \frac{N+1}{N} \quad (5)$$

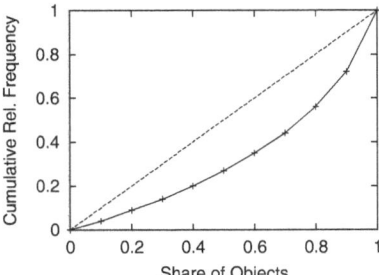

Fig. 1. Lorenz curve

Like the Herfindahl coefficient, the Gini coefficient is minimal ($GC = 0$) when the frequencies are equally distributed, and maximal ($GC = \frac{N-1}{N}$), when only one certain unit of observation occurs at all. Often the normalized Gini coefficient $GC^* = \frac{N}{N-1} \cdot GC$ is used in order to get values ranging from 0 to 1.

3.3 Comparison

The difference between the two coefficients is that the Gini coefficient solely reflects relative concentrations, whereas the Herfindahl coefficient also expresses the absolute concentration of a frequency distribution.

Thereby, absolute concentration means that the observed incidences are distributed over a small set O of possible units of observation, e.g. if an information space only contains a few information objects. Thus, the absolute concentration can be high, although the frequencies are equally distributed. Whereas relative concentration means, that a big part of the observed incidences is distributed over a small part of all possible units of observation. This implies an unequal frequency distribution.

4 Location-Dependency

In this section, we describe our metrics for the location-dependency of a single information object and that of a complete information space. In both metrics we assume that the coverage area of the information system, i.e. the area where the system can be accessed, is separated into a set S of equally sized squares, which do not overlap each other.

4.1 Definitions

As explained above, the location-dependency of an information object should describe, how strong the requests for this object are concentrated on certain locations. Thus, we define the location-dependency of a single information object as follows:

Definition 1. *Let i be an information object and $f : S \rightarrow \mathbb{N}$ a frequency distribution that assigns to each square $s \in S$ the frequency with that the information object i has been accessed in the square s. Then, the normalized Gini coefficient of the distribution f is called the location-dependency of the information object i.*

We chose the Gini coefficient for the definition of the location-dependency because we want our metric to reflect solely the distribution of the information accesses over the grid. The Herfindahl coefficient would also reflect the number of considered squares.

In order to make estimations about the benefit of location-awareness for a mobile data management mechanism, it is not enough to analyze only one single information object. Therefore the whole information space has to be considered. Thus, we also define the location-dependency of an information space:

Definition 2. *Let IS be an information space and $f : IS \to \mathbb{N}$ a frequency distribution that assigns to each object in the information space the number of requests that occurred for this object during the observation period. Let furthermore $l : IS \to [0, 1]$ be a function that assigns to each object in the information space its location-dependency. Then, the location-dependency L of the information space is defined as the weighted average of the location-dependencies of the objects in the information space:*

$$L = \sum_{i \in IS} \frac{f(i)}{\sum_{j \in IS} f(j)} \cdot l(i) \qquad (6)$$

According to our definition, the location-dependency of an information space lies between 0 and 1, where 0 indicates a location-independent information space and 1 indicates an inherently location-dependent information space.

4.2 Discussion

Unfortunately, a high location-dependency is only a necessary condition for a beneficial effect of a data management mechanism's location-awareness. That it is not a sufficient condition can be seen from the access matrix given in Figure 2. The matrix shows how often each of the five considered information objects $i_1, \ldots, i_5$ are accessed in each of the five considered squares $s_1, \ldots, s_5$. The value given in row m and column n states how often the information object i_n is accessed in square s_m.

	i_1	i_2	i_3	i_4	i_5
s_1	0	0	0	0	0
s_2	0	0	0	0	0
s_3	10	10	10	10	10
s_4	0	0	0	0	0
s_5	0	0	0	0	0

Fig. 2. Strongly location-dependent access matrix

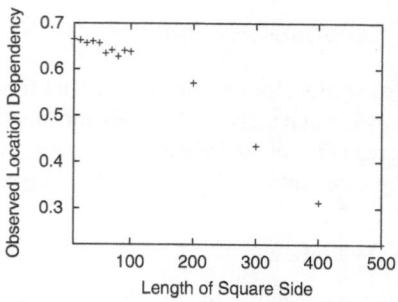

Fig. 3. Location-dependencies observed with different grid resolutions

Since each information object is only accessed in one square, we have a strongly location-dependent information access ($L = 1$). However, a location-aware mobile data management mechanism can not profit from this location-dependency, because all objects are accessed in the same square s_3. For example, a location-aware filter mechanism could not filter out any information objects, if the user is located in square s_3. A location-aware caching strategy also can not profit from knowledge about the user's position, since in all squares each information object is of the same value. Hence, a strong location-dependency is only a hint that exploiting location-awareness might be beneficial.

So far we have not made any assumption about the size of the squares in the set S. However, it has to be reasonably chosen. If the squares are too big, i.e. the resolution of the grid is too small, inequalities in the frequency distributions are hardly recognized. The worst case is, if the whole coverage area of the information system consists of just one square. Then, inequalities can not be observed at all. Because, if only one square is considered, the access frequencies are trivially the same in all squares.

In Figure 3 this effect is illustrated. It shows the observed location-dependencies for a certain information object, which we got using different grid sizes. Although we always used the same, quite location-dependent, access pattern, we got different values for the observed location-dependency. As expected, the observed location-dependency decreases with a decreasing grid resolution, i.e. an increasing length of the squares' sides. For a length of a square side of up to $100m$, what corresponds to 100 squares in the considered $1000m$ x $1000m$ coverage area, we get acceptable values for the location-dependency. For bigger squares the accuracy becomes poor and the observed location-dependency is noticeably lower than the actual one.

However, the square size should not be too small either because smaller squares obviously imply higher storage costs and computational costs for the determination of the location-dependency. Moreover, if the squares are smaller, more requests for an information object have to be observed to make a general statement on how the requests to the object are distributed over the squares. This means that the observation periods have to be longer.

5   Focus

In this section, we introduce the focus of a mobile information system as a second characteristic that has a strong impact on the efficiency of location-aware mobile data management mechanisms.

5.1   Definitions

As shown in the previous section, a high location-dependency alone is not enough to guarantee a high benefit from considering a user's location. In addition, it is also important that at each location, only a specific part of all available information is accessed. This means, that we also have to consider, how the information requests originating from each location are distributed over the information space. Therefore, we define the focus as follows:

Definition 3. *Let $s \in S$ be a square and let $f : IS \to \mathbb{N}$ be a frequency distribution that assigns to each information object in IS the frequency with that it has been accessed*

within the square s during the observation period. Then, the focus within the square s
is the Herfindahl coefficient of the distribution f.

This time we chose the Herfindahl coefficient for our definition, since now the absolute number of information objects requested in a square is also important, not only the inequality in the frequency distribution. For example, the number of objects that should be cached or prefetched for a certain square grows with the absolute number of objects that the average user requests there.

For a good estimation of a location-aware mechanism's performance, all squares have to be considered. The contribution of each square should correlate with the number of requests originating from this square, because the focus in a square with only a few requests will have less impact on the performance of a location-aware mechanism than that of a square with many requests will have. Thus, we define the the focus of an information system as the weighted average of the focuses observed in each square $s \in S$:

Definition 4. *Let $fr : S \to \mathbb{N}$ be a frequency distribution that assigns to each square $s \in S$ the number of information requests that occurred there during the observation period. Furthermore, let $fo : S \to [\frac{1}{N}, 1]$ be a function that assigns to each square $s \in S$ the focus observed there. Then, the focus F of the according information system is defined as follows:*

$$F = \sum_{s \in S} \frac{fr(s)}{\sum_{i \in S} fr(i)} \cdot fo(s) \qquad (7)$$

With this definition values between $\frac{1}{N}$, where N is the number of objects in the information space, and 1 are possible for the focus of an information system. The value $\frac{1}{N}$ indicates that within each considered square all objects of the information space are accessed with the same frequency. A value of 1 for the focus indicates that within each square only one information object is requested.

5.2   Discussion

Similar to the location-dependency, a high focus alone is also no guarantee for a beneficial use of a location-aware mechanism. For an example, consider the access matrix in Figure 4. With this access pattern, we will get a high focus, since in each square only one information object is accessed. However, a location-aware mechanism can not profit from the knowledge of a user's position, since information object i_3 is preferred at any location, what can easily be observed without any location-awareness.

Note, that the location-dependency is 0 for the access pattern given in Figure 4. In fact, if both the location-dependency and the focus are high (see Figure 5) the benefit from using location-awareness will also be high. Because then, at each location many requests are concentrated on a small part of the available information objects (high focus) and the preferred information objects will differ between different locations (high location-dependency). Thus, location-awareness can then be used to identify the objects which are specifically preferred at each location. Without considering a user's location, all information objects could seem to be of equal popularity. This would be, for example, the case with the access pattern given in Figure 5.

	i_1	i_2	i_3	i_4	i_5
s_1	0	0	10	0	0
s_2	0	0	10	0	0
s_3	0	0	10	0	0
s_4	0	0	10	0	0
s_5	0	0	10	0	0

Fig. 4. Highly focused access matrix

	i_1	i_2	i_3	i_4	i_5
s_1	10	0	0	0	0
s_2	0	10	0	0	0
s_3	0	0	10	0	0
s_4	0	0	0	10	0
s_5	0	0	0	0	10

Fig. 5. Highly focused and strongly location-dependent access matrix

6   Evaluation

In this section, we determine the location-dependency and the focus of some typical access patterns in order to give an idea about the meaning of concrete values obtained with the metrics. We also analyse how important parameters of the access patterns influence these values. Finally, we use a location-aware hoarding mechanism as an example to show how the location-dependency and the focus may effect the efficiency of a location-aware mobile data management mechanism.

6.1   Location-Dependency

For the analysis of the location-dependency, we assume that the considered information system covers a quadratic area of $1000m$ x $1000m$. For the determination of the location-dependency the coverage area was separated into 1600 squares measuring $25m$ x $25m$. We always considered the requests for one single information object. In order to specify the origin of an information request we used a coordinate system with a resolution of 1 meter, with $(0, 0)$ representing the upper left corner of the coverage area and $(1000, 1000)$ the lower right corner.

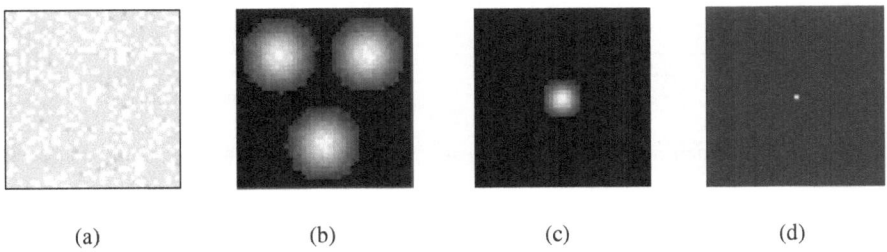

(a)   (b)   (c)   (d)

Fig. 6. The considered access patterns

Location-Independent Access. For the simulation of a location-independent access, we assumed that the coordinates from which the requests for the considered information object originate are both equally distributed over the range from 0 to 1000. In Figure 6(a) a graphical representation of this access pattern is given. It shows the number of requests for the considered information object that occurred in each square during the observation period. The brighter a square is depicted the higher is the number of requests originating from this square. Since in all squares approximately the same number of requests occurred, the brightness of the squares does not differ much. And, as expected, we get a very low value for the location-dependency: 0.023.

Hot Spot Access. For the simulation of the wide spectrum of location-dependent information accesses, we used an access pattern, which we call the hot spot access pattern. In this pattern, the requests for the considered information object primarily originate from one or more hot spots. If there is more than one hot spot, we first randomly select one of the hot spots as the origin of a request. Each hot spot is selected with the same probability. Afterwards, the coordinates, where the information object was requested, are determined. Therefore, we assume that the x- and y-coordinate are distributed according to a gaussian distribution around the coordinates of the selected hot spot.

Thus, this pattern has two important parameters, which allow to simulate information accesses with quite different location-dependencies. These two parameters are the number of hot spots and the standard deviation σ of the gaussian distribution. Figures 6(b) and 6(c) show examples for the hot spot access pattern, with three hot spots and a standard deviation of 100, and with one hot spot and a standard deviation of 50. The plots in Figure 7 show the location-dependencies we get with one and three hot spots for different values of σ.

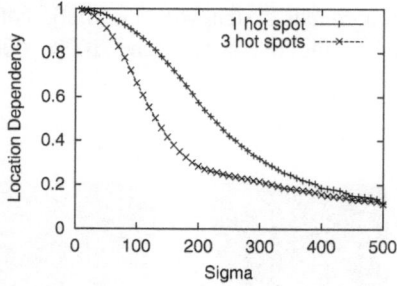
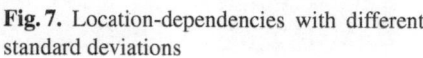
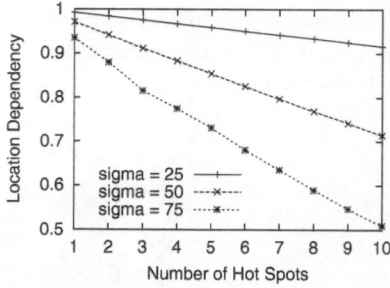

Fig. 7. Location-dependencies with different standard deviations

Fig. 8. Effect of the number of hot spots on the location-dependency

In Figure 8 the location-dependencies are shown, which we got for different numbers of hot spots. The hot spots were evenly distributed over the information system's coverage area. The three plots show the results for standard deviations of $\sigma = 25$, $\sigma = 50$, and $\sigma = 75$.

Inherently Location-Dependent Access. With an inherently location-dependent information access, all requests for a certain information object are done at the same location, thus they all originate from the same square (see Figure 6(d)) and, not surprisingly, the location-dependency is 1.

6.2 Focus

In the previous section, we had to consider the whole coverage area of the information system but only one information object to make statements about the location-dependency. In contrast, we now have to consider only one square of the coverage area, but all information objects requested there. Analogously to our examinations of the location-dependency, we now determine the focus of typical access patterns and analyze the effect of the patterns' parameters on the focus.

Uniform Distribution. As mentioned above, the focus does not only reflect inequalities in the frequency distributions of the information requests but also the absolute number of locally accessed information objects. Hence, we can get high focuses, although the local access frequencies are equally distributed, if only a small number N of different objects is locally requested at all. For a uniform distribution the focus F will be $F = \frac{1}{N}$.

Zipf-like Distribution. In [2] it has been shown that the distribution of the requests over single web pages follows a Zipf-like distribution. In such a distribution the relative probability of a request for the i'th most popular page is proportional to $\frac{1}{i^\alpha}$. The observed value of α varies between the different considered traces, ranging from 0.64 to 0.83.

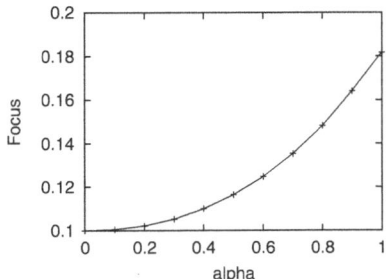 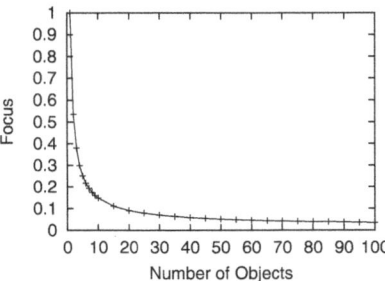

Fig. 9. Focus depending on the parameter α of a Zipf-like distribution

Fig. 10. Focus depending on the total number of different objects accessed locally

We assumed such a distribution for the information requests observed within the considered square. We experimented with different values of the parameter α. A higher value for α means a stronger concentration on the most popular information objects. Figure 9 shows the focus depending on the parameter α. We assumed that a total of 10 different objects is accessed in the considered square.

As mentioned above, the focus also reflects an absolute concentration, i.e. the total number of different objects that are accessed in a square is reflected in our metric. Figure 10 shows how this number influences the focus. To get these results, we simulated a Zipf-like distributed information access and varied the total number of locally requested objects. The parameter α of the Zipf-like distribution was this time set to a fixed value of 0.8.

6.3   An Example Application

In this section, we illustrate the effects that the location-dependency and the focus might have on location-aware mobile data management mechanisms using a hoarding mechanism, we proposed in [8], as an example.

This mechanism aims to allow the users of a mobile information system to access information during disconnections, i.e. when no network is available. Therefore, the mechanism tries to predict the information objects a user will probably access during a future disconnection and transfers them to the user's device as long as there is a connection. Thus, the information is already locally available, when the user accesses it during the disconnection. In order to predict the objects that a certain user will probably need, the mechanism uses knowledge about the objects' popularities at each location and predictions of the user's future movement (for details see [8]).

A measure often used to rate the efficiency of hoarding and caching mechanisms is the hit ratio. This is the ratio between the number of requests that can be answered with information objects that are stored locally and the total number of requests a user makes.

If there is no location-dependency and no focus, i.e. if all objects are requested everywhere with the same probability, the hoarding mechanism will not be able to benefit from its location-awareness. In this worst case scenario, the user will only get an average hit ratio of $\frac{m}{n}$, where m is number of information objects that can be stored locally on the user's device and n is the total number of objects in the whole information space. Usually, m will be much smaller than n. If m is bigger than n or equal to n, we will trivially get a hit ratio of one.

Finally, let us consider an example with a location-dependency of 1 and a high focus. In this example, at each square s a specific set of information objects O_s is requested, where $O_i \cap O_j = \emptyset$, if $i \neq j$ and $|O_s| = \frac{n}{|S|}$. n is again the total number of objects in the information space and S the set of all squares within the information systems' coverage area. Furthermore, let us assume that a user located in square s accesses all objects in O_s with the same probability. Then, we will get an average hit ratio of $\min(\frac{m \cdot |S|}{n \cdot k}, 1)$, where k is the number of squares the user will visit during the disconnection. Figure 11 shows the hit ratios for this and the previous example depending on the number of squares a user visits during a disconnection. The parameters were chosen as follows: $|S| = 100$, $m = 100$, $n = 5000$.

7   Conclusion

In this paper, we claimed that there are not only inherently location-dependent and location-independent information systems, but that there is a complete spectrum of

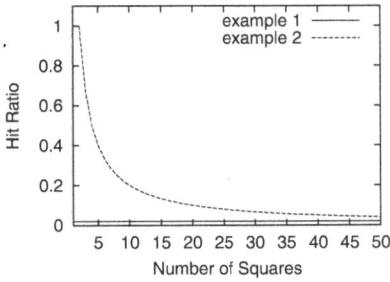

Fig. 11. Hit ratios of a location-aware hoarding mechanism

systems differing in their degree of location-dependency. We supported this claim by giving examples of information systems with different location-dependencies.

Furthermore, we proposed a metric, which can be used to measure the location-dependency of an information system and which helps to better understand what location-dependent information is. We explained why considering only the location-dependency of an information system is not enough in order to estimate the benefit a location-aware data management mechanism can get from its location-awareness. Additionally, we proposed a second metric, the focus, which together with the metric for the location-dependency finally provides a good means for this estimation.

In the evaluation part of the paper, we analyzed the location-dependency and the focus of various access patterns to give an idea of the values that can be expected with typical access patterns. Finally, we used a location-aware hoarding mechanism to show how the location-dependency and the focus can effect the efficiency of a mobile location-aware data management mechanism.

For the near future, we plan to combine the two metrics into only one metric completely characterizing an information space. To do so, we first have to further analyze the proposed metrics with a simulation tool which allows to directly manipulate the location-dependency and the focus of a considered information space. With such a tool we could more exactly analyze how the proposed metrics effect different typical location-aware mechanisms. This will also help us to further refine our metrics such that they also distinguish access patterns, which are, with our current metrics, considered as equivalent.

Finally, we plan to adjust our metrics to other fields of application, e.g. the measurement of coverage in a wireless ad-hoc sensor network (see Section 2). In contrast to existing metrics, our metric would measure the expected coverage for an average object, not the maximal or minimal possible coverage. Another field of application for our metrics are mobile ad-hoc networks. There, inequalities in the average spatial distribution of the network nodes are crucial for the delay and the throughput of the network.

References

1. A. B. Atkinson. *Wealth, Income and Inequality*. Oxford University Press, Incorporated, second edition, 1981.

2. L. Breslau, P. Cao, F. Li, G. Philips, and S. Shenker. Web caching and zipf-like distributions: Evidence and implications. In *Proceedings of the IEEE INFOCOM '99*, pages 126–134, New York City, NY, USA, March 1999.

3. H. Chang, C. Tait, N. Cohen, M. Shapiro, S. Mastrianni, R. Floyd, B. Housel, and D. Lindquist. Web browsing in a wireless environment: Disconnected and asynchronous operation in artour web express. In *Proceedings of the Third Annual ACM/IEEE International Conference on Mobile Computing and Networking (MobiCom '97)*, pages 260–269, Budapest, Hungary, September 1997.

4. J. Chim, M. Green, R. Lau, H. Leong, and A. Si. On caching and prefetching of virtual objects in distributed virtual environments. In *Proceedings of the Sixth ACM International Conference on Multimedia (ACM MULTIMEDIA '98)*, pages 171–180, Bristol, UK, 1998. http://www.acm.org/sigmm/MM98/electronic_proceedings/chim/index.html.

5. S. Dar, M. Franklin, B. Jónsson, D. Srivastava, and M. Tan. Semantic data caching and replacement. In *Proceedings of the 22nd International Conference on Very Large Data Bases*, Bombay, India, 1996.

6. N. Davies, K. Cheverst, K. Mitchell, and A. Friday. Caches in the air: Disseminating information in the guide system. In *Proceedings of the Second IEEE Workshop on Mobile Computing Systems and Applications (WMCSA '99)*, pages 11–19, New Orleans, LA, USA, February 1999.

7. F. Hohl, U. Kubach, A. Leonhardi, K. Rothermel, and M. Schwehm. Next century challenges: Nexus - an open global infrastructure for spatial-aware applications. In *Proceedings of the Fifth Annual ACM/IEEE International Conference on Mobile Computing and Networking (MobiCom '99)*, pages 249–255, Seattle, WA, USA, August 1999.

8. U. Kubach and K. Rothermel. Exploiting location information for infostation-based hoarding. In *Proceedings of the Seventh Annual ACM SIGMOBILE International Conference on Mobile Computing and Networking (MobiCom '01)*, pages 15–27, Rome, Italy, July 2001.

9. S. Meguerdichian, F. Koushanfar, M. Potkonjak, and M. B. Srivastava. Coverage problems in wireless ad-hoc sensor networks. In *Proceedings of Twentieth Annual Joint Conference of the IEEE Computer and Communications Societies (Infocom 2001)*, volume 3, pages 1380–1387, Anchorage, AK, USA, April 2001.

10. S. Meguerdichian, F. Koushanfar, G. Qu, and M. Potkonjak. Exposure in wireless ad-hoc sensor networks. In *Proceedings of the Seventh Annual ACM SIGMOBILE International Conference on Mobile Computing and Networking (MobiCom '01)*, pages 139–150, Rome, Italy, July 2001.

11. Q. Ren and M. Dunham. Using semantic caching to manage location dependent data in mobile computing. In *Proceedings of the Sixth Annual ACM SIGMOBILE International Conference on Mobile Computing and Networking (MobiCom 2000)*, pages 210–221, Boston, MA, USA, August 2000.

12. A. Sen. *On Economic Inequality*. Clarendon Paperbacks, second edition, 1993.

13. T. Ye, H.-A. Jacobsen, and R. Katz. Mobile awareness in a wide area wireless network of info-stations. In *Proceedings of the Fourth Annual ACM/IEEE International Conference on Mobile Computing and Networking (MobiCom '98)*, pages 109–120, Dallas, TX, USA, 1998.

iCAMS: A Mobile Communication Tool Using Location and Schedule Information

Yasuto Nakanishi, Kazunari Takahashi, Takayuki Tsuji, and Katsuya Hakozaki

Graduate School of Information Systems, Univ. of Electro-Communications,
1-5-1 Chofugaoka, Chofu-City, Tokyo, Japan 182-8585
{naka, tkazu, taka, hako}@hako.is.uec.ac.jp
http://www.hako.is.uec.ac.jp

Abstract. We have developed a communication support system that estimates the situation of a person using schedule information and location information provided by a PHS (Personal Handy Phone System). From the lessons provided by our prior studies and inventions, we developed a new mobile communication tool for cellular phones that uses location information and schedule information. This is a kind of dynamic telephone book, which we have named iCAMS. We performed user studies for eight weeks in Tokyo with a group of students and with a group of small-office workers. By analyzing the communication logs, questionnaires and interviews we conducted with the users, we evaluated our system.

1 Introduction

The spread of cellular phones, email, and mobile computers has freed us from the restrictions of time and place. Although we spend every day enjoying the merits that these tools bring, new problems have arisen. Today, there is a wide range of technologies, access methods, and devices for communication. People often have several phone numbers and several email addresses, and switch among them according to their situation. Because the modes of communication have become so varied, it is difficult to make appropriate use of these various media and addresses when sending messages to people who use all these modes of communication. Even if a message is delivered, the addressee might not read or listen to it. In a mobile environment in particular, it is difficult to know where a person is, and what media he or she is using or can be contacted by, since it can change hour by hour. Although cellular phones and email free us from the restrictions of time and place, they take away the context of communication that fixed times and places provid.

Many researchers have been exploring how to provide awareness information to help distributed collaborators communicate smoothly. Especially with the increasingly widespread use of mobile phones, researchers have been studying how to give meaningful awareness information to anyone, anywhere, in order to facilitate communication among persons in a mobile environment [1-7].

F. Mattern and M. Naghshineh (Eds.): Pervasive 2002, LNCS 2414, pp. 239–252, 2002.

We developed and tested such a messaging service, which uses location information and scheduling information [8, 9]. Our system, which is called "CAMS" selects the most suitable telephone number or email address, and redirects each incoming message dynamically, according to the callee's communication context (location, schedule, and available media). It also documents the communication contexts in an HTML file. By obtaining a callee's communication context (location, schedule, available media), a caller is able to choose the best time to call and/or select the best media to use (Figure 1).

With the lessons provided from trials using the CAMS system, we have developed a new mobile communication tool, and have conducted user studies. In this paper, we describe this new mobile communication tool, "iCAMS," which uses location information and schedule information to connect callers with callees. Location and schedule alone might not be sufficient information for determining user context. It may involve details such as immediate task or background information such as family or office situations. Although it might be necessary to consider more information in determining the context of communications, our system handled only location and schedule information.

2   System

2.1   Problems with Prior System: CAMS

Each CAMS user carried a PDA and a PHS (Personal Handyphone System), which was a kind of cellular phone. The system redirects incoming messages with estimated communication contexts using location information obtained from the PHS and schedule information. The users can also share communication contexts on the WWW (World Wide Web).

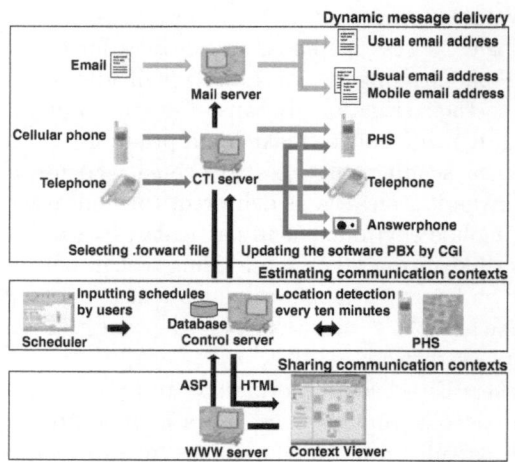

Fig. 1. The system architecture of CAMS

The CAMS users pointed out that the dynamic message redirection was convenient for the callee; however, the flaw in the system was that the caller didn't know where the message would be redirected. This seemed to make them feel uneasy a little. There were also instances when CAMS redirected a telephone call, even when the caller just needed to talk with the callee on the telephone. They pointed out that callees could adapt the system to their needs, but the system was not always so adaptable to the callers' needs.

The CAMS users preferred sharing communication contexts on a WWW server to the dynamic message redirection because they could select the media or the timing of communication by oneself. However, they found it especially bothersome to make a call after viewing the shared communication contexts in the mobile environment. In order to view the context over the WWW, they had to connect the PHS and the PDA and then remove the PHS from the PDA in order to make the call. They wanted to see improvement of the interface, and that they wanted to be able to use the functions of CAMS with one cellular phone.

2.2   iCAMS

In response to these comments, we developed a new mobile communication tool, "iCAMS," as a dynamic address book on the WWW for cellular phones. It is a client/server system that seeks to make CHTML (Compact HTML) enabled mobile phones an integrated part of the communication network of companies (Figure 2).

The intended users are friends who would be willing to share location or other sensitive information with each other. Each user carries a PHS, which is a type of cellular phone that can be used to read or write email, and that contains a WWW browser, which also works as a location sensor.

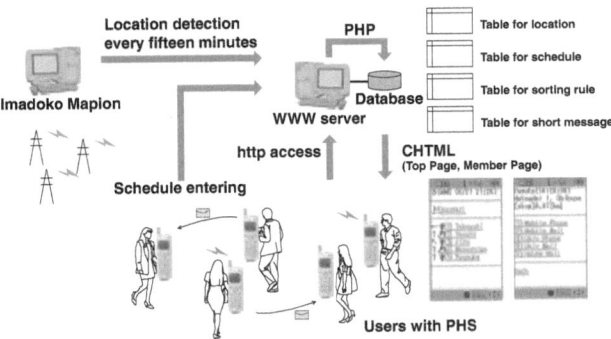

Fig. 2. The system architecture of iCAMS

To obtain location information on each user, our system posts a cgi message every fifteen minutes to a map database service on the WWW called "Imadoko

Mapion" [10]. Together with this service, we can utilize the location detection service of PHS that NTT DoCoMo offers through the WWW (Figure 3). This service performs localization through the cell phone network, with a precision of the location detected of within about 100m. When a cgi message with the telephone number of the PHS and the password is posted, "Imadoko Mapion" returns a HTML file that contains the location information of the PHS on a map (Figure 3). The user is then able to obtain the location name, the longitude and latitude where the PHS carrier is, from the HTML file which has been returned. The detected location information for each user is stored to the table for location in our database.

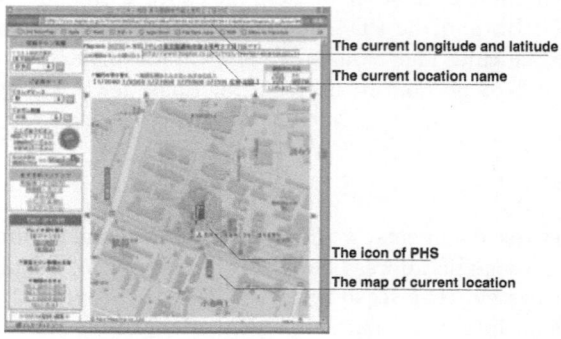

Fig. 3. An example of "Imadoko Mapion"

Whenever a user accesses our WWW server, a PHP script queries the database and returns a CHTML file called "Top Page." The file sorts other users in their nearest order of position, using the latest location information (Figure 4). It also shows whether they are moving or not, and that in which direction they are located in relation to the user.

When a user accesses the Top Page in order to communicate with another user, he/she might find out that another user is nearer. The Top Page might let him/her know where the nearest user is. The major purpose of CAMS was to facilitate smooth telecommunication for its users in a mobile environment. The added purpose of iCAMS is to give users an opportunity to meet in the mobile environment.

When clicking an entry in the Top Page, another PHP script queries the database and returns a CHTML file called "Member Page". This file contains more detailed awareness information on the user, and options for contacting him/her (Figure 4). This page shows the user's name, the time at which the latest location information was detected, the location name where the person is, if he/she is moving/staying and his/her distance from the user.

The Member Page also displays a list of telephone numbers and email addresses for the selected user, which are sorted by registered rules concerning locations and schedules. The system uses two kinds of rules, those regarding locations, and those regarding types of schedules. If a person is in his/her house, the phone number of the house might be listed first. If that person is in his/her office, the phone number of the office would be listed first. If that person is in a meeting in his/her office, that person's office email address would be listed first. The schedule information entered is not shown.

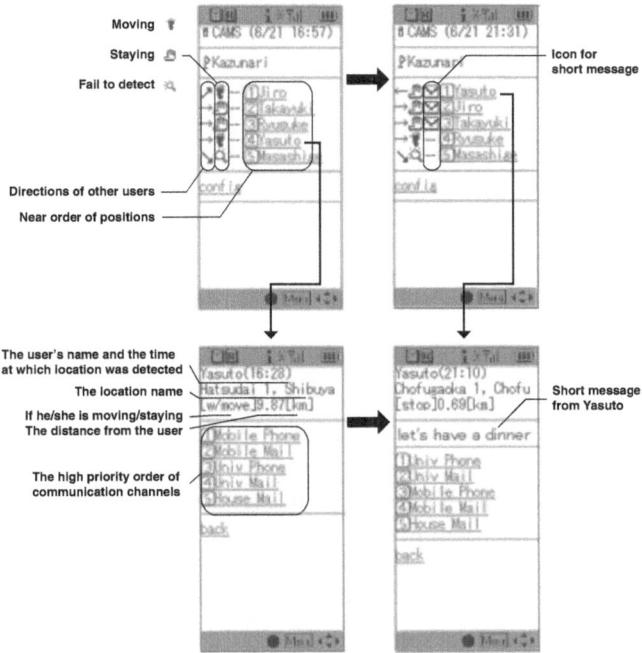

Fig. 4. Examples of the Top Page and the Member Page

Each user chooses and registers rules about which means of communication are appropriate under what conditions (Figure 5). In comparing a user's current location with the locations described in his/her rules for sorting of telephone numbers and email addresses, we compared the location name that "Imadoko Mapion" had returned to the location name in the rules as character strings. If the latest location information for a user is listed in the rules for locations, the rule is changed to the one for the latest location. When no schedule has been inputted, that rule is applied. If a schedule has been input, the rule is updated to the one for the schedule. The caller can find out which channel is the best for the callee in the Member page. This will solve the problem that the users of CAMS experienced, feeling uneasy that they weren't be able to find out where their calls would be redirected.

Even if the email address of a mobile phone is listed first and the mobile phone number is listed second in the sorted list, the caller can select the mobile phone number and make a call to the callee in case contact is urgently required. This will solve the problem of CAMS redirecting a telephone call even when the caller needs to talk with the callee on the telephone. The sorted list of communication channels in our system will help callers and callees find a point of mutual agreement between the caller's demands and the callee's wishes.

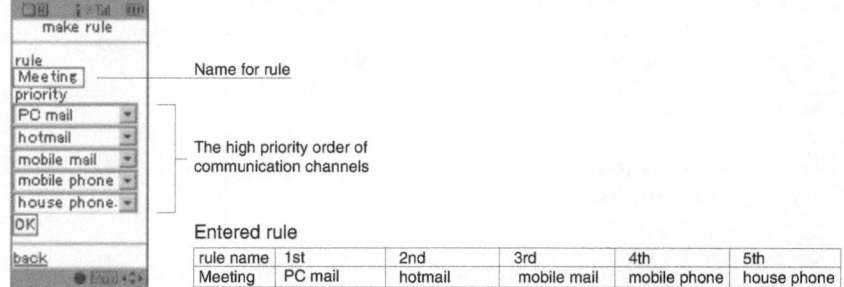

rule name	1st	2nd	3rd	4th	5th
Meeting	PC mail	hotmail	mobile mail	mobile phone	house phone

Fig. 5. Entering a rule for a schedule with the mobile phone

We prepared another PHP script for the registration of rules for sorting telephone numbers or email addresses regarding schedules. A user inputs a rule by entering a name for the rule and deciding a priority sequence of telephone numbers and email addresses for the rule (Figure 5). The entered rule is stored to the table for rule in the database. In the present system, users carry PHSs for several days before they begin to use the system, and we inform them location names obtained from the PHS. We register rules regarding locations according to the user's criteria.

In order to reduce the labor of inputting schedule information, we prepared another PHP script for each user according to the registered rules (Figure 6). A user inputs a schedule by selecting the day, the time, and the schedule contents. The user can input a schedule using the WWW browser in the cellular phone. The entered schedule is stored to the table for schedule in the database.

Because telephone numbers are written with the "tel" tag of CHTML, users can make a call directly by clicking a telephone number. Because email addresses are written with the "mailto" tag of CHTML, users can begin to write an email by clicking an email address (Figure 7). By writing the sorted list as a CHTML file, iCAMS users can make use of the function of the shared communication contexts in CAMS, just by using their cellular phone. This will improve the interface and will also solve the operability problem of the CAMS system.

We prepared a function called "short message" for sending a message to more than one member simultaneously. An icon is shown beside the name of the sender in the Top Page of the addressee when a message is entered. In clicking the entry for the sender, the message is shown in the Member Page of the sender (Figure

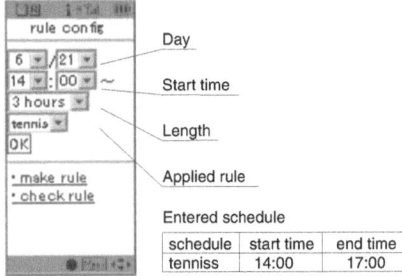

Fig. 6. Entering a schedule with the mobile phone

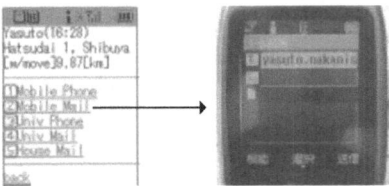

Fig. 7. Writing an email via the Member page

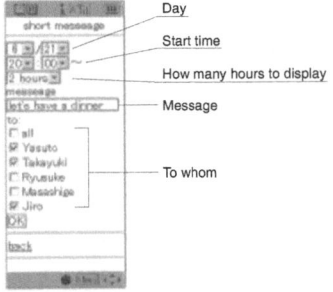

Fig. 8. Entering a short message with the mobile phone

4). The sender enters the message, the start time to display, and the length of time to display, and checks the members to send the message (Figure 8). The entered message is stored to the table for short messages in the database. Users may exchange light messages with persons they do not need to send emails to.

3   User Studies

We conducted user studies for eight weeks (24/09/2002-18/11/2002) in Tokyo with a group of students and with a group of small-office workers. All of the members of the two groups were dispersed spatially. The ratio of men and women and the number of members were the same in the two groups. The group of students was comprised of eight men and two women. They were undergraduate students and were 21 years old or 22 years old. They all belonged to a university tennis club, and often had to practice tennis and attend meetings after school or at intervals between attending lectures. They made and received telephone calls with cellular phones mainly. They sent and received emails using cellular phones more often than using personal computers. The group of small-office workers was also comprised of eight men and two women. They were editors or designers publishing a magazine collaboratively. They were between the late 20's and early 30's. They both worked together and were friends. Each had his/her own office, and they sometimes held meetings in one of the offices. They made and received telephone calls with cellular phones and with regular phones. They sent and received emails with personal computers more often than with cellular phones.

During the first five weeks of the study, we increased the number of functions of the system every week. Throughout the rest of the period studied, the users used all functions (Table 1). We analyzed the logs in the server at the end of the first five weeks, and administered questionnaires and interviews to all of the users at the end of the study. In the first week, the Top Page and the Member page were fixed. In the second week, the Top Page sorted the users' name with location information and the Member Page displayed the location name of the users, but the telephone numbers and email addresses were fixed. In the third week, the Top Page sorted the users' name with location information, and the Member Page sorted the telephone numbers and email addresses only with location information. In the fourth week, the Top Page sorted the users' name with location information, and the Member Page sorted the telephone numbers and email addresses with location information and schedule information. In the fifth week, the function of short messages was added. During the rest of the study period, we stopped taking a log of the WWW server and the users used all functions. We expected that when we stopped taking the log, the users wouldn't feel so conscious of participating in an experiment, and would feel freer to give us their honest opinions in the questionnaires and the interviews.

Table 1. Experimental conditions for each week

	sorting the users' name with location in the Top Page	sorting the address list with location in the Member Page	sorting the address list with schedule in the Member Page	the function of short message	taking of the log
1st week	x	x	x	x	O
2nd week	O	x	x	x	O
3rd week	O	O	x	x	O
4th week	O	O	O	x	O
5th week	O	O	O	O	O
6th week	O	O	O	O	x
7th week	O	O	O	O	x
8th week	O	O	O	O	x

Table 2 shows the number of times the user accessed the Top Page and the Member Page, the number of emails made, the number of telephone calls made and the number of short messages made through our system in the first five

Table 2. Numbers of times users accessed the WWW server and used the communications in the first five weeks

	access to the Top Page	access to the Member Page	email	telephone	short message
the group of students	1093	2664	567	171	8
the group of workers	670	1218	72	73	3

weeks. All of the numbers in all categories were higher for the group of students
than for the group of small-office workers. This fact might come from that the
students are heavier users of cellular phones than the small-office workers. Users
accessed the Member Page more often than they accessed the Top Page. This
fact shows that the users viewed the Member Page of some members more times.
Both the numbers of time users accessed to the Top Page and the number of
times users accessed the Member Page were more than the total of the number
of emails and telephone calls combined. This might show that the users didn't
use iCAMS only to decide what communication channel to use. We had thought
that users would access the Top Page in order to communication with another
user a particular aim and might communicate with other users at the time they
accessed the Top Page. The users told us that there were many such instances
when in fact they did check the Top Page in order to initiate communication,
however, they also told us that they accessed to the Top Page even when they
didn't have a particular aim in mind. The users seemed to have accessed the
Top Page first, and then the Member Page of somebody more often than they
seemed to have accessed the Top Page in order to access the Member Page of
a specified member. Our system seemed to aid members not only in contact a
specified callee with specified media, but also in providing an opportunity for
communication. They seemed to have used iCAMS not only for confirming the
situation of other members, but also for obtaining the situation of the whole
group. The users told us that they most often accessed the Top Page when they
had a little free time, for example they were waiting for a train. In Japan, people
often write or read emails with cellular phones in trains or at stations. They also
browse old emails as a way to spend free time. The users in our study seem to
have accepted our system as part of such activities.

We will now summarize the results of our questionnaire mainly from now
(Table 3). The responses were five grade evaluations. However, for a simple
analysis, we state here the number of people who evaluate each item highly as
the fourth grade or as the fifth grade.

Thirteen users out of the twenty total stated that they appreciated the sorted
list of the telephone numbers and email addresses using location information and
schedule information. Sixteen users stated that they appreciated that obtaining
information on the callee's situation in order help select which medium of communication to use.

Ten users told us that the sorted address book showed them whether they
could make a call, and enable them to make it without hesitation. These results
might mean that iCAMS could solve the problem of the users of CAMS feeling

Table 3. The results of questionnaire

	the group of students (10 people)	the group of small-office workers (10 people)
Telephone numbers and email addresses sorted using location information and schedule information was nice.	4	9
Knowing the callee's situation helped you to select the communication medium.	7	9
The sorted address book showed whether you could make a call and enabled you make it without hesitation.	4	6
You were not troubled with what form of medium you should use among some telephone numbers and some email addresses.	4	9
You had used telephone numbers or email addresses those weren't listed first for giving priority to your situations.	6	7
Having to input a schedule into this system was bothersome.	8	9
You wen to meet another user with a specified aim because you found out his/her location.	1	4
You went to meet other users without a specified aim because you learned their locations.	2	3
There were advantages to obtaining each other's location information.	6	8
The list of members sorted in their nearest order of position was useful.	4	9
The list of members sorted in their nearest order of position was enough.	0	0
Learning the location names where others were was enough.	2	4
Knowing each other's locations on a map is desirable.	8	6

uneasy about where their calls would be redirected. Thirteen users responded that when the callee had several phone numbers and email addresses, they were not troubled with what form of medium they should use among them in our system.

Thirteen users stated that they had used telephone numbers or email addresses those weren't listed as a first choice of contact in order to give priority to their situations over those of callees. This might mean that iCAMS has solved the problem of the dynamical message redirection in CAMS, and could help callers and callees to find a point of mutual agreement between the caller's demands and the callee's wishes.

Seventeen users stated that having to input a schedule into this system was bothersome. Only three users among the group of students inputted their schedules, and none of the small-office workers inputted their schedule. The bother of making inputs seemed to have exceeded the merits of the address book that can sort with priority using schedules. We need to improve the interface to resolve this problem or to import information in schedulers that users usually use such as an MS Outlook calendar.

One person among the group of the students went to meet another user with a specified aim after they found out his/her location using our system, while four of the group of small-office workers went to meet another user with a specified aim. These instances might mean that iCAMS changed the medium of communication from electronic or phone to an in-person meeting. The number of those went to meet other users without a specified aim after they learned their locations was two from the group of students, and three from the group of small-office workers. These instances would mean that iCAMS gave these users chances to meet in-person.

Fourteen users stated that there were advantages to obtaining each other's location information. Thirteen users agreed that the list of members sorted in their nearest order of position in the Top Page was useful. However, most of the users wanted to see improvement of the interface for sharing location information. Nobody stated that only the list of members sorted in their nearest order of position was sufficient. Six users stated that learning the location names where others were was sufficient. The other fourteen users wanted to know each other's locations on a map. Many users in both groups told us that they wanted to know who and who were together. This means we would have to show the context of a group, although our current system mainly shows personal context information. Mapping location information on a map would show easily who is with who. Such information about users being together might facilitate more meetings between users. We plan to add such a function in the future.

All of the users wanted the precision of the location information to be improved, because the current range of precision is about 100m. In particular, the students wanted to know where on campus others were. Even though they learned that another user was on campus with the name of location where he/she was, the obtained location name was same even if he/she was in the classroom or in the tennis court. They wanted to obtain more detailed location; whether he/she was in the tennis court or the classroom. On the other hand, the small-office workers did not have the same problem, because most of them do their work in the downtown area of Tokyo, and they, unlike the students, moved about within a 10km area where each location has a specific name (in this group, one location name corresponded to one house or one office). Currently, we have been able to use the cellular phones that GPS has built in Japan from December 2001. We will develop a new system using such cellular phones.

For the one user among the group of the small-office workers who works mainly in Yokohama (which is approximately 30km from Tokyo), the list of members sorted in their nearest order of position seemed to make almost no sense. Even if other users moved around within Tokyo, there seems to have been few changes in his list. For the user who works at a remote place and has little chance to meet others in person, the status of others (whether they were at their office or home or in the middle of a meeting) seems to have been more important than detailed location information. This particular user suggested to us in the interview that we should show the status of each user using an icon in the Top Page.

The intended users are to be friends who would share location or other sensitive information with each other. However, we have to consider the problem of privacy, because location information is acquired automatically and periodically.

Table 4. Sensitivity of the users to revealing personal information

	to a close person		to a person who is not so close	
	the group of students	the group of small-office workers	the group of students	the group of small-office workers
the list of members sorted in their nearest order of position	8	10	5	9
location name	9	9	1	5
distance	7	10	1	6
movig or staying	7	10	4	10
longitude and latitude via PHS	5	9	0	4
location information via GPS	5	7	0	2
the sorted address book	7	8	4	2

Actually, some users had worries about sharing position information before they began to use our system. After the studies were over, we asked them what kind of information made from location information they would be willing to reveal and Table 4 shows their answers.

Most of the users didn't mind if persons who were close to saw the list and location name, distance, and state of movement. However, the number of persons increased who minded revealing the longitude and latitude of where they were. When we show only the location name via "Imadoko mapion", the precision of the location information decreases, because others can't figure out at which point the target person is in the map shown. They seem to think it undesirable to show their own locations in order to increase the precision of the location information.

Two-thirds of the users didn't want to show even a location name to persons who were not so close to. Most of the group of students especially did not want to reveal their location to persons who weren't so close to. The ratio of decrease in the number of users who agreed to show their location seemed to depend on human relations more than on the kinds of information revealed or on the precision of information. Most users wanted to be able to control which personal information about them gets revealed, according to human relations. We need to a function that allows them this control in the future.

However, we learned after the study was over that most of the users valued sharing location information. Most users said that our system was not only convenient to use but also pleasant to use. Our system would suit for friends or families more than the people who are busy colleagues at the same company.

4   Discussion

Interfaces have been studied that help mobile telephone users answer the question "What is the best way to make contact right now?" [1-6]. The idea of integrating schedule and location information into a communication tool might not be new. Some projects have explored this with PDAs, using MS Outlook for schedule info and relying on the user to update their location information. Our aim is not only to share location information as awareness information and to help users to select an appropriate communication channel, but also to give users a chance to meet in-person.

One novel aspect of our project is the on-screen interface for the cellular phone using CHTML. We sort online users by proximity, then prioritize the available channels of communication for each user based on the activity the user is doing according to their schedule and location. [1- 3] are also web-based applications for the cellular phone. However they facilitate the ongoing interaction between two people.

Another novel aspect of this project is the use of NTT DoCoMo's location detection service. Some projects also use location information as awareness information. However, the location information was limited to information in a building [6], or involved users entering location information by manual operation [2, 3, 5]. In our projects, location information is within the city spaces, and is acquired automatically.

Work in the active badges [6] requires a specialized environment and the possession of specialized equipment. In contrast, our system uses conventional and very popular technologies, and explores how these technologies can be used for new purposes. We think that these two approaches have important differences in terms of where they might be used, and also in terms of who might use them (e.g., highly-paid engineers and researchers in high-tech workplaces, vs. a broad diversity of people in common, everyday settings). The formal similarity of the two research programs is contradicted by their very different settings – that is, the settings are very different in terms of where they might occur, how much preparation would be required in these environments, the social classes involved in usage, and the work/life activities that could be affected. In other projects users often forgot to update their information manually [5], and we had the same problem with the users in our system. However, location information is acquired automatically and periodically in our system. That is advantageous in being able to update awareness information on a user at any time; however we need pay attention to privacy issues in revealing location information.

The intended users are to be friends who would share location or other sensitive information with each other. However, we have to consider the problem of privacy, because location information is acquired automatically and periodically. Actually, some users had worries about sharing position information before they began to use iCAMS. After the study was over, however, sharing location information had become popular among most of the users. There seems to have been occasional instances when users didn't want to reveal their position to someone else. In the present system, users can hide their location information by turning off the PHS. Users also want to be able to change the level of detail of information that is revealed, according to who wants to access them. Future work for our project might involve both increasing the sensitivity and range of locations that the system can display to the user, as well as researching ways that the user can more finely select what is revealed to the outside world.

sectionConclusion We developed a new mobile communication tool for cellular phones that uses location information and schedule information using the lessons learned from our prior invention, which sorts the member list and the address list in the cellular phone. We conducted user studies for eight weeks in order to evaluate this system. We are planning to develop a new system from the lessons we've learned, one for a cellular phone built in a GPS and a Java virtual machine.

References

1. Bergqvist, J. and Ljungberg, ComCenter: a Person Oriented Approach to Mobile Communication, CHI2000 Extended Abstract, pp.123-124 (2000).
2. Schmidt, A., Takaluoma, A. and Mantyjarvi, J., Contex-Aware Telephony Over WAP, Journal of Personal Technologies, Vol.4, No.4, pp.225-229 (2000).
3. Pedersen, E.R., Calls.calm: Enabling Caller and Callee to Collaborate, CHI2001 Extended Abstract, pp.235-236 (2001).
4. Tang, J., Yankelovich, N., Begole, J., Van Kleek, M., Li, F. and Bhalodia, J., ConNexus to Awarenex: Extending awareness to mobile users, Proceedings of CHI2001, pp.221-228 (2001).
5. Milewski, E., A and Smith, M, T, Providing Presence Cues to Telephone Users, Proceedings of CSCW2000 (2000).
6. Want, R., Schlit, B., Adams, N., Gold, R., Petersen, K., Ellis, J., Goldberg, D. and Weiser, M., An Overview of the Parctab Ubiquitous Computing Experiment, IEEE Personal Communications, Vol. 2, No. 6, pp.28-43 (1995).
7. Watanabe, S., Kakuta, J., Mitsukoka, M. and Okuyama, S., A Field Experiment on the Communication of Awareness-Support Memos among Friends with Mobile Phones, Proceedings of CSCW2000 (2000).
8. Nakanishi, Y., Tusji, T., Ohyama, M. and Hakozaki, K., Context Aware Messaging Service: a Dynamical Messaging Delivery using Location Information and Schedule Information, Journal of Personal Technologies, Vol.4, No.4, pp.221-224 (2000).
9. Nakanishi, Y., Kitaoka, N., Ohyama, M. and Hakozaki, K., Estimating Communication Context through Location Information and Schedule Information - A Study with Home Office Workers, ACM Conference on Human Factors in Computing Systems (CHI2002) Extended Abstracts, pp.642-643 (2002).
10. http://imadoko.mapion.co.jp

Browser State Repository Service

Henry Song, Hao-hua Chu, Nayeem Islam, Shoji Kurakake, and Masaji Katagiri

DoCoMo Communications Laboratories USA, Inc.
181 Metro Drive, Suite 300, San Jose, CA 95110 USA
{csyus, haochu, nayeem, Kurakake, katagiri}@docomolabs-usa.com

Abstract. We introduce *browser state repository (BSR) service* that allows a user to save and restore multiple independent *snapshots* of web sessions on a browser. At a later time, the user can retrieve any saved snapshot on a potentially different browser on a different device to continue any one of the chosen saved session in any order. The web session snapshot captures a complete browser running state, including the last page that appears on the browser, document object state, script state, values that a user enters in forms on the last page, browser history for back and forward pages, and cookies. BSR service consists of a *browser plug-in* that takes browser session snapshots, and a *repository server* that stores snapshots securely for each user. The main contribution of BSR service is *that it decouples association between browser state and a device*, in favor of *association between browser state and its user*.

1   Introduction

Most of the web applications today are *session-oriented*. Most websites require a client browser to first establish a web session and obtain a session ID. This session ID can then be used by a website to track and identify the client browser as it moves between different web pages within the website domain. During a web session, a client browser may accumulate *session state* (e.g., session identifier, shopping cart, form inputs, and etc.) that is needed to interact with the web application servers over the stateless HTTP protocol. The session state on a browser can appear in cookies, document objects, and script objects (e.g., JavaScript or VBScript). When the client browser exits a website, the active web session is closed and some of client-side session state is un-recoverable.

This session-oriented model places a limitation such that *for the duration of a web session*, a user cannot switch devices; otherwise, she might lose her active web session and would need to restart it on a new device. Consider a case when she is running an active web session on a stationary device (a desktop PC), but an alternative mobile device (a Pocket PC with wireless access) is available to her. She would like to continue browsing but on a mobile device. However, she could not do so without losing her active web session on the stationary device. She is restricted to the location of her stationary device. Consider another case when she is running an active web session on a mobile device (a Pocket PC) with a small screen, but an alternative stationary device (a desktop PC) with a large screen is available to her. She would like to use the large screen of the stationary device, but again she could not

F. Mattern and M. Naghshineh (Eds.): Pervasive 2002, LNCS 2414, pp. 253-266, 2002.

do so without losing her active web session on the mobile device. She is tied to the device that she starts a web session with. To address these limitations, we propose a *browser state repository (BSR) service* that enables her to seamlessly or effortlessly migrate an active web session to any device that is accessible or convenient to her. A unique aspect of the BSR service is that it is based on a *client-side approach* that has many advantages over existing server-based approach that can also address this limitation. These advantages are described in details in section 1.3.

1.1   Motivation Scenario

We will use the following scenario to further illustrate motivations of our work:
> *Jane was shopping for window draperies online. Jane started by browsing an online drapery store using her office PC while she was in her office. When she was asked about sizes of her windows, Jane did not know them, so she stopped browsing the online drapery store. After work, Jane went home to get measurements on the sizes of windows. She returned to the online drapery store using her home PC, and continued shopping for her window draperies. After making her selections, she was not sure if the online drapery store offered good prices, so she visited a local store carrying her Pocket PC (or PDA) with wireless network access. On her way to the drapery aisle, she saw some beautiful wallpaper that fits perfectly with her selected drapery. Unfortunately, the local store did not carry enough stock. She used her Pocket PC to do an online search for the home page of the company that produces this wallpaper. Then she continued to the drapery aisle. She compared the prices of the online drapery store with prices found at the local store. After she checked that the online prices were good bargains, she ordered draperies from the online drapery store using her Pocket PC. When she got home, she used her home PC to retrieve the homepage of the wallpaper company. The homepage provided her with contact information, and she placed her order over the phone.*

We make two observations in Jane's scenario.
- *Multiple Devices*: Jane had multiple computing devices (office PC, home PC, and Pocket PC), and she used different devices to perform her online task(s) at different locations.
- *Multiple Tasks*: Jane had multiple ongoing tasks (ordering drapery, and finding contact information of wallpaper company), and she performed them in *discontinuously incremental steps* rather than one continuous step from the start to the finish.

Jane would like *flexibility* to use whatever device is accessible or convenient to her to continue her unfinished online tasks in an effortless or seamless manner. In Jane's scenario, she returned to the online drapery store on her home PC, and she would like to continue exactly where she left off on her office PC without having to start over again. To provide this flexibility, it requires the decoupling of an association between browser state to a device, in favor of an association between a browser state and its user and independent of any device. The BSR service is designed based on this new association. It enables Jane to seamlessly migrate her browser state with her across different devices. The mechanism used in BSR service is simple but effective. Before Jane switches out of her current device, she preserves her active web session by saving a snapshot of her current browser state on a trusted BSR repository server. When she finds a new device at a later time, Jane retrieves the browser state snapshot from the BSR repository server and restores it on the new device. Then Jane can continue with her online activity on the new device. The session migration becomes seamless and effortless for her.

An additional benefit of using BSR service is that it helps people to keep track of their ongoing tasks and do them incrementally in any order convenient to them. In Jane's scenario, she switched from one online task (ordering drapery) to another online task (finding contact information of wallpaper company). Using BSR service, Jane could juggle between several online tasks. Each online task-in-progress would be preserved in a browser state snapshot on a BSR repository server, and Jane could freely stop and continue any task at any time from any device.

1.2   Client-Based Approach vs. Server-Based Approach

To accommodate users like Jane who are mobile and have a need to do tasks on different devices at different locations, a service like BSR is needed to allow Jane to save and retrieve intermediate session state. There are two possible approaches in the design for this repository service called *client-based approach* and *server-based approach*. BSR service chooses the *client-based* approach that captures session state through a browser state snapshot, and stores this snapshot on a repository server (not on the web application server). In the client-based approach, web application servers can be made completely unaware of a session migration occurring on the client browsers and devices. The reason is that *HTTP protocol is stateless*. A client browser on a different device can connect to a web application server to continue an ongoing web session, as long as the client browser can present the same browser state (e.g., cookies, hidden forms, and etc.) to the web application server.

In the server-based approach, intermediate session state is captured and saved on the web application servers. It is used by many of advanced websites such as amazon.com or expedia.com. These websites typically require Jane to first register with them to obtain unique sign-on names and passwords. By identifying her sign-on name, websites can track Jane on different devices and retrieve any saved intermediate session state for Jane. However, there are limitations in server-based approach when comparing to client-based approach:

- *Incomplete Browser State*: the session state that the web application server can save is limited to information that a client browser sends to the web application server, which is unlikely to be the complete browser state. For example, it cannot capture browser state such as the last web page (if retrieved from browser cache) that Jane navigates to, values of document objects on the last web page, values of scripting objects, values of forms that Jane filled but had yet submitted to the website, and the browser history for back pages and forward pages. Consider the amount of efforts needed for Jane to switch from a PC to a Pocket PC using the server-based approach. She would need to re-construct the browser state on Pocket PC browser by signing-on with the online drapery store, navigating her Pocket PC browser to the page where she left off on the PC browser, and re-entering any empty forms. This can be considerable effort. In contrast, the client-based approach can capture and restore complete client browser state, making session migration across devices transparent and effortless.

- *Privacy*: Jane may not trust websites to keep track of her intermediate session state, which may involve personal information that she does not want to share with websites before she makes her final purchase (e.g., the website may intentionally increase the price of items in her shopping cart). She would prefer a trusted third party, such as a BSR service provider, to maintain her intermediate session state.
- *Decentralized Storage*: In server-side approach, each web application server maintains its own intermediate session state with Jane. If Jane has multiple incomplete online activities on multiple websites, Jane needs to remember all these websites. This can be a problem when the number of such websites becomes large. In comparison, the client-based approach such as BSR service uses a centralized repository server to store all her incomplete online activities. Jane would only need to remember the location of the centralized repository server to retrieve all saved sessions.

In addition to the limitations described above, there is one more reason that the client-based approach is better than the server-based approach. That is only small portions of web websites are advanced like amazon.com and expedia.com that allow users to preserve intermediate session state on the web application servers. Some websites do not identify and track users, because they do not want to burden users with the registration and sign-on processes until they are ready to make purchases. These websites will not save any intermediate session state for Jane. This means that after Jane closed the session with the website, she would lose her shopping cart and would need to start over again on a new device.   Some websites store intermediate session state on browser-side cookies. Since cookies are not access-able to browsers on other devices, Jane would again need to start over again on a new device. On the other hand, the client-based approach is applicable to any websites regardless of whether or how they perverse intermediate session state for users.

There can be a possible redundancy in a snapshot if the BSR service is used in conjunction with a website that keeps track of the session state using the server-based approach. That is, a portion of the snapshot may be redundant information that a website also maintains, and that portion of snapshot can be retrieved directly from the website rather than from the snapshot. BSR service must be properly used to ensure that the session state stored the snapshot is consistent with the session state on the website. After a user takes a snapshot of a browser session, the user should not continue with browser session because further actions can create inconsistency between the session snapshot and the data on website.

We identify several design requirements for BSR service so that it can be easily deployed to current WWW. BSR service should require little or no modifications to the existing web application servers to support, and it should require little or no modifications to browser internal. A key assumption in BSR service is that websites do not set short time-out policies that automatically close a session-in-progress with a client browser after the session-in-progress is preserved and becomes inactive.

The rest of the paper is organized as follows: Section 2 discusses related work; Section 3 describes design of BSR service; Section 4 explains its implementation; and Section 5 draws conclusion and future work.

2   Related Work

BSR service is similar to bookmark concept on existing web browsers (it is also called "Favorites" on Microsoft Internet Explorer). Bookmarks allow a user to save URLs to web pages, so that she can quickly come back to these pages at a later time. An alternative solution to BSR service is to simply synchronize bookmarks across devices, but this solution is insufficient. Bookmarks can only work on static web pages which do not accumulate any runtime browser state with users, e.g., these static web pages do not contain client-side scripting and cookies to track web sessions. In comparison, BSR service can work on both static and dynamic web pages. As a matter of fact, most web services available today on Internet make extensive use of browsers' capabilities to support client-side scripting and cookies to track sessions.

Application-layer mobility in SIP [1] shares a similar goal with BSR service. SIP provides session mobility that allows a user to change terminals while maintaining the same running media session. It provides personal mobility that allows other people to address a single user located at different terminals. It also provides service mobility that allows a user to change terminals while maintaining access to services. However, SIP is targeted toward telecommunication services, such as video conferencing, voice over IP, and instant messages, whereas BSR service is targeted specifically to provide session mobility for web applications.

Service hand-off [3] in the Iceberg project and Universal Inbox [4] describe architectures that can support personal mobility and service mobility for a user who may want to switch between heterogeneous access networks or between heterogeneous access devices. They deal with issues such as data transformation into different formats that can be accepted by heterogeneous devices, storage and processing for redirecting messages to preferred device that users designate, and heterogeneous device name translation and mapping which may be based on naming schemes other than IP. Like SIP, service hand-off and Universal Inbox are targeted toward telecommunication services.

Mobile People Project [5] utilizes Personal Proxy to route communication to a mobile user, independently of the user's location and applications she is currently using. Personal Proxy shares a similar concept of personal data storage as our BSR repository server. In addition to personal data storage, Personal Proxy also plays an active role in tracking mobile people's whereabouts, routing application communications, and transforming communication protocol to the preferred devices designated by mobile people. In comparison, BSR service provides a centralized personal data storage specifically for browser session state preservation and migration, rather than a broader range of personal data.

Roma Project [6] provides centralized personal metadata service to locate and synchronize current version of personal files and to ensure the availability of the personal files across different repositories. It is similar to BSR service in the sense that BSR service is synchronizing the browser state across devices, whereas Roma is synchronizing personal data in general. In Roma architecture, the data can be distributed across multiple devices and repositories. We believe, for the specific purpose of synchronizing web session states, it is more appropriate to keep them in a centralized proxy server for ease of security and data management.

Aglet [8] and Roam system [9] are mobile agent systems that enable an application to migrate at runtime from one device to another device. This runtime migration involves taking a snapshot of application running state on a source device, transferring and restoring the snapshot on a target device. This snapshot approach is also used in our BSR service. However, BSR service differs from these agent-based systems in that it is focused on web applications, and more importantly, it introduces the concept of a personal repository where a user to store multiple snapshots and retrieve them anytime on any device.

3   Design

The design of the BSR service is based on decoupling the association between web sessions (browser state) and a device, in favor of a new association between web sessions (browser state) and its user as shown in Fig. 1. The benefits of this new association are that (1) it allows a user to switch devices in the middle of an active web session without losing it and having to restart it on a new device, and (2) it allows a user to keep track of multiple active web sessions and freely save and continue any active web sessions at any time from any device.

The architecture of BSR service is shown in Fig. 2. It contains 2 modules: *BSR plug-in* and *BSR repository server*. The BSR plug-in is like any browser plug-in that can be downloaded and installed from the Internet. BSR plug-in exposes user interface (UI) as a bar in a browser for users to perform two basic operations:

- Take a snapshot of the current browser state and store the browser state snapshot on BSR repository server in a secure manner.
- Retrieve a saved snapshot from BSR repository server to a browser in a secure manner.

BSR repository server provides an always-connected storage for authenticated users to store browser state snapshots. In addition, BSR repository server is also a web host that serves the web document displayed in the BSR plug-in bar shown in Fig. 4. By using two basic operations provided by BSR service, Jane can easily migrate an active web session between different devices as shown in Fig. 2. Prior to leaving the source device (device A), Jane takes a snapshot of the browser state and stores the snapshot on BSR repository server. After the target device (device B) becomes accessible to her, Jane retrieves the browser state snapshot from BSR repository server, and restores it on the browser on the target device (device B).

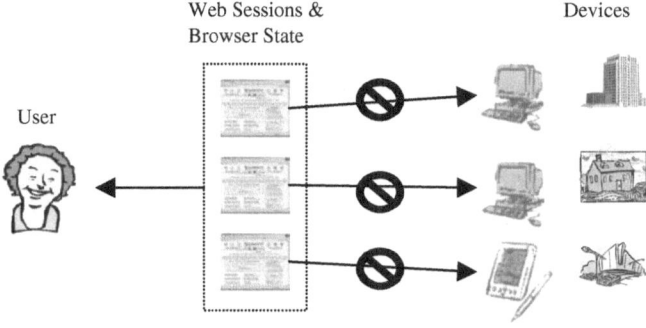

Fig. 1. Design principle of BSR service

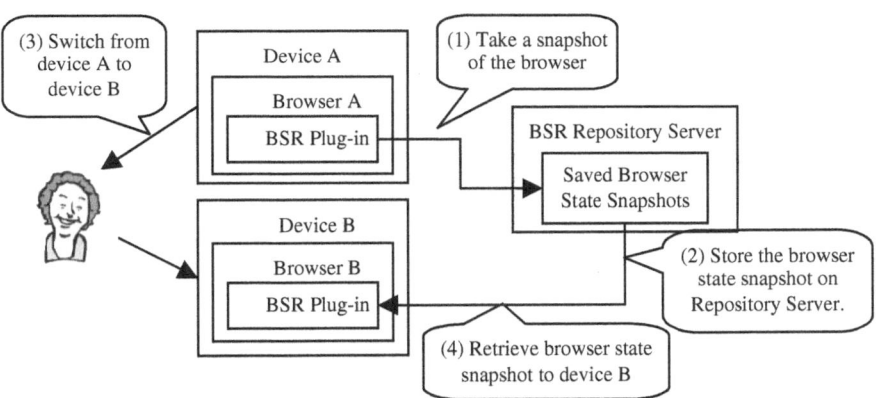

Fig. 2. Web session preservation architecture

For the remainder of this section, we provide a detailed step-by-step description of BSR service corresponding to the example in Fig. 2. Jane first needs to activate the BSR plug-in when she starts her web browser on device A as shown in step (1) of Fig. 3. The steps needed to activate a browser plug-in differ from one web browser to another web browser. In Microsoft Internet Explorer (referred to as IE for remainder of this paper), BSR plug-in can be found and activated by this sequence of browser menus: View -> Explorer Bar -> BSR. By activating BSR plug-in, a horizontal explorer bar appears at the bottom of a browser window as shown in Fig. 4. The horizontal explorer bar displays a HTML page downloaded from her designated BSR repository server. We assume that Jane has already downloaded and installed BSR plug-in in her browser, and she has created an account with BSR repository server.

From the horizontal explorer bar (referred to as BSR plug-in bar for remainder of this paper), Jane can sign on with BSR repository server by entering her user ID and password and then click on sign-on button. Then BSR plug-in downloads a list of saved browser state snapshots from the BSR repository server. Jane has an option to select either a commercial BSR service provider, or she can setup her own BSR repository.

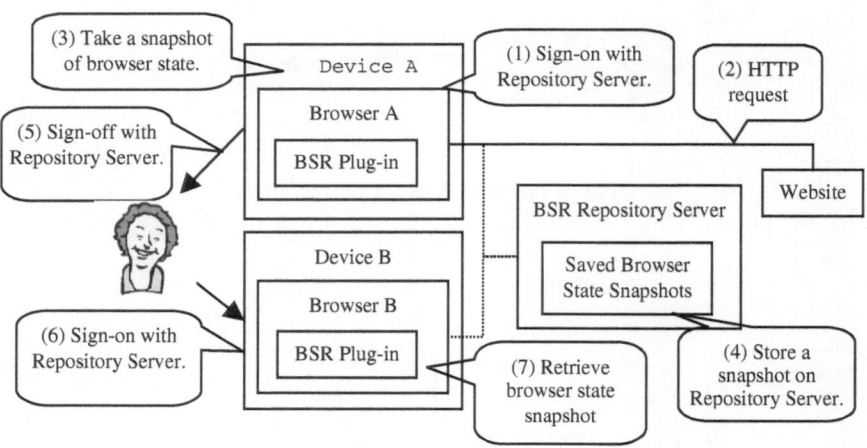

Fig. 3. A step-by-step description of browser session migration from device A to device B

Jane can now browse a website on browser/device A as shown in steps (2) of Fig. **3**. After some time, Jane wants to leave device A and switches to device B. As shown in step (3) of Fig. 3, Jane first clicks on the snapshot button in BSR plug-in bar. Then BSR plug-in takes a snapshot of the browser session. The snapshot includes the document objects (DOM) of the current web page on the browser, scripting objects (JavaScript or VBScript), URL history, and cookies corresponding to the domain of website(s) serving the current web page. The snapshot is stored securely on BSR repository server in step (4) of Fig. 3 via a SSL connection. Jane can assign a unique *session name* to the snapshot as an index in the list of stored snapshots. If Jane does not give a session name, a default one is generated which is the website hostname. Jane also has an option to protect this snapshot with a *session password*. The session password offers additional protection in the case that Jane carelessly leaves her device open to other people without signing off with BSR service.

Before Jane closes browser on device A, she clicks on sign-off button in the browser plug-in bar. It disconnects the browser from BSR service as shown in step (5) of Fig. 3. Jane is now ready to use device B. As shown in step (6) of Fig. 3, Jane first signs on with BSR service on device B. The BSR plug-in downloads a list of saved snapshots from BSR repository server. Jane selects the session that she was last working on device A. Then the BSR plug-in downloads the snapshot and restores it on the browser as shown in step (7) of Fig. 3. Now Jane has successfully migrate her session from device A to device B. In the previous example, we make an assumption that the web session in the snapshot has not been time-out by the web application server when Jane restores it. However, this may not be true for some web application servers that set short time-out policies when they detect user inactivity. We propose the following solutions to help keep alive any saved web session:

- The BSR service can periodically ping the web application server (e.g., by refreshing the web page stored the browser state snapshot) in order to keep alive a saved web session. It requires no modification to the web application servers, but it has undesirable effect of generating additional network and server loads.

- The BSR service can probe the websites for the session time-out value, and inform its user to retrieve the browser state snapshot before time-out occurs. In addition, the websites can be modified to allow the BSR service to set time-out policies on saved web sessions.

4   Implementation

We describe how to implement for the BSR plug-in, in IE version 5.0 and later, to capture and restore a browser state snapshot. The data structure for the browser state snapshot includes the following components – document object state, cookies, script state, and browser history.

4.1   Document Object State

W3C defines a standard specification for document objects called DOM (document object model) [2]. DOM describes a logical structure of web documents and standard interfaces for access and manipulation of web documents by scripting. When a browser parses a HTML page, it creates a DOM structure to represent the structure of a HTML page. Each node in the DOM structure represents a DOM element, which in turn may represent a particular HTML tag or a HTML element defined in DOM specification, such as <HEAD>, <BODY>, <SCRIPT>, <FRAME> and etc. Each DOM element has a certain set of properties that describe its presentation and behavior in the browser. In a dynamic HTML page, properties of DOM elements and DOM structure are its runtime state that can be changed by scripting when a user interacts with the page. As a result, there is a need to capture this DOM runtime state of a web page in a browser state snapshot.

We have designed two possible methods for capturing and restoring document objects called *URL reloading method* and *content reloading method*. The URL reloading method captures the DOM runtime state of a web page in a browser and the URL(s) to this web page. The DOM runtime state is extracted by traversing each node in the DOM structure of a web page and serializing its values and properties. When a snapshot is restored, the web page is first loaded from the original website specified in the saved URL(s). After the web page is completely downloaded in the browser, the retrieved DOM runtime state is de-serialized and applied to the web page. For example, when Jane enters her name "Jane" on a text field of a HTML page from the online drapery store (e.g., <INPUT NAME= "name" TYPE="text">), the browser assigns the value property of the DOM text field element to "Jane". When Jane takes a snapshot, BSR plug-in serializes the value property of the text field DOM element and stored it on the BSR proxy. In a restoration, BSR plug-in reloads the HTML page from the online drapery store, which initializes the text field to be empty. Then it de-serializes the DOM runtime state, applies it to DOM text element, and restores the text field back to "Jane".

The content reloading method captures the content of the web page as displayed on the browser. In a later restoration, it retrieves the content from the BSR repository server to the browser rather than reloading it from the original website. There is no need to serialize the DOM runtime state because the saved content contains modified runtime state. Consider the same example which Jane enters her name "Jane" on a text field of a HTML page from the online drapery store. When Jane takes a snapshot, the browser plug-in saves the modified HTML page including her input name, e.g., <INPUT NAME="name" TYPE="text" VALUE="Jane">.

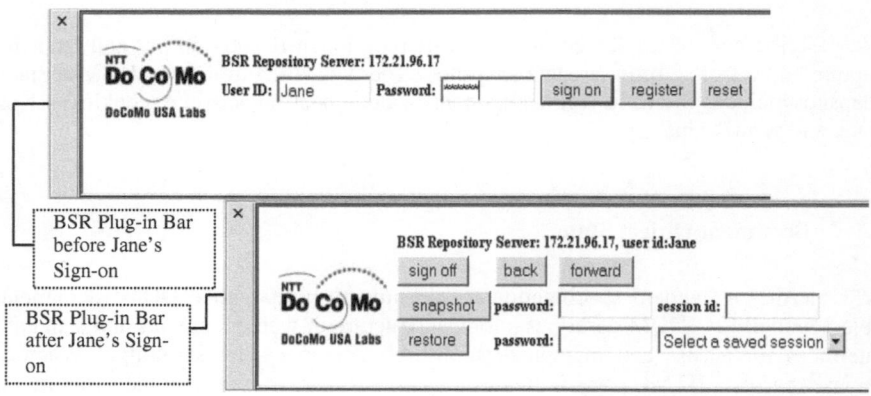

Fig. 4. BSR plug-in bar before Jane's sign-on and after Jane's sign-on

The main difference between these two methods is that when a snapshot is restored, URL reloading method generates request(s) to the original website, whereas the content reloading method does not generate any request(s) to the original website. However, the URL reloading method works better for web pages that are updated periodically. When Jane restores a snapshot, URL reloading method can retrieve the most updated content from the website; whereas the content reloading method may display content that have expired already.

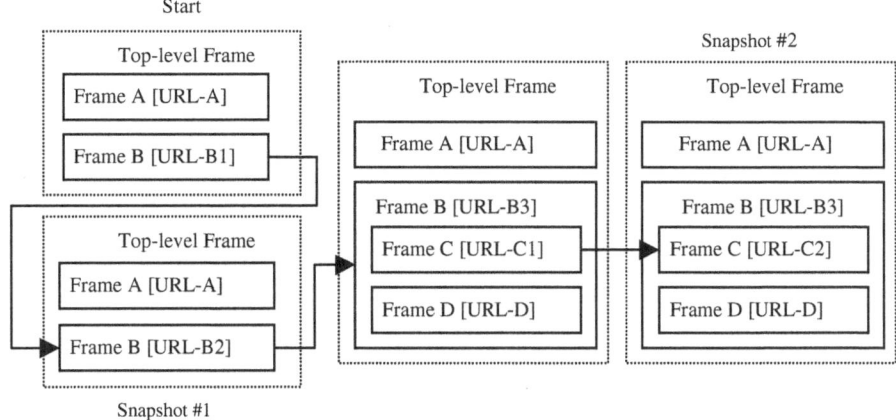

Fig. 5. Keeping track of URLs to frames in a web page

4.2   Tracking Frames

A web page can contain multiple frames and sub-frames. Content in each frame or sub-frame can be changed at runtime by scripting to load from a different URL/website. Since frames are also document objects, tracking runtime state of frame and sub-frame is a necessary part of capturing the overall DOM runtime state of a web page. Consider the example in Fig. 5. The web page contains two frames – an index frame A and a content frame B. When the web page is first downloaded to a browser, URL-B1 is loaded in frame B. When Jane clicks on a link in frame A, frame B is reloaded with URL-B2. At this time, Jane takes snapshot #1 of the browser state. In URL reloading method, the browser plug-in creates a *frame-URL-mapping table* for snapshot #1. In this table, each entry contains the name of a frame and the *modified URL* to the frame (Frame B, URL-B2). When the user restores snapshot #1, browser plug-in traps all HTTP requests and checks if frames that HTTP requests are originated from match entries in frame-URL mapping table. In case of a match, the URL in the HTTP request (URL-B1 for frame B) is replaced with the modified URL (URL-B2 for frame B) recorded in the frame-URL mapping entry. A more complex example is shown in snapshot #2 of Fig. 5, where frame B is loaded with yet another frameset (URL-B3) that contains two sub-frames C and D. Frame C is first loaded with URL-C1, and reloaded with URL-C2. In snapshot #2, the frame-URL-mapping table contains two entries where URLs have been modified {(Frame B, URL-B3), (Frame C, URL-C2)}. When a user restores snapshot #2, HTTP requests for URL-B1 and URL-C2 are replaced with URL-B3 and URL-C2.

4.3   Cookie State

Cookies are information stored on client browsers for specific domains of web documents. Cookies can be used for many different purposes, such as keeping track of a web session with a web application server, storing intermediate session state, identifying the user between sessions, and etc. When a snapshot is taken on a source browser (device) and restored on a different target browser (device), the snapshot may not work on the target browser if cookies are not identical between two browsers. For example, a web application server may set a cookie containing the session ID on the client browser. If the session ID cookie were not set on the target browser, the web application server would not be able to identify the session from HTTP requests sent by the target browser.

A simple but inefficient method is to save all cookies on the source browser in a browser snapshot, regardless of domains of cookies, and transfer them to the BSR repository server. When the browser snapshot is restored, all cookies are transferred to the target browser. This is inefficient because only the cookies that match domains of the websites in the snapshot are used on the target browser. A more efficient method is to identify domains of websites in the snapshot, and transfer only the yet expired cookies matching those domains.

4.4   Script State

Popular client-side script languages embedded or referenced in HTML pages are JavaScript and VBScript. Client-side scripts run on a client browser, and they can be used to manipulate the appearance and content of a dynamic HTML page without having to go to a web application server. Like many programming languages, client-side scripts contain script methods and script variables. The scope of a script variable can be either local (declared within a script method) or global (declared outside of any script methods). In a browser state snapshot, BSR plug-in does not need to save local script variables because they are reinitialized each time a script method is invoked. In addition, BSR plug-in makes sure that it does not take a snapshot in the middle of script method execution where the execution stack is not empty. It waits to take a snapshot until the script engine of the browser has completed executing script methods. However, BSR plug-in does need to save global script variables because their values remain valid until a browser navigates to other web pages.

BSR plug-in does not need to save any script methods, because script methods do not change and they are reloaded into a browser when a snapshot is restored.

4.5   History State

Browser history state can be an essential part of a browser state snapshot, which is used to support the back and forward buttons found on almost all browsers. BSR plug-in maintains its own *history state* for a browser state snapshot, and it is implemented independently from the browser history mechanism.

According to the HTTP 1.1 protocol specification [7], browser history should show exactly what a user saw at the time that the web page was displayed in the browser. It differs from browser cache mechanism in that if a web page retrieved from history has expired, browser should still display the expired web page as is in the browser. On the other hand, if the same expired web page is retrieved from the browser cache mechanism, browser should re-fetch the updated copy from the original website. This means that BSR plug-in needs to save the actual content of historical web pages in a snapshot. Saving URLs of historical web pages is insufficient. As a result, only the content reloading method described in section 4.1 can be applied to saving history state but not the URL reloading method.

BSR plug-in starts recording browser history when it is activated in a browser, so history of a snapshot can only go back to the web page where the BSR plug-in is activated. However, the amount of data for history state can be huge at the end of a long browsing period. It can become too big to be transferred to and from a BSR repository server in a snapshot or a restoration. To address this issue, BSR plug-in has an adjustable limit on the size of the most recent history that it keeps track of. If the history state grows beyond its limit, older web pages are thrown out. In addition, for mobile devices where bandwidth is a premium, history state is downloaded only as needed from repository server to a mobile device when the user clicks on the back button. A user can also disable history state tracking completely on BSR plug-in

4.6   Performance

The performance of the a snapshot or restore operation depends on several factors – the type of network connectivity, the size of snapshot which is proportional to complexity of web pages and the size of the history, and the load on the BSR server. We have conducted preliminary testing of snapshot and restore operations on various popular web sites (such as amazon.com and yahoo.com). The size of the snapshot is approximately equal to the size of the downloaded HTML page(s). If history tracking is enabled, the size of the snapshot is approximately the sum of all HTML pages recorded in history. We have found that on average, a snapshot or a restore operation, with history tracking disabled, takes less than 5 seconds on a desktop PC with a fast WLAN/LAN connection, with the majority of the time in the serialization and de-serialization of the DOM tree.

5   Conclusion and Future Work

In this paper, we describe BSR service that brings browser session mobility to users. BSR enables a user to switch devices in the middle of an active browser session. In addition, it enables a user to preserve multiple active web sessions for restoration at a later time on any device with a built-in browser support. We believe that as browser-based applications gain popularity, browser session mobility provided by BSR service can help browser-based applications to better integrate with people's mobile work and life styles.

We expect such browser session mobility to occur not only among devices and/or browsers with similar capabilities, but also across heterogeneous device platforms with different browsers/micro-browsers (HTML, cHMTL, WML, and etc.) and hardware capabilities (screen sizes, input methods, and etc.).   This introduces additional challenges.   For examples, the same web page can require different presentations (HTML, cHTML, or WML) on different browsers, so snapshot on one device platform may require transformation before it can be restored on another device platform.   Browsers and micro-browsers also differ on cookie management, the type of scripting language, and history management.   BSR service for multiplatform applications is our future work.

References

1. Henning Schulzrinne and Elin Wedlund, "Application-Layer Mobility Using SIP", *Mobile Computing and Communications Review*, Volume 4, Number 3, pp. 47–57, July 2000.
2. W3C, "Document Object Model (DOM) Level 1 Specification (Second Edition)", W3C Working Draft, September 2000.
    http://www.w3.org/TR/2000/WD-DOM-Level-1-20000929
3. Anthony Joseph, B. R. Badrinath, and Randy Katz, "A Case for Services over Cascaded Networks", *First ACM/IEEE International Conference on Wireless and Mobile Multimedia (WoWMoM'98)*, October 30, 1998.
4. Bhaskaran Raman, Randy H. Katz, and Anthony D. Joseph, "Universal Inbox: Providing Extensible Personal Mobility and Service Mobility in an Integrated Communication Network", *Workshop on Mobile Computing Systems and Applications (WMSCA'00)*, December 2000.
5. P. Maniatis, M. Roussopoulos, E. Swierk, K. Lai, G. Appenzeller, X. Zhao, and Mary Baker, "The Mobile People Architecture". *ACM Mobile Computing and Communications Review*, Volume 3, Number 3, July 1999.
6. Edward Swierk, Emre Kicman, Nathan Williams, Takashi Fukushima, Hideki Yoshida, Vince Laviano, Mary Baker, "The Roma Personal Metadata Service", *Proceedings of the IEEE Workshop on Mobile Computing Systems and Applications*, December 2000.
7. R. Fielding, J. Gettys, J. Mogul, H. Frystyk, L. Masinter, P. Leach, T. Berners-Lee, "Hypertext Transfer Protocol – HTTP/1.1", RFC 2616, June 1999.
8. D. B. Lange, M. Oshima, "Mobile Agents with Java: The Aglet API", *World Wide Web Journal*, 1998.
9. Hao-hua Chu, Henry Song, Candy Wong, and Shoji Kurakake, "Seamless Applications over Roam System", *UbiTools'01 (Part of UbiComp'01)*, September 2001, http://choices.cs.uiuc.edu/UbiTools01/.

Annotation by Transformation for the Automatic Generation of Content Customization Metadata

Masahiro Hori, Kouichi Ono, Teruo Koyanagi, and Mari Abe

IBM Tokyo Research Laboratory
1623-14 Shimotsuruma, Yamato-shi
Kanagawa-ken, 242-8502, Japan
{horim, onono, teruok, maria}@jp.ibm.com

Abstract. Users are increasingly accessing the Internet from mobile devices as well as conventional desktop computers. However, it is not reasonable to expect content authors to create different data presentations for each device type, but the content source should be reused across multiple delivery contexts whenever possible. The objective of this research is to develop a supporting tool for the presentation customization that follows after the content specialization in device-independent authoring. This paper presents a tool that automatically generates content customization metadata on the basis of users' editing operations toward the desired results of the customization. A prototype of the metadata generator was developed for the generation of page-clipping annotations to be used for an annotation-based transcoding system.

1 Introduction

As more and more Web-enabled personal devices are becoming available for connecting to the Internet, the same Web content needs to be rendered differently on different client devices, taking account of their physical and performance constraints such as screen size, memory size, and connection bandwidth. For example, a large full-color image may be reduced with regard to size and color depth, removing unimportant portions of the content. Such device adaptation is exploited for Web documents delivered over HTTP, and results in better presentation and faster delivery to the client device. Device adaptation is thus crucial for transparent Web access under different delivery context, which may depend on client capabilities, network connectivity, or user preferences [2,3,9,22]. It must be noted that the Web documents refer not only to existing HTML pages, but also to XML/XHTML documents that may be generated from a content source.

It is not reasonable to expect content authors to create multiple versions of the same Web content, each specialized for a different delivery context. Such multiple authoring encounters the *M times N problem*, namely, an application composed on M pages to be accessed via N devices requires M x N authoring steps, and results in M x N presentation pages that must be maintained [18]. The

F. Mattern and M. Naghshineh (Eds.): Pervasive 2002, LNCS 2414, pp. 267–281, 2002.

key challenge here is to enable Web content to be delivered through a variety of access mechanisms with minimum effort.

The objective of this research is to develop technologies to be used for presentation customization in a device-independent authoring environment. The customization requires additional data about the ways of modifying the presentation, and it is assumed in this study that metadata or annotations[1] are exploited by a runtime engine for the content customization at content delivery time. An annotation is a remark attached to a document, and declares properties that qualify a particular portion of a target document. In addition, annotations may indicate structural changes for the annotated portion of a target document. In order to clarify the distinction of these two roles, we call the former *assertional annotations* and the latter *transformational annotations*. It is important to note that this distinction is not exclusive, because every annotation is intrinsically an assertion.

It is straightforward for annotation authors to indicate a location to be annotated and create an assertion as annotation content. This is an approach that we call *annotation by assertion*, and is adopted by existing annotation editors [1,5, 8,12,14,20,23]. On the other hand, for transformational annotations, it is easier for the authors to modify a target document toward the desired results of the customization, rather than to indicate the ways of modifications declaratively as assertional annotations. This is a basic idea behind an approach what we call *annotation by transformation*, which was originally proposed for the automatic generation of document transformation rules [15]. In this paper, we present a tool that follows the annotation by transformation approach, and automatically generates content customization metadata on the basis of users' editing operations for the content customization.

In the next section, we present a perspective on device independent authoring, and briefly explains the usefulness of the annotation by transformation approach for supporting content customization. Section 3 presents an annotation-based content adaptation system, and introduces a content customization language for annotation-based document clipping. Section 4 explains approaches to metadata authoring taking account of the distinction between assertional and transformational annotation. Finally, we present our prototype of the metadata generation tool, and emphasize the complementary roles between the annotation by transformation and annotation by assertion approaches.

2   A Perspective on Device Independent Authoring

The ultimate goal of device-independent authoring is to realize a *single authoring* environment [18], in which presentation suitable for each delivery context is generated from an abstract interaction model that does not depend on any particular delivery context. The single authoring, which is illustrated in Fig. 1, consists of not only the content *specialization* process, but also a content *customization* process.

---

[1]   Metadata and annotations are used interchangeably in this study.

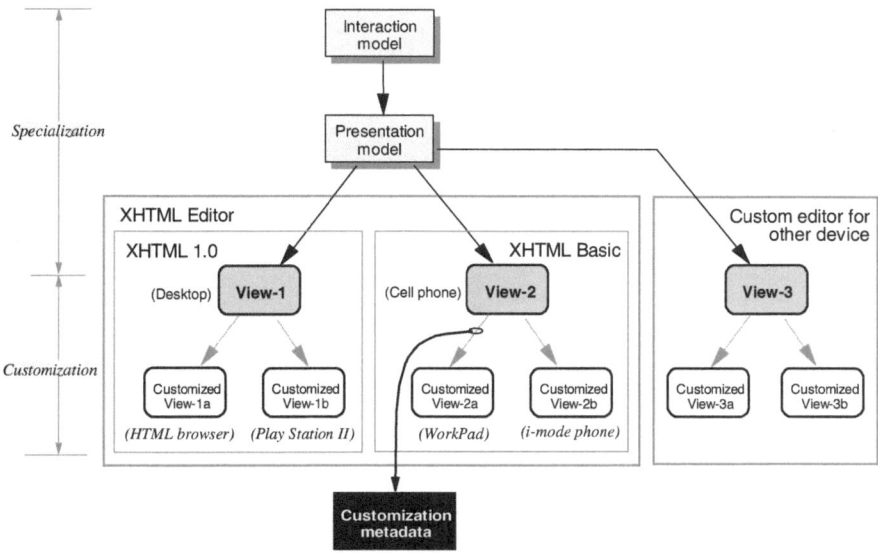

Specialization

Customization

Fig. 1. Illustration of content customization in the single-authoring process

In the authoring for the specialization, content authors need to think abstractly by going back and forth between the device-neutral interaction model and a presentation model specialized for a class of document representation. The specialization thus entails document type conversion from the schema for an interaction model toward another schema for a presentation model. The customization, on the other hand, is characterized as document modification under the same document type or schema.

The content customization allows content authors to further elaborate the specialized presentation to be more specific to the delivery context, in order to meet criteria such as ease of use and quality of presentation [6]. Authoring for the customization is largely concerned with details of the presentation, which depend on individual device capabilities and users' preferences. In such a situation, it is desirable for content authors to create an expected result interactively with an example, rather than working abstractly without any concrete example.

Learning an abstract language and writing programs are not easy tasks for most people. However, if a person knows how to perform a task to be executed by a computer, perhaps the person's knowledge can somehow be exploited for the creation of a program to perform the task. This is the motivation behind *programming by example* [17], which is also called *programming by demonstration* [4]. Programming by example would be a natural approach to creating the presentation customization metadata for page designers or novice programmers, because users need only work with examples of how to transform a page at hand, and are given automatically generated metadata that can replicates the same transformation. Fig. 2 illustrates the idea of generating content customization

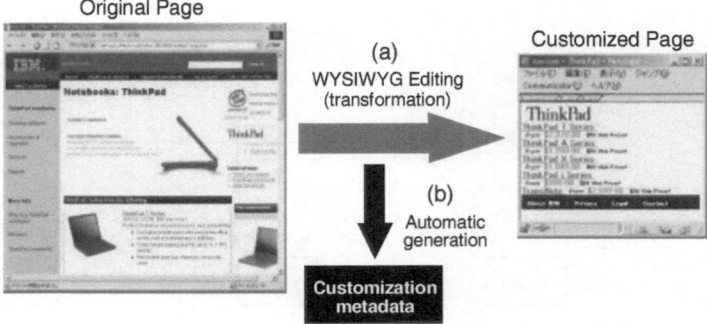

Fig. 2. Illustration of annotation generation by example

metadata (e.g., page-clipping annotation) as results of WYSIWYG editing of an
original page.

On the basis of the idea of the metadata generation from users' operation
histories, we have already developed a prototype of a metadata generation tool
XSLbyDemo [21,15] as an add-on module for a commercially available WYSI-
WYG HTML editor. The XSLbyDemo generates document transformation rules
for the XSL Transformation Language (XSLT) [28]. However, the XSLbyDemo
could only function within the HTML editor, even though the generation functions were designed to be independent of any specific editor programs.

The metadata generator presented in this paper was developed as a plug-in
module for an open tool integration platform [25], so that arbitrary editors can
be chosen for the content customization authoring. In addition, the metadata
generator is designed as a tool framework to be customized for a class of metadata
languages. In the next section, annotation-based page clipping is introduced and
an example of the page-clipping annotation is explained. The metadata generator
specialized for the page-clipping annotation language is then presented in the
section 4.

3   Annotation-Based Content Adaptation

Web-content metadata or annotations have a variety of applications [16], which
can be categorized into three types: discovery, qualification, and adaptation of
Web contents. The primary focus of this research is on the Web-content adaptation, and the role of annotation is to characterize ways of content adaptation
rather than to describe individual contents themselves.

3.1   Annotation Framework

Annotations can be embedded into a Web document as an inline annotation.
Inline annotations are often created as comments or extra attributes of document
elements. Because of its simplicity, inline annotation has been often adopted as a

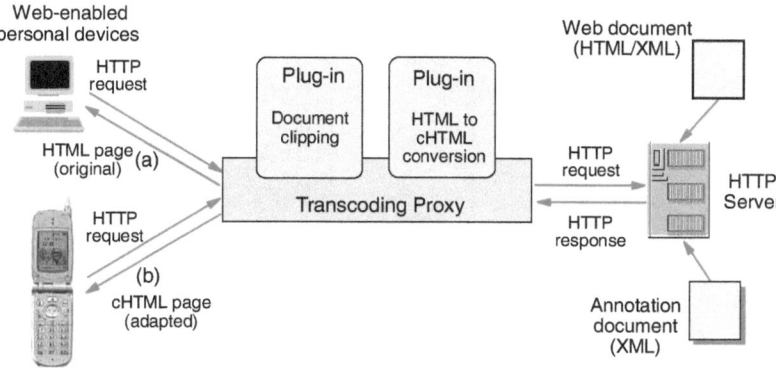

Fig. 3. Overview of an annotation-based transcoding

way of associating annotation with HTML documents [19,22,10]. An advantage of the inline annotation approach is the ease of annotation maintenance without the bookkeeping task of associating annotations with their target document. The inline approach, however, requires annotators to have document ownership, because annotated documents need to be modified whenever inline annotations are created or revised. Furthermore, mixing of contents and metadata is not desirable with regard to the design consideration that separates content from presentation.

The external annotation approach [11,14,20,24], on the other hand, does not suffer from the issues related to the document ownership. Moreover, the most important point of the external annotation is that this approach facilitates the sharing and reuse of annotations across Web documents. Since an external annotation points to a portion of a Web document, the annotation can be shared by Web documents that have the same document fragment.

Annotations provide additional information about Web contents, so that an adaptation engine can make better decisions on the content re-purposing. The role of annotations is to provide explicit semantics that can be understood by a content adaptation engine [13]. Fig. 3 depicts an overview of an annotation-based transcoding process. Upon receipt of a request from a client, a Web document is retrieved from a content server. Taking account of the capabilities of the client specified in the HTTP request header, a transcoding proxy selects one or more transcoding modules, which are indicated as plug-ins in Fig. 3. When a selected transcoding module requires an annotation document, an annotation file is also retrieved from a content server, which may or may not be the same server that retrieved the Web document. The transcoding module may simply return the original document, if a client agent has the rendering capabilities compatible with ordinary desktop computers [Fig. 3 (a)]. Alternatively, the original document may be returned with modification, so that the original content can fit into a small screen device [Fig. 3 (b)]. The decisions about the content adaptation are made taking account of the client capabilities specified in the HTTP request header.

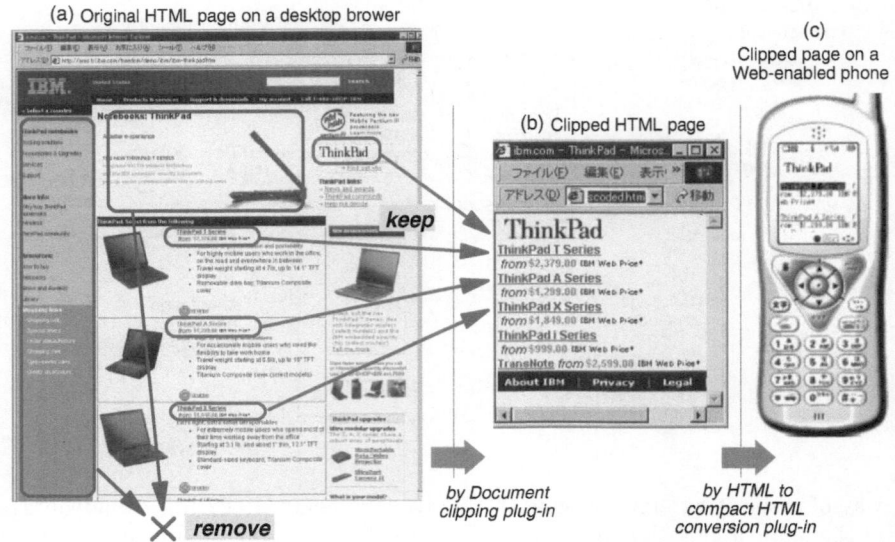

Fig. 4. Illustration of content-adaptation process

3.2   Annotation-Based HTML-Page Clipping

Fig. 4 shows a catalog page of notebook computers. The catalog page contains a lot of information, such as details of the product specification, a search field, and numerous links to other areas of the site that might be of interest to the user. However, it may be necessary to deliver portions of this page for users to access through a Web-enabled phone rather than a desktop browser. In such a case, the images and nested HTML tables prepared for a nicely laid out page are a hindrance rather than help. The sheer amount of information becomes unwieldy in the small display, and potentially expensive depending on the user's wireless service.

Content adaptation as illustrated in Fig. 4 can be done, for example, by using an annotation-based page-clipping engine [24]. At content delivery time, the page-clipping engine may modify the original document with reference to page-clipping annotations and client profiles sent over HTTP. The main idea in the page-clipping annotation language is the notion of a clipping state. By using <keep> and <remove> elements in the annotation descriptions, users can specify the clipping state to indicate whether the content being processed should be preserved or removed.

As a simple example, an HTML page and its clipped results are shown in Fig. 5. In this example, the header and the first paragraph are preserved as shown in Fig. 5(a). The table element is modified by deleting the third column and the second row. The cell-padding attribute of the table is increased, so that each table cell can be provided with margin space [Fig. 5(b)]. In addition, the whole of the second paragraph is removed as shown in Fig. 5(c).

Original Page   Clipped Page

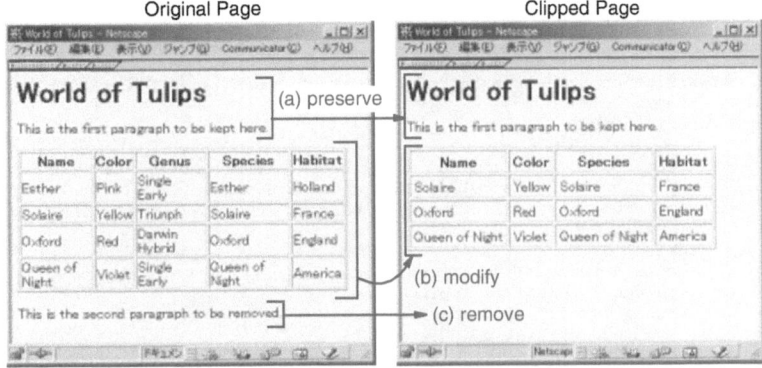

Fig. 5. Simple example of an HTML page and its clipped result

Fig. 6 shows an example of annotation document that allows the page clipping illustrated in Fig. 5. The <description> element prescribes a unit of an annotation statement in the annotation language. The target attribute is set to an XPath expression [27], and identifies the node on which the annotation will be applied, and the take-effect attribute indicates whether the annotation is applied before or after the target node. By specifying target="/HTML[1]/BODY[1]/*[1]" as in Fig. 6(a), the clipping state is activated after the first element after the first <BODY> element, which in this case is an <H1>. The <keep> element in Fig. 6(a) indicates that all the document elements encountered are preserved, until otherwise instructed by another annotation statement.

The clipping state is changed to 'remove' just before the second <P> element [Fig. 6(c)], and changed back to 'keep' after the <P> element [Fig. 6(d)]. As results, the second paragraph element indicated by "/HTML[1]/BODY[1]/P[2]" is removed while preserving the elements just before and after the removed element.

Since HTML tables can often be complex elements to clip, the annotation language provides special-purpose elements to make table clipping easier. The <row> and <column> elements allow user to clip rows and columns without relying on complicated XPath expressions. The table-clipping elements are used in the description shown in Fig. 6(b). This description sets the clipping state to 'keep' just before the first table element, and also changes the value of cellpadding attribute to 4 by using the <insertattribute> element. The name attribute of <insertattribute> can be specified with an arbitrary name of an attribute available for a target document.

In addition, the description element [Fig. 6(b)] declares that the third column, which is indicated by the index value of the <column> element, is discarded, while the remaining columns are preserved. Note here that the wildcard character to indicate multiple columns (index="*"). If a wildcard is specified, all rows (or columns) will be affected, except for those specifically indicated by a separate <row> (or <column>) element. So, all rows but the second will be preserved for the target table.

```
<?xml version='1.0' ?>
<annot version="2.0">
  <!-- (a) Set the default clipping state to 'keep' -->
  <description take-effect="before" target="/HTML[1]/BODY[1]/*[1]">
    <keep/>
  </description>

  <!-- (b) Remove a column and a row of the first table, -->
  <!--     and change a cellpadding attribute value      -->
  <description take-effect="before" target="/HTML[1]/BODY[1]/TABLE[1]">
    <keep/>
    <table>
      <column index="3" clipping="remove"/>
      <column index="*" clipping="keep"/>
      <row index="2" clipping="remove"/>
      <row index="*" clipping="keep"/>
    </table>
    <insertattribute name="cellpadding" value="4"/>
  </description>

  <!-- (c) Set the clipping state to 'remove' -->
  <description take-effect="before" target="/HTML[1]/BODY[1]/P[2]">
    <remove/>
  </description>
  <!-- (d) Set the clipping state back to 'keep' -->
  <description take-effect="after" target="/HTML[1]/BODY[1]/P[2]">
    <keep/>
  </description>
</annot>
```

Fig. 6. Example of a page-clipping annotation document

4   Automatic Generation of Annotation

In the page-clipping annotation language, annotation elements such as `<keep>`, `<remove>`, and `<insertattribute>` impose modifications of a target document, and can be regarded as the transformational annotations. On the other hand, in the other annotation languages, such as a page-splitting annotation language [11], the language provides assertional elements like `<role>` for specifying a role for an annotated element (as one of the values like proper content, advertisement, or decoration), and `<importance>` for specifying the priority of an annotated element in relation to the other elements in the same page.

Every annotation is an assertion in essence, but some annotations may further imply document transformation *with regard to the procedural semantics of a runtime engine*. Taking account of this distinction between annotations, it is possible to think about two approaches to annotation authoring: annotation by assertion, and annotation by transformation.

4.1   Annotation by Assertion

Existing annotation editors support the creation of assertional annotations by providing views easy to indicate a location to be annotated (e.g., an HTML browser view [1,5,8,20,23] and a WYSIWYG HTML editor [12,14]). We have

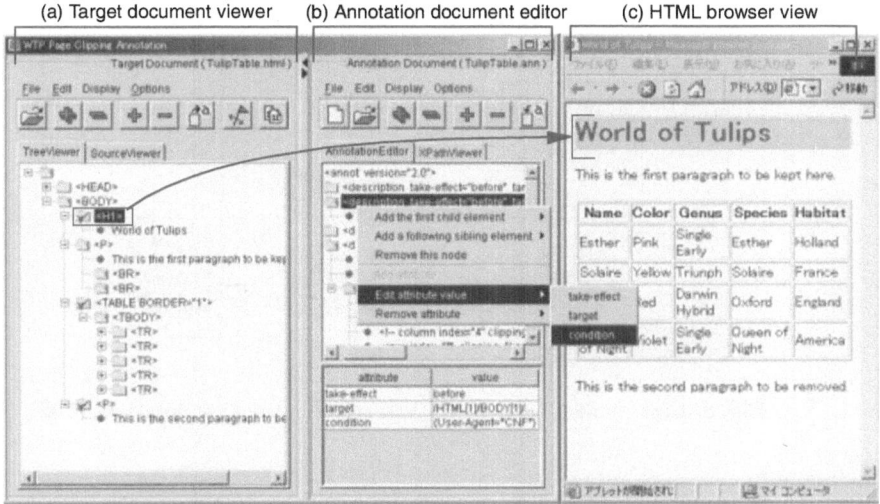

(a) Target document viewer   (b) Annotation document editor   (c) HTML browser view

Fig. 7. Sample display of the authoring tool for assertional annotations

developed an annotation editor, which follows the annotation by assertion approach [1]. Fig. 7 shows a screen of the annotation editor.

This annotation editor can be customized for different annotation vocabularies as long as the language is compliant with a prescribed schema for a class of annotation languages [1]. With this annotation editor, a user can select a node to be annotated in either a target document viewer [Fig. 7(a)] or an HTML browser view [Fig. 7(c)] if the target document is an HTML page. The node selection in either view synchronizes with the other, and a user can create an annotation description by using a popup menu that appears when the user clicks the right mouse button. The details of the annotation content can be edited by using the menu-based annotation document editor [Fig. 7(b)], which can provide context-sensitive help based on the annotation language schema at hand. Since the page-clipping annotation language mentioned earlier is compliant with the prescribed schema, the annotation document shown in Fig. 6, for example, can be created with this annotation editor.

In this way, the annotation editor allows users to indicate a position to be annotated and declare properties as the content of an annotation. The HTML browser view, for example, is very helpful to indicate the position. However, the HTML page or target document remains unchanged, even when annotations specify transformation (e.g., insertion and removal) of the annotated portions.

4.2   Annotation by Transformation

By using a WYSIWYG editor, it is easy for the users to modify a target document toward the desired results of the customization. The annotation by transformation is an approach to generating the content customization metadata au-

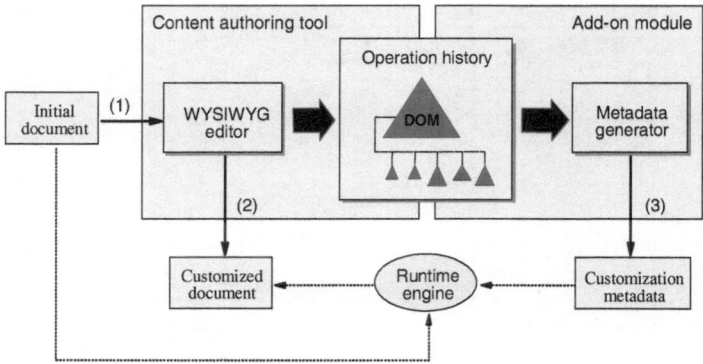

Fig. 8. Environment for generating transformational annotations

tomatically from the user's editing operations for the customization. An environment for generating transformational annotations is depicted in Fig. 8. With this metadata generation tool, first a user opens a document to be customized (e.g., an HTML file). The user then edits the document by using the full capabilities of the WYSIWYG authoring tool. The user's actions are recorded into an operation history while working in a recording mode. The user, however, does not have to care about the recording process behind the scenes. When the editing is finished, the user will have a customized document. At the same time, the metadata generator creates the customization metadata that can be used by a runtime engine to replicate the transformation from the initial document to the customized document.

Since users can perform a given editing task in different ways, it is necessary to prescribe a set of edit operations to be recorded in the user interface. The essential mechanism of the metadata generation depends solely on the Document Object Model (DOM) [7], and can be used for not only as arbitrary metadata format when a custom generator is provided, but also with an arbitrary content authoring tool as long as the editor adopts the DOM as an internal document model.

Fig. 9 shows a screenshot of the prototype environment that generates page-clipping annotations on the basis of users' editing operations with a WYSIWYG HTML editor [Fig. 9(a)]. The area of the WYSIWYG editor can be changed by selecting one of the sizes available from the page size button [Fig. 9(b)], for example, reducing the area in the screen shot 1/4 VGA size.

Fig. 9(c) shows the generated metadata, which is an XML document in the page-clipping annotation language. The look and feel of this WYSIWYG editor is exactly the same as in the original HTML editor, except for the small window for the generated annotations. A recording button [Fig. 9(d)] is used to start and stop recording a user's operations. While the toggle button remains depressed, the user's operations are recorded. When the button returns to its normal state, the recording stops. A page-clipping annotation document is created from the recorded operations by pressing the create annotation button [Fig. 9(e)].

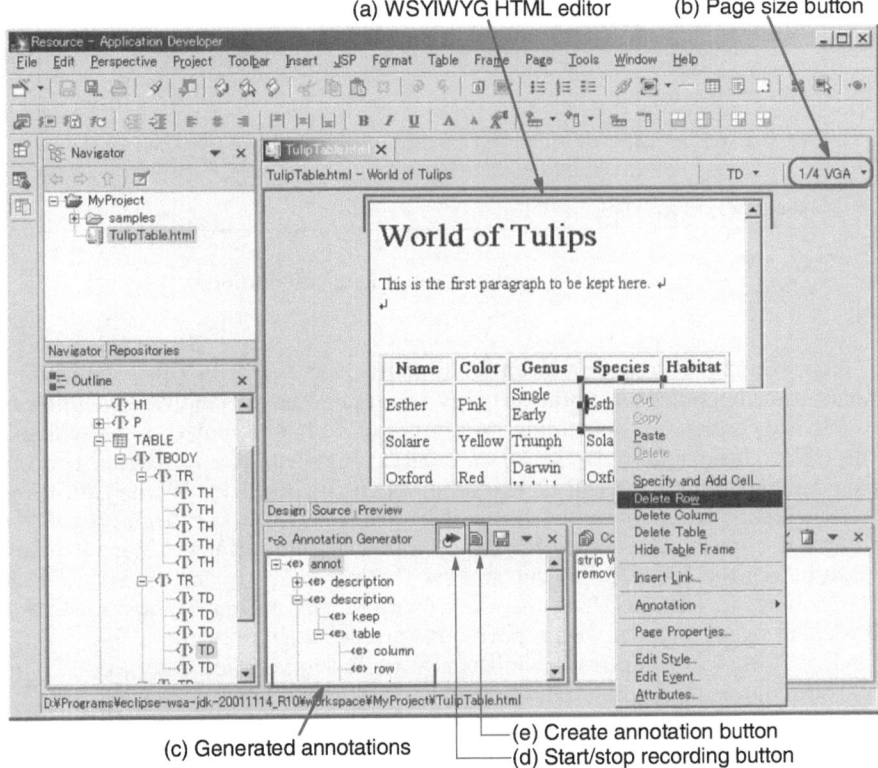

Fig. 9. Sample display from the environment for generating transformational annotations

Transformations for the page-clipping shown in Fig. 5 includes the removal of a table column and a row, and change of a table attribute value [Fig. 6(b)], as well as the removal of a paragraph [Fig. 6(c)]. All of the transformations can be done easily by using the HTML editor in a WYSIWYG manner. For example, a table row can be removed by selecting a cell in the row to be deleted and then choosing the menu item "Delete Row" from a popup menu (Fig. 9) that appears after clicking the right mouse button. The `cellpadding` attribute of the table can also be changed from a dialog window for the table attribute settings.

All the above-mentioned editing operations actually modify the target document by means of DOM manipulation operations, and can be created as transformational annotations. However, intrinsically assertional annotations cannot be created on the basis of the DOM manipulation operations, but need to be declared explicitly as assertions. Annotation editors that follow the annotation by assertion approach play a complementary role in such situations instead of the metadata generator for the transformational annotations.

```
<description take-effect="before" target="/HTML[1]/BODY[1]/H1[1]"
    condition="(User-Agent=*CNF*)">
  <remove/>
</description>

<description take-effect="after" target="/HTML[1]/BODY[1]/H1[1]"
    condition="(User-Agent=*CNF*)">
  <keep/>
</description>
```

Fig. 10. Examples of conditional annotations

An example of assertional annotation is given below in the page-clipping annotation language. Annotations may be applied conditionally depending on a delivery context, or depending on profiles of a client agent to be specified in the HTTP header field. By adding a `condition` attribute to a `<description>` element, the annotation will be executed only if the condition is true [26]. When it is necessary to remove the header element (`<H1>`) at the beginning of the sample document, the user needs to add a pair of annotations to start a remove state before the first H1 element and set the state back to 'keep' just after the H1 element. In addition, it is necessary for these annotations to be applied only if a client agent is given with a particular profile attribute value.

Fig. 10 shows examples of conditionally applied annotation descriptions. That is, the removal of the first H1 element is done only when a substring 'CNF' is included as a device type in the HTTP header. The asterisk ('*') here is a wild-card expression for the string matching. In addition, more complex conditions can be given by using conjunctive and/or disjunctive expression.

4.3 Metadata Generation Procedure

The DOM (Document Object Model) [7] defines a logical structure for documents, and provides ways of accessing and manipulating XML or well-formed HTML documents. The DOM represents a document as a hierarchy of node objects, which may be associated with attributes. DOM trees are changed by node insertion (*insertBefore, replaceChild, appendChild*), node removal (*removeChild*), or attribute change (*setAttribute, removeAttribute*). Since the DOM provides standardized, general-purpose operations for the transformation of DOM trees, it is reasonable to record users' operations as a sequence of DOM manipulation operations rather than as primitive operations such as a sequence of keystrokes and mouse clicks. Therefore, along with the notion of current node, which may be either a leaf node or an entire subtree, the model of WYSIWYG editing adopted here consists of the four basic operations: *insert* to add a subtree, *remove* to delete a subtree, *modify* to change a node, and *copy* to replicate a subtree. The rationale behind this editing model was already reported in another article [15].

Since the WYSIWYG editing for page clipping is typically done by removing the unnecessary portion of the original document, it is assumed that the clipping state of a created annotation document begins with 'keep' as shown in Fig. 6(a).

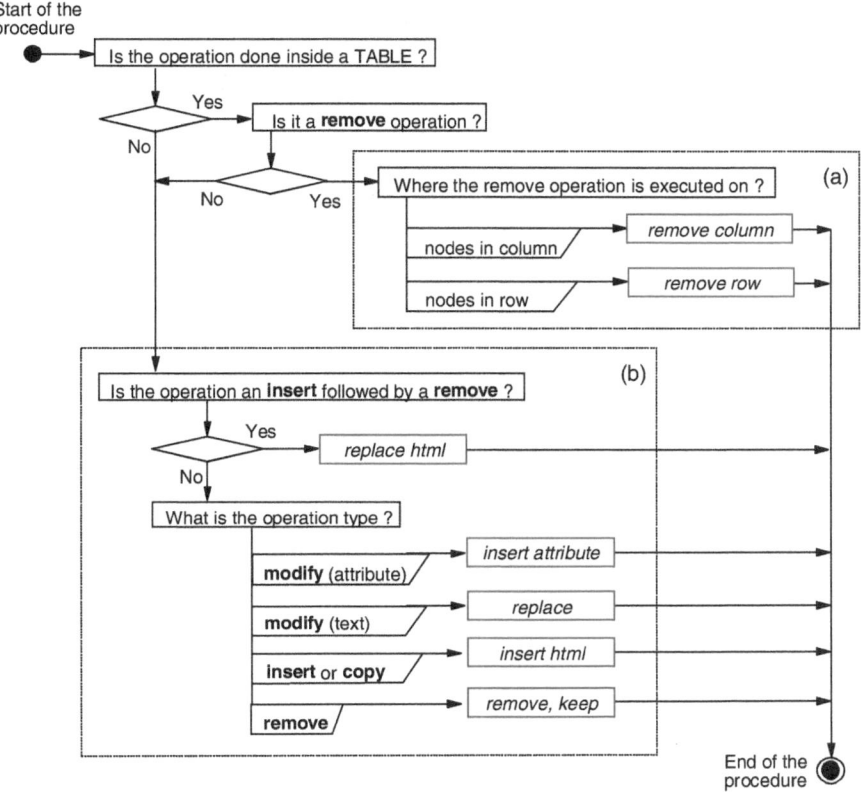

Fig. 11. Procedure for the generation of page-clipping annotations

Individual annotations are then created by examining the sequence of the user's recorded operations from the beginning to the end.

Fig. 11 depicts the procedure of annotation generation. In the figure, the terms in bold face are the primitive operations prescribed in the model of editing, while the terms in italics correspond to the elements of the page-clipping annotation language.

The generation procedure consists of the two sub-processes: one is to deal with manipulation with in a HTML table element, and the other is for other cases. When a focal node is a descendant of a TABLE element and its operation is removal [Fig. 11(a)], the generator creates annotations for removing the table rows or columns by using the <column> and <row> elements. In other cases [Fig. 11(b)], clipping annotations are created taking account of the distinction of the primitive operations (i.e., insert, remove, modify, and copy),

5   Concluding Remarks

Annotation by transformation is an innovative approach to helping users with the generation of content customization metadata, because it allows the users to create a customized result interactively with an example, rather than abstractly without any concrete example. Since users of the metadata generator do not have to learn the annotation language at all, this approach is particularly suitable for page designers or novice programmers who are not necessarily familiar with annotation languages. However, it is also useful for skilled designers too, because the environment allows users to concentrate on customizing Web pages without paying any attention to the metadata authoring task.

It has been pointed out that the success of an environment for programming by example depends far more on the user experience of interacting with the environment than the induction algorithms used to create the user's programs [4]. The advantage of annotation by transformation in this regard is that it does not rely on any particular editors, including a conventional WYSIWYG HTML editor, rather than a special editor tailored for the annotation generation. Further research will be needed to investigate the applicability of this technique. The main problem is to determine to what extent users are willing to accept the annotation by transformation environment that guesses what they are doing, and that occasionally might make inadequate or inappropriate generalizations. However, the approach proposed in this paper would be an important step towards the realization of a full-fledged support tool for a device-agnostic authoring environment.

References

1. Abe, M. and Hori, M.: A visual approach to authoring XPath expressions. *Proceedings of Extreme Markup Languages 2001*, pp. 1–14 Montréal, Canada (2001).
2. Bickmore, T. W. and Schilit, B. N.: Digestor: Device-independent access to the World Wide Web. *Proceedings of the 6th International World Wide Web Conference*, Santa Clara, CA (1997).
3. Butler, M. H., Current Technologies for Device Independence. Technical Report HPL-2001-83, Hewlett-Packard Company (2001).
4. Sypher, A.: Introduction: Bringing programming to end users. In Allen Sypher *et al.* (Eds.): *Watch What I Do: Programming by Demonstration.* pp. 1–11, The MIT Press, Cambridge, MA (1993).
5. Denoue, L. and Vignollet, L.: An annotation tool for Web browsers and its applications to information retrieval. *Proceedings of the 6th Conference on Content-Based Multimedia Information Access (RIAO 2000)*, Paris, France (2000).
6. Device Independence Principles. *W3C Working Draft*, http://www.w3.org/TR/di-princ/ (2001).
7. Document Object Model (DOM) Level 1 Specification Version 1.0. *W3C Recommendation*, http://www.w3.org/TR/REC-DOM-Level-1/ (1998).
8. Erdmann, M., Maedche, A., Schnurr, H.-P., and Staab, S.: From manual to semi-automatic semantic annotation: about ontology-based text annotation tools. *Proceedings of the COLING 2000 Workshop on Semantic Annotation and Intelligent Content*, Luxembourg (2000).

9. Fox, A. and Brewer, E. A.: Reducing WWW latency and bandwidth requirements by real-time distillation. *Proceedings of the 5th International World Wide Web Conference*, Paris, France (1996).

10. Heflin, J. and Hendler, J.: Semantic interoperability on the Web. *Proceedings of Extreme Markup Languages 2000*, pp. 111–120 (2000).

11. Hori, M., Kondo, G., Ono, K., Hirose, S., and Singhal, S.: Annotation-based Web content transcoding. *Proceedings of the 9th International World Wide Web Conference (WWW9)*, pp. 197–211, Amsterdam, Netherlands (2000).

12. Hori, M., Ono, K., Kondo, G., and Singhal, S.: Authoring tool for Web content transcoding. *Markup Languages: Theory & Practice*, **2**(1): 81–106 (2000).

13. Hori, M.: Semantic annotation for Web content adaptation. In D. Fensel, J. Hendler, H. Lieberman, and W. Whalster (Eds), *Spinning the Semantic Web*, pp. 542–573, MIT Press, Boston, MA (2002).

14. Kahan, J. and Koivunen, M.-R.: Annotea: an open RDF infrastructure for shared Web annotations. *Proceedings of the 10th International World Wide Web Conference (WWW10)*, pp. 623–632, Hong Kong (2001).

15. Koyanagi, T., Ono, K., and Hori, M.: Demonstrational Interface for XSLT Stylesheet Generation. *Markup Languages: Theory & Practice*, **2**(2): 133–152 (2001).

16. Lassila, O.: Web metadata: a matter of semantics. *IEEE Internet Computing*, **2**(4): 30–37 (1998).

17. Lieberman, H. (Ed.): *Your Wish is My Command: Programming by example*. Morgan Kaufmann Publishers, San Francisco (2001).

18. Maes, S. H. and Raman, T. V.: Position paper for the W3C/WAPWorkshop on the Multi-modal Web, W3C, http://www.w3.org/2000/09/Papers/IBM.html (2000).

19. Mea, V. D., Beltrami, C. A., Roberto, V., and Brunato, D.: HTML generation and semantic markup for telepathology. *Proceedings of the 5th International World Wide Web Conference (WWW5)*, pp. 1085–1094, Paris, France (1996).

20. Nagao, K., Shirai, Y., and Kevin, S.: Semantic annotation and transcoding: making Web content more accessible. *IEEE Multimedia*, **8**(2): 69–81 (2001).

21. Ono, K., Koyanagi, T., Abe, M. and Hori, M.: XSLT Stylesheet Generation by Example with WYSIWYG Editing. *Proceedings of the International Symposium on Applications and the Internet (SAINT 2002)*, pp. 150–159 (2002).

22. Rousseau, J. F., Macias, A. G., de Lima, J. V., and Duda, A.: User adaptable multimedia presentations for the World Wide Web. *Proceedings of the 8th International World Wide Web Conference*, pp. 195–212, Toronto, Canada (1999).

23. Sakairi, T. and Takagi, H.: An annotation editor for nonvisual Web access. *Proceedings of the 9th International Conference on Human-Computer Interaction (HCI International 2001)*, pp. 982–985, New Orleans, LA (2001).

24. Spinks, R., Topol, B., Seekamp, C., and Ims, S.: Document clipping with annotation. developerWorks, IBM Corp. http://www.ibm.com/developerworks/ibm/library/ibm-clip/ (2001).

25. Eclipse Platform. *eclipse.org Consortium*, http://www.eclipse.org/ (2002).

26. WebSphere Transcoding Publisher Version 4.0 Developer's Guide. IBM Corp. (2001).

27. XML Path Language (XPath) Version 1.0. *W3C Recommendation*, http://www.w3.org/TR/xpath (1999).

28. XSL Transformations (XSLT) Version 1.0. *W3C Recommendation*, http://www.w3.org/TR/xslt (1999).

SCAN: A Dynamic, Scalable, and Efficient Content Distribution Network

Yan Chen, Randy H. Katz, and John D. Kubiatowicz

Computer Science Division, University of California at Berkeley

Abstract. We present *SCAN*, the Scalable Content Access Network. SCAN combines dynamic replica placement with a self-organizing application-level multicast tree to meet client QoS and server resource constraints. It utilizes an underlying distributed object routing and location system (DOLR) as an essential component. Simulation results on both flash-crowd-like synthetic workloads and real Web server traces show that SCAN deploys close to an optimal number of replicas, achieves good load balance, and incurs a small delay and bandwidth penalty for update multicast relative to static replica placement on IP multicast. We envision that SCAN could enhance a number of different applications, such as content distribution and peer-to-peer file sharing.

1 Introduction

Exponential growth in processor performance, storage capacity, and network bandwidth is changing our view of computing. Our focus has shifted away from centralized, hand-choreographed systems to global-scale, distributed, self-organizing complexes – composed of thousands or millions of elements. Unfortunately, large pervasive systems are likely to have frequent component failures and be easily partitioned by slow or failed network links. Thus, use of local resources is extremely important – both for performance *and* availability. Further, pervasive streaming applications must tune their communication structure to avoid excess resource usage. To achieve both local access *and* efficient communication, we require flexibility in the placement of data replicas and multicast nodes.

One approach for achieving this flexibility while retaining strong properties of the data is to partition the system into two tiers of replicas[9] – a small, durable *primary* tier and a large, soft-state, *second-tier*. The primary tier could represent a Web server (for Web content delivery), the Byzantine inner ring of a storage system [3,16], or a streaming media provider. The important aspect of the primary tier is that it must hold the most up-to-date copy of data and be responsible for serializing and committing updates. We will treat the primary tier as a black box, called simply "the data source". The second-tier becomes soft-state and will be the focus of this paper. Examples of second-tiers include content-distribution networks (CDNs), file system caches, or web proxy caches.

Because second-tier replicas (or just "replicas") are soft-state, we can dynamically grow and shrink their numbers to meet constraints of the system. We may, for instance, wish to achieve a Quality of Service (QoS) guarantee that bounds

F. Mattern and M. Naghshineh (Eds.): Pervasive 2002, LNCS 2414, pp. 282–296, 2002.
© Springer-Verlag Berlin Heidelberg 2002

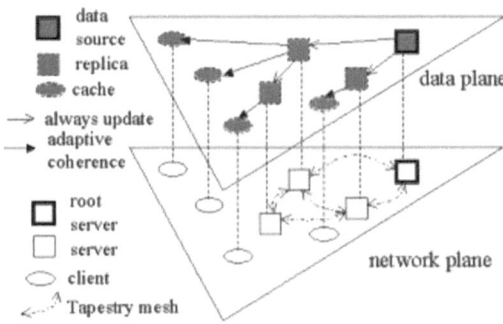

Fig. 1. Architecture of a SCAN system

the maximum network latency between each client and replicas of the data that it is accessing. Since replicas consume resources, we will seek to generate as few replicas as possible to meet this constraint. As a consequence, popular data items may warrant hundreds or thousands of replicas, while unpopular items may require no replicas.

One difficult aspect of unconstrained replication is ensuring that content does not become stale. Slightly relaxed consistency, such as in the Web [10], OceanStore [16], or Coda [14], allows delay between the commitment of updates at the data source and the propagation of updates to replicas. None-the-less, update propagation must still occur in a timely manner. The potentially large number of replicas rules out direct, point-to-point delivery of updates to replicas. In fact, the extremely fluid nature of the second tier suggests a need to self-organize replicas into a multicast tree; we call such a tree a *dissemination tree* (d-tree). Since interior nodes must forward updates to child nodes, we will seek to control the *load* placed on such nodes by restricting the fanout of the tree.

The challenge of second-tier replication is to provide good QoS to clients while retaining *efficient* and *balanced* resource consumption of the underlying infrastructure. To tackle this challenge, we propose a self-organizing soft-state replication system called SCAN: the *Scalable Content Access Network*. Figure 1 illustrates a SCAN system. There are two classes of physical nodes shown in the network-plane of this diagram: *SCAN servers* (squares) and *clients* (circles). We assume that SCAN servers are placed in Internet Data Centers (IDC) of major ISPs with good connectivity to the backbone. Each SCAN server may contain replicas for a variety of data items. One novel aspect of the SCAN system is that it assumes SCAN servers participate in a distributed routing and location (DOLR) system, called Tapestry [11]. Tapestry permits clients to locate nearby replicas without global communication.

There are three types of data illustrated in Figure 1: Data *sources* and *replicas* are the primary topic of this paper and reside on SCAN servers. *Caches* are the images of data that reside on clients and are beyond our scope[1] Our goal is to

---

[1] Caches may be kept coherent in a variety of ways (for instance [22]).

translate client requests for data into replica management activities. We make the following contributions:

- We provide algorithms that dynamically place a minimal number of replicas while meeting client QoS and server capacity constraints.
- We self-organize these replicas into d-tree with small delay and bandwidth consumption for update dissemination.

The important intuition here is that the presence of the DOLR system enables simultaneous placement of replicas and construction of a dissemination tree without contacting the data source. As a result, each node in a d-tree must maintain state only for its parent and direct children.

The rest of the paper is organized as follows: Section 2 presents related work. Section 3 introduces the Tapestry DOLR while Section 4 describes SCAN algorithms. Results are given in Section 5, while discussion is given in Section 6. We conclude with Section 7.

2   Related Work

Much previous work on replica placement involves *static* placement of replicas – assuming that clients' distribution and access patterns are known in advance[20, 12]. These techniques ignore server capacity constraints and assume explicit knowledge of the global IP network topology. Content Distribution Networks (CDNs) use DNS-based redirection to route clients' requests [1,7,25]. Such centralized services do not record locations for each replica. Thus CDNs often place more replicas than necessary, consuming excess storage and update bandwidth.

Inter-domain IP multicast is not widely available; it is also not ideal for Internet distribution [8]. Application-Level Multicast (ALM) builds data distribution trees on top of an efficient overlay network of unicast connections [8,4,6,13,28]. Most ALM systems utilize a central node to maintain state for all existing children [6,13,19,4] and are not very scalable; some replicate the root to help with this problem [13].

Similar to SCAN, Bayeux [28] is an overlay multicast system build on the Tapestry [11] DOLR; however, unlike SCAN, Bayeux filters all "join" requests through a replicated set of root nodes. Scribe [24] provides a scalable, event-notification multicast system built on top of the Pastry [23] DOLR. It provides mechanisms for dynamic, decentralized construction of multicast groups, but does not include mechanisms for replica placement.

3   Distributed Object Location and Routing

Peer-to-peer researchers have begun to explore distributed object location and routing (DOLR) services [11,21,26,23]. DOLR systems offer a distributed framework in which objects that are named by opaque strings can be located quickly. Since information about objects is distributed throughout the system, messages are routed to objects by passing them from node to node until they reach their

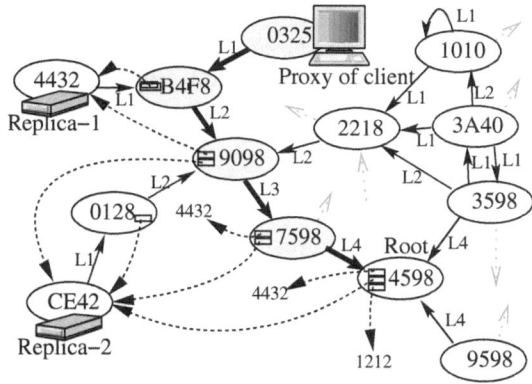

Fig. 2. *The Tapestry Infrastructure:* Nodes route to nodes one digit at a time: e.g. 0325 → B4F8 → 9098 → 7598 → 4598. Objects are associated with a particular "root" node (e.g. 4598). Servers *publish* replicas by sending messages toward root, leaving back-pointers (dotted arrows). Clients route directly to replicas by sending messages toward root until encountering a pointer (e.g. 0325 → B4F8 → 4432)

destination. Some DOLR systems have the important property of *routing locality*. This means two things: First, that messages destined for a given object will be routed over a minimal overlay path to their destination; and Second, when two or more objects have the same name, messages destined for that name will be routed to a *close* object rather than a distant object.

SCAN is built on top of Tapestry [11], a DOLR that exhibits routing locality. Tapestry is an IP overlay network that uses a distributed, fault-tolerant architecture to track the location of objects in the network. In Figure 1, the SCAN servers double as Tapestry nodes. Each client talks to its nearby Tapestry node (*the proxy*) to send object requests. In practice, the proxy node can be located through bootstrap mechanisms.

Figure 2 shows a portion of Tapestry. Each Tapesty node has neighbor links to other Tapestry nodes, forming a routing mesh. Neighbor links provide a route from every node to every other node through incremental, single-digit resolution of the destination. Nodes are added and removed from Tapestry via dynamic membership algorithms [11]. The neighbor-link construction process ensures routing locality.

Each replica is associated with a *Tapestry root node* through a deterministic mapping function[2]. To advertise a replica, the SCAN server storing the object sends a publish message toward the Tapestry location root for that object, depositing *location pointers* each hop. Figure 2 shows two replicas and the Tapestry root for an object. Each object location message is routed toward the object's root until it encounters a pointer, at which point it routes directly to the object.

---

[2] This root is for location purposes only and has nothing to do with the data source.

The Essential Property: The important property of Tapestry for SCAN is that
it provides routing locality, *e.g.* for Figure 2, Client (0325) will find Replica-1
instead of Replica-2. Any DOLR that provides this property will be compatible
with the SCAN algorithms in the next section.

4   SCAN Replica Management Algorithms

The presence of an underlying DOLR with routing locality can be exploited
to perform simultaneous replica placement and tree construction. Every SCAN
server is a member of the DOLR. Hence, new replicas are published into the
DOLR. Further, each client directs its requests to its proxy SCAN server; this
proxy server interacts with other SCAN servers to deliver content to the client.

Although we use the DOLR to locate replicas during tree building, we otherwise communicate through IP. In particular, we use IP between nodes in a d-tree
– parents and children keep track of one another. Further, when a client makes
a request that results in placement of a new replica, the client's proxy keeps a
cached pointer to this new replica. This permits direct routing of requests from
the proxy to the replica. Cached pointers are soft state since we can always use
the DOLR to locate replicas.

4.1   Goals for Replica Placement

Replica placement attempts to satisfy both *client latency* and *server load* constraints. *Client latency* refers to the round-trip time required for a client to read
information from the SCAN system. We keep this within a pre-specified limit.
Server load refers to the communication volume handled by a given server. We
assume that the load is directly related to the number of clients it handles and
number of d-tree children it serves. We keep the load below a specified maximum. Our goal is to meet these constraints while minimizing the number of
deployed replicas, keeping the d-tree balanced, and generating as little traffic
during update as possible. Our success will be explored in Section 5.

4.2   Dynamic Placement

Our dynamic placement algorithm proceeds in two phases: *replica search* and
replica placement. The replica search phase attempts to find an existing replica
that meets the client latency constraint without being overloaded. If this is
successful, we place a link in the client and cache it at the client's proxy server.
If not, we proceed to the replica placement phase to place a new replica.

Replica search uses the DOLR to contact a replica "close" to the client proxy;
call this the *entry* replica. The locality property of the DOLR ensures that the
entry replica is a reasonable candidate to communicate with the client. Further,
since the d-tree is connected, the entry replica can contact all other replicas. We
can thus imagine three search variants: *Singular* (consider only the entry replica),
Localized (consider the parent, children, and siblings of the entry replica), and
Exhaustive (consider all replicas). For a given variant, we check each of the

---

procedure DynamicReplicaPlacement_Smart(c, o)

1 c sends *JOIN* request to o through DOLR, reaches entry server s

2 From s, request forwarded to children (sc), parent (p), and siblings (ss)

3 Each family member t with $rc_t > 0$ collects $dist_{IP}(c, t)$.
   Each t returns rc_t and $dist_{IP}(c, t)$ to s.

4 At s: Choose family member t with biggest rc_t and $dist_{IP}(t, c) \leq d_c$.
   Send result to c.

5 **if** *c gets positive result t* **then**

6   c chooses t as parent
   else

7   c sends *PLACEMENT* request to o through DOLR, reaches entry server s
       Request collects $IP_{s'}$, $dist_{IP}(c, s')$ and $rc_{s'}$ for each server s' on the path.

8   At s, choose s' on path with $rc_{s'} > 0$ *and* largest $dist_{IP}(t, c) \leq d_c$

9   s puts a replica on s' and becomes its parent, s' becomes parent for c

10   s' publishes replica in DOLR
   end

Algorithm 1: Smart Replica Placement. Notation: Object o. Client c with latency constraint d_c. Entry Server s. Every server s' has remaining capacity $rc_{s'}$ (additional children it can handle). IP distance, $dist_{IP}(\text{x,y})$, collected by sending "pings"

included replicas and select one that meets our constraints; if none meet the constraint, we proceed to place a new replica.

We restrict replica placement to servers visited by the DOLR routing protocol when sending a message from the client's proxy to the entry replica. We can locate these servers without knowledge of global IP topology. The locality properties of the DOLR suggest that these are good places for replicas. We consider two placement strategies: *Eager* places the replica as close to the client as possible and *Lazy* places the replica as far from the client as possible. If all servers that meet the latency constraint are overloaded, we replace an old replica; if the entry server is overloaded, we disconnect the oldest link among its d-trees.

Dynamic Techniques: We can now combine some of the above options for search and placement to generate dynamic replica management algorithms. Two that we would like to highlight are as follows (see [5] for more information[3]).

- *Naive Placement:* A simple combination utilizes *Singular* search and *Eager* placement. This heuristic generates minimal search and placement traffic.
- *Smart Placement:* A more sophisticated algorithm is shown in Algorithm 5. This algorithm utilizes *Localized* search and *Lazy* placement.

The tradeoff between these approaches is that the latter one consumes more "join" traffic, but constructs a tree with fewer replicas, less update delay, and lower bandwidth consumption. We evaluate this in Section 5.

---

[3] In fact, in [5] we show how overlay distance can be used to estimate IP distance and reduce ping traffic for part of the algorithm.

Static Comparisons: The dynamic methods above are unlikely to be optimal in terms of the number of replicas deployed, since clients are added sequentially and with limited knowledge of the network topology. For comparison, we will consider two static placement methods in Section 5:

- *IP Static:* Each data source has global.IP topology knowledge and is given complete knowledge of all clients. It runs an optimal algorithm [5] to place replicas. It is assumed that the d-tree is formed from IP multicast.
- *Overlay Static:* Again the data source knows the identities of all clients at once. However, it utilizes the overlay distances instead of IP distances to place replicas and build the d-tree.

The first of these is a "guaranteed-not-to-exceed" optimal placement. We expect that it will consume the least total number of replicas and lowest multicast traffic. The second algorithm explores the best that we could expect to achieve gathering all topology information from the DOLR system.

4.3   Soft State Tree Management

Soft-state infrastructures have the potential to be extremely robust, precisely because they can be easily reconfigured to adapt to circumstances. For SCAN we target two types of adaptation: fault recovery and performance tuning.

To achieve fault resilience, the data source sends periodic *heartbeat* messages through the d-tree. Members know the frequency of these heartbeats and can react when they have not seen one for a sufficiently long time. In such a situation, the replica initiates a *rejoin* process – similar to the replica search phase above – to find a new parent. Further, each member periodically sends a *refresh* message to its parent. If the parent does not get the refresh message within a certain threshold, it invalidates the child's entry. With such soft-state group management, any SCAN server may crash without significantly affecting overall CDN performance.

Performance tuning consists of pruning and re-balancing the d-tree. Replicas at the leaves are pruned when they have seen insufficient client traffic. To balance the d-tree, each member periodically rejoins the tree to find a new parent.

5   Evaluation

In this section, we evaluate the performance of the SCAN dynamic replica management algorithms. What we will show is that:

- For realistic workloads, SCAN places close to an optimal number of replicas, while providing good load balance, low delay, and reasonable update bandwidth consumption relative to static replica placement on IP multicast.
- SCAN outperforms the existing DNS-redirection based CDNs on both replication and update bandwidth consumption.
- The performance of SCAN is relatively insensitive to the SCAN server deployment, client/server ratio, and server density.
- The capacity constraint is quite effective at balancing load.

5.1   Experimental Setup

We utilize an event-driven simulation of SCAN. This includes a packet-level network simulator (with a static version of the Tapestry DOLR) and a replica management framework. The soft-state replica layer is driven from simulated clients running workloads.

Our goal is to evaluate the replica schemes of Section 4.2. These strategies are dynamic naive placement (*od_naive*), dynamic smart placement (*od_smart*), overlay static placement (*overlay_s*), and static placement on IP network (*IP_s*). We compare the efficacy of these four schemes via three classes of metrics:

- *Quality of Replica Placement*: Includes number of deployed replicas and degree of load distribution, measured by the ratio of the standard deviation vs. the mean of the number of client children for each replica server.
- *Multicast Performance*: We measure the relative delay penalty (RDP) and the bandwidth consumption which is computed by summing the number of bytes multiplied by the transmission time over every link in the network. For example, the bandwidth consumption for 1K bytes transmitted in two links (one has 10 ms, the other 20 ms latency) is $1KB \times (10+20)ms = 0.03(KB.sec)$.
- *Tree Construction Traffic:* We count both the number of application-level messages sent and the bandwidth consumption for deploying replicas and constructing d-tree.

In addition, we quantify the effectiveness of capacity constraints by computing the *maximal load* with or without constraints. The maximal load is defined as the maximal number of client cache children on any SCAN server. Sensitivity analysis are carried out for various client/server ratios and server densities.

Network Setup. We use the GT-ITM transit-stub model to generate five 5000-node topologies [27]. The results are averaged over the experiments on the five topologies. A packet-level, priority-queue based event manager is implemented to simulate the network latency. The simulator models the propagation delay of physical links, but does not model bandwidth limitations, queuing delays, or packet losses.

We utilize two strategies for placing SCAN servers. One selects all SCAN servers at random (labeled *random SCAN*). The other preferentially chooses transit and gateway nodes (labeled *backbone SCAN*). This latter approach mimics the strategy of placing SCAN servers strategically in the network.

To compare with a DNS-redirection-based Web content distribution network (CDN), we simulate typical behavior of such a system. We assume that every client request is redirected to the closest CDN server, which will cache a copy of the requested information for the client. This means that popular objects may be cached in every CDN server. We assume that content servers are allowed to send updates to replicas via IP multicast.

290   Y. Chen, R.H. Katz, and J.D. Kubiatowicz

Table 1. Statistics of Web site access logs used for simulation

Web site	Period	# Requests total - simulated	# Clients	# Client groups total - simulated	# Objects simulated
MSNBC	10-11 am, 8/2/99	1604944 - 1377620	139890	16369 - 4000	4186
NASA	All day, 7/1/95	64398 - 64398	5177	1842 - 1842	3258

Workloads. To evaluate the replication schemes, we use both a synthetic work-load and access logs collected from real Web servers. These workloads are a first step toward exploring more general uses of SCAN.

Our synthetic workload is a simplified approximation of *flash crowds*. Flash crowds are unpredictable, event-driven traffic surges that swamp servers and disrupt site services. For our simulation, all the clients (not servers) make requests to a given hot object in random order.

Our trace-driven simulation includes a large and popular commercial news site, MSNBC [17], as well as traces from NASA Kennedy Space Center [18]. Table 1 shows the detailed trace information. We use the access logs in the following way. We group the Web clients based on BGP prefixes [15] using the BGP tables from a BBNPlanet (Genuity) router [2]. For the NASA traces, since most entries in the traces contain host names, we group the clients based on their domains, which we define as the last two parts of the host names (e.g., a1.b1.com and a2.b1.com belong to the same domain). Given the maximal topology we can simulate is 5000 (limited by machine memory), we simulate all the clients groups for NASA and 4000 top client groups (cover 86.1% of requests) for MSNBC. Since the clients are unlikely to be on transit nodes nor on server nodes, we map them randomly to the rest of nodes in the topology.

5.2   Results for Synthetic Workload

We start by examining the synthetic, flash crowd workload. 500 nodes are chosen to be SCAN servers with either "random" or "backbone" approach. Remaining nodes are clients and access some hot object in a random order. We randomly choose one non-transit SCAN server to be the data source and set as 50KB the size of the hot object. Further, we assume the latency constraint is 50ms and the load capacity is 200 clients/server.

Comparison between Strategies. Figure 3 shows the number of replicas placed and the load distribution on these servers. *Od_smart* approach uses only about 30% to 60% of the servers used by *od_naive*, is even better than *overlay_s*, and is very close to the optimal case: *IP_s*. Also note that *od_smart* has better load distribution than *od_naive* and *overlay_s*, close to *IP_s* for both *random* and *backbone SCAN*.

Relative Delay Penalty (RDP) is the ratio of the overlay delay between the root and any member in d-tree vs. the unicast delay between them [6]. In Figure 4, *od_smart* has better RDP than *od_naive*, and 85% of *od_smart* RDPs between any member server and the root pairs are within 4. Figure 5 contrasts the bandwidth consumption of various replica placement techniques with the optimal IP

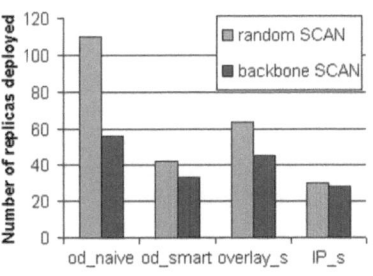 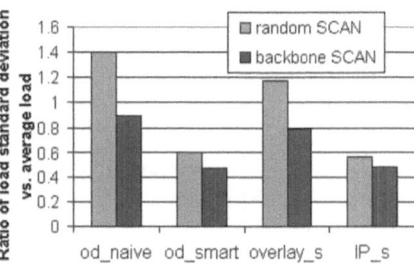

Fig. 3. Number of replicas deployed (left) and load distribution on selected servers (right) (500 SCAN servers)

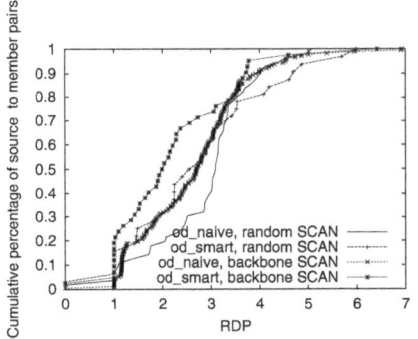 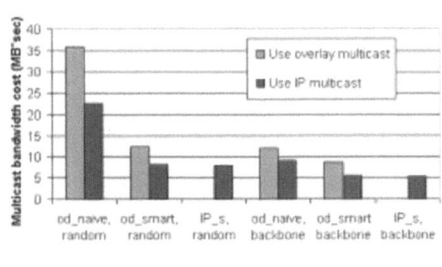

Fig. 4. Cumulative distribution of RDP (500 SCAN servers)   **Fig. 5.** Bandwidth consumption of 1MB update multicast (500 SCAN servers)

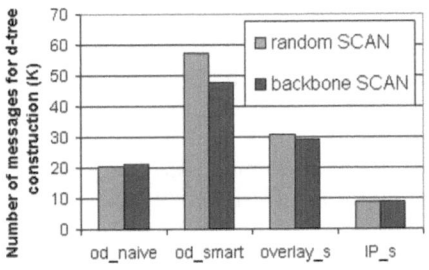 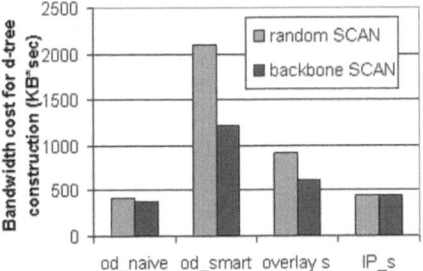

Fig. 6. Number of application-level messages (left) and total bandwidth consumed (right) for d-tree construction (500 SCAN servers)

static placement. The results are very encouraging: the bandwidth consumption of *od_smart* is quite close to *IP_s* and is much less than that of *od_naive*.

The performance above is achieved at the cost of d-tree construction (Figure 6). However, for both *random* and *backbone SCAN*, *od_smart* approach produces less than three times of the messages of *od_naive* and less than six times of that for optimal case: IP_s. Meanwhile, od_naive uses almost the same amount of bandwidth as IP_s while od_smart uses about three to five times that of IP_s.

In short, the smart dynamic algorithm has performance that is close to the ideal case (static placement with IP multicast). It places close to an optimal number of replicas, provides better load distribution, and less delay and multicast bandwidth consumption than the naive approach – at the price of three to five times as much tree construction traffic. Since d-tree construction is a much less frequent than data access and update this is a good tradeoff.

Due to the limited number and/or distribution of servers, there may exist some clients who cannot be covered when facing the QoS and capacity requirements. In this case, our algorithm can provide hints as where to place more servers. Note that experiments show that the naive scheme has many more uncovered clients than the smart one, due to the nature of its unbalanced load. Thus, we remove it from consideration for the rest of synthetic workload study.

Comparison with a CDN. As an additional comparison, we contrast the overlay smart approach with a DNS-redirection-based CDN. Compared with a traditional CDN, the overlay smart approach uses a fraction of the number of replicas (6-8%) and less than 10% of bandwidth for disseminating updates.

Effectiveness of Distributed Load Balancing. We study how the capacity constraint helps load balancing with three client populations: 100, 1000 and 4500. The former two are randomly selected from 4500 clients. Figure 7 shows that lack of capacity constraints (labeled $w/o\ LB$) leads to hot spot or congestion: some servers will take on about 2-13 times their maximum load. Performance with load balancing is labeled as $w/\ LB$ for contrast.

Performance Sensitivity to Client/Server Ratio. We further evaluate SCAN with the three client populations Figure 8 shows the number of replicas deployed. When the number of clients is small, $w/\ LB$ and $w/o\ LB$ do not differ much because no server exceeds the constraint. The number of replicas required for od_smart is consistently less than that of $overlay\_s$ and within the bound of 1.5 for IP_s. As before, we also simulate other metrics, such as load distribution, delay and bandwidth penalty for update multicast under various client/server ratios. The trends are similar, that is, "od_smart" is always better than "overlay_s", and very close to "IP_s". We omit the graphs to save the space.

Performance Sensitivity to Server Density. Next, we increase the density of SCAN servers. We randomly choose 2500 out of the 5000 nodes to be SCAN servers and measure the resulting performance. Obviously, this configuration can support better QoS for clients and require less capacity for servers. Hence, we set the latency constraint to be 30 ms and capacity constraint 50 clients/server. The number of clients vary from 100 to 2500.

With very dense SCAN servers, our od_smart still uses less replicas than $overlay\_s$, although they are quite close. IP_s only needs about half of the replicas,

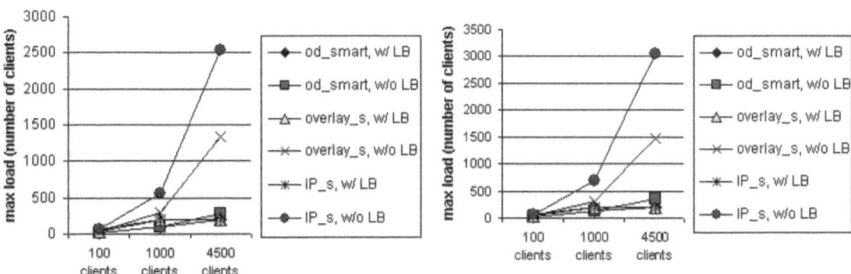

Fig. 7. Maximal load measured with and without load balancing constraints (LB) for various numbers of clients (left: 500 random servers, right: 500 backbone servers)

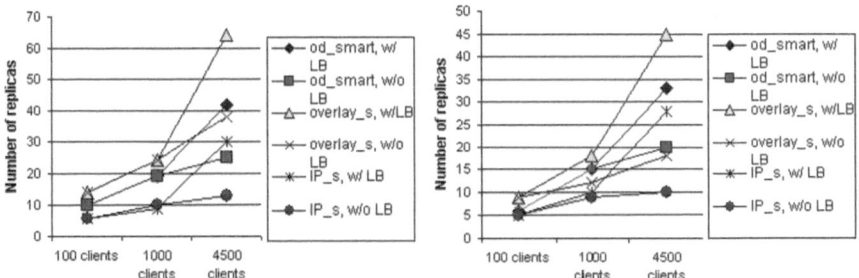

Fig. 8. Number of replicas deployed with and without load balancing constraints (LB) for various numbers of clients (left: 500 random servers, right: 500 backbone servers)

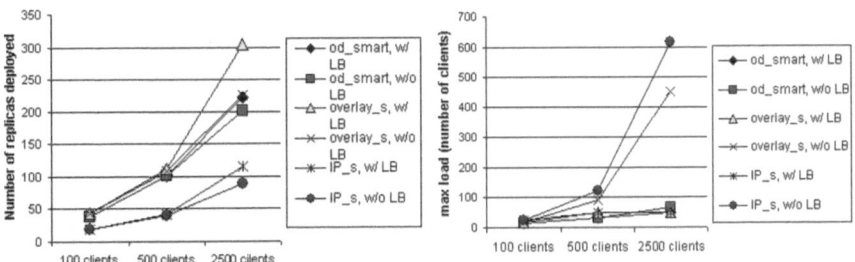

Fig. 9. Number of replicas deployed (left) and maximal load (right) on 2500 random SCAN servers with and without the load balancing constraint (LB)

as in Figure 9. In addition, we notice that the load balancing is still effective. That is, overloaded machines or congestion cannot be avoided simply by adding more servers while neglecting careful design.

Due to space limitations, we cannot show all simulation results here. In summary, *od_smart* performs well with various SCAN server deployments, various client/server ratios, and various server densities. The capacity constraint based distributed load balancing is effective.

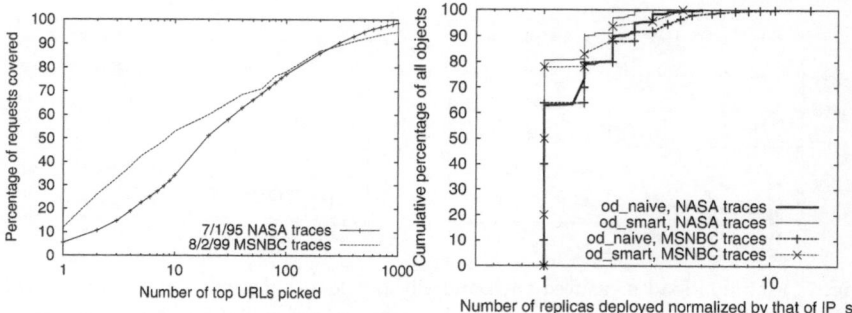

Fig. 10. Simulation with NASA and MSNBC traces on 100 backbone SCAN servers. (a) Percentage of requests covered by different number of top URLs (left); (b) the CDF of replica number deployed with *od_naive* and *od_smart* normalized by the number of replicas using IP_s (right)

5.3   Results for Web Traces Workload

Next, we explore the behavior of SCAN for Web traces with documents of widely varying popularity. Figure 10.a characterizes the request distribution for the two traces used (note that the x-axis is logarithmic.). This figure reveals that the request number for different URLs is quite unevenly distributed for both traces.

For each URL in the traces, we compute the number of replicas generated with *od_naive*, *od_smart*, and IP_s. Then we normalize the replica numbers of *od_naive* and *od_smart* by dividing them with the replica number of IP_s. We plot the CDF of such ratios for both NASA and MSNBC in Figure 10.b. The lower percentage part of the CDF curves are overlapped and close to 1. The reasons are most of the URLs have very few requests, and we only simulate a limited period, thus the number of replicas deployed by the three methods are very small and similar. However, *od_smart* and *od_naive* differ significantly for popular objects, exhibited in the higher percentage part. *Od_smart* is very close to IP_s, for all objects, the ratio is less than 2.7 for NASA and 4.1 for MSNBC, while the ratio for *od_naive* can go as high as 5.0 and 15.0, respectively.

In addition, we contrast the bandwidth consumption for disseminating updates. Given an update of unit size, for each URL, we compute the bandwidth consumed by using (1) overlay multicast on *od_naive* tree, (2) overlay multicast on *od_smart* tree, and (3) IP multicast on IP_s tree. Again, we have metric (1) and (2) normalized by (3), and plot the CDF of the ratios. The curves are quite similar to Figure 10.b. So we omit them in the interest of space.

In conclusion, although *od_smart* and *od_naive* perform similarly for infrequent or cold objects, *od_smart* outperforms dramatically over *od_naive* for hot objects which dominate overall requests.

6   Discussion

How does the distortion of topology through Tapestry affect replica placement? Notice that the overlay distance through Tapestry, on average, is about 2-3 times more than the IP distance. Our simulations in Sec. 5, shed some light on the resulting penalty: $Overlay\_s$ applies exactly the same algorithm as IP_s for replica placement, but uses the static Tapestry-level topology instead of IP-level topology. Simulation results show that $overlay\_s$ places 1.5 - 2 times more replicas than IP_s. For similar reasons, od_smart outperforms $overlay\_s$. The reason is that od_smart uses "ping" messages to get the real IP distance between clients and servers. This observation also explains why od_smart gets similar performance to IP_s. One could imagine scaling overlay latency by an expected "stretch" factor to estimate real IP distance – thereby reducing ping probe traffic.

We start this paper with a general discussion of second-tier, soft-state replication. How does the SCAN system fulfill that role? Our initial results are encouraging precisely because they demonstrate that a dynamic, self-organizing system can utilize resources *conservatively* in the *interior* of the network to meet client QoS requirements. This is in marked contrast to many peer-to-peer replication systems that cache information only at clients, or CDN systems that cache data widely. The underlying DOLR (Tapestry) provides the essential mechanism for distributing information about resources and topology. The exploitation of such an infrastructure is a powerful organizing principle for future systems.

7   Conclusions

The importance of adaptive replica placement and update dissemination is growing as distribution systems become pervasive and global. In this paper, we present SCAN, a scalable, soft-state replica management framework built on top of a distributed object location and routing framework (DOLR) with locality. SCAN generates replicas on demand and self-organizes them into an application-level multicast tree, while respecting client QoS and server capacity constraints. An event-driven simulation of SCAN shows that SCAN places close to an optimal number of replicas, while providing good load distribution, low delay, and small multicast bandwidth consumption compared with static replica placement on IP multicast. Further, SCAN outperforms existing DNS-redirection based CDNs in terms of replication and update cost. SCAN shows great promise as an essential component of global-scale peer-to-peer infrastructures.

Acknowledgments. We graciously acknowledge sponsorship and grants from DARPA (grant N66061-99-2-8913), NSF career award #ANI-9985250, California Micro Grant #01-042, Ericsson, Nokia, Siemens, Sprint, NTTDoCoMo and HRL laboratories. We thank Lili Qiu for providing anonymized MSNBC traces, thank Albert Wang for help with implementation, and thank Johnny Lam and Lakshminarayanan Subramanian for reviewing the draft of the paper.

References

1. Akamai Technologies Inc. http://www.akamai.com.
2. BBNPlanet. telnet://ner-routes.bbnplanet.net.
3. M. Castro and B. Liskov. Proactive recovery in a byzantine-fault-tolerant system. In *Proc. of USENIX Symp. on OSDI*, 2000.

296 Y. Chen, R.H. Katz, and J.D. Kubiatowicz

4. Y. Chawathe, S. McCanne, and E. Brewer. RMX: Reliable multicast for heterogeneous networks. In *Proceedings of IEEE INFOCOM*, 2000.
5. Y. Chen et al. Dynamic replica placement for scalable content delivery. In *Proc. of 1st International Workshop on Peer-to-Peer Systems (IPTPS)*, Mar. 2002.
6. Y. Chu, S. Rao, and H. Zhang. A case for end system multicast. In *Proceedings of ACM SIGMETRICS*, June 2000.
7. Digital Island Inc. http://www.digitalisland.com.
8. P. Francis. Yoid: Extending the Internet multicast architecture. Technical report, ICIR, http://www.icir.org/yoid/docs/yoidArch.ps, April, 2000.
9. J. Gray, P. Helland, P. O'Neil, and D. Shasha. The dangers of replication and a solution. In *Proc. of ACM SIGMOD Conf.*, pages 173–182, 1996.
10. James Gwertzman and Margo Seltzer. World-Wide Web Cache Consistency, 1996.
11. K. Hildrum, J. Kubiatowicz, S. Rao, and B. Zhao. Distributed data location in a dynamic network. In *Proc. of ACM SPAA*, 2002.
12. S. Jamin, C. Jin, A. Kurc, D. Raz, and Y. Shavitt. Constrained mirror placement on the Internet. In *Proceedings of IEEE Infocom*, 2001.
13. J. Jannotti et al. Overcast: Reliable multicasting with an overlay network. In *Proceedings of OSDI*, 2000.
14. J. Kistler and M. Satyanarayanan. Disconnected operation in the Coda file system. *ACM Transactions on Computer Systems*, 10(1):3–25, February 1992.
15. B. Krishnamurthy and J. Wang. On network-aware clustering of Web clients. In *Proc. of SIGCOMM*, 2000.
16. John Kubiatowicz et al. Oceanstore: An architecture for global-scale persistent storage. In *Proceeedings of 9th ASPLOS*, 2000.
17. MSNBC. http://www.msnbc.com.
18. NASA server traces. http://ita.ee.lbl.gov/html/contrib/NASA-HTTP.html.
19. D. Pendarakis, S. Shi, D. Verma, and M. Waldvogel. ALMI: An application level multicast infrastructure. In *Proceedings of 3rd USITS*, 2001.
20. L. Qiu, V. N. Padmanabhan, and G. Voelker. On the placement of Web server replicas. In *Proceedings of IEEE Infocom*, 2001.
21. S. Ratnasamy, P. Francis, M. Handley, R. Karp, and S. Shenker. A scalable content-addressable network. In *Proceedings of ACM SIGCOMM*, 2001.
22. P. Rodriguez and S. Sibal. SPREAD: Scaleable platform for reliable and efficient automated distribution. In *Proceedings of WWW*, 2000.
23. A. Rowstron and P. Druschel. Pastry: Scalable, distributed object location and routing for large-scale peer-to-peer systems. In *Proc. of Middleware 2001*.
24. A. Rowstron, A-M. Kermarrec, M. Castro, and P. Druschel. SCRIBE: The design of a large-scale event notification infrastructure. In *Proceedings of NGC*, 2001.
25. Speedera Inc. http://www.speedera.com.
26. I. Stoica et al. Chord: A scalable peer-to-peer lookup service for Internet applications. In *Proceedings of ACM SIGCOMM*, 2001.
27. E. Zegura, K. Calvert, and S. Bhattacharjee. How to model an Internetwork. In *Proceedings of IEEE INFOCOM*, 1996.
28. S. Q. Zhuang et al. Bayeux: An architecture for scalable and fault-tolerant wide-area data dissemination. In *Proceedings of ACM NOSSDAV*, 2001.

Author Index

Lecture Notes in Computer Science

For information about Vols. 1–2341
please contact your bookseller or Springer-Verlag

Vol. 2383: M.S. Lew, N. Sebe, J.P. Eakins (Eds.), Image and Video Retrieval. Proceedings, 2002. XII, 388 pages. 2002.

Vol. 2384: L. Batten, J. Seberry (Eds.), Information Security and Privacy. Proceedings, 2002. XII, 514 pages. 2002.

Vol. 2385: J. Calmet, B. Benhamou, O. Caprotti, L. Henocque, V. Sorge (Eds.). Artificial Intelligence, Automated Reasoning, and Symbolic Computation. Proceedings, 2002. XI, 343 pages. 2002. (Subseries LNAI).

Vol. 2386: E.A. Boiten, B. Möller (Eds.), Mathematics of Program Construction. Proceedings, 2002. X, 263 pages. 2002.

Vol. 2387: O.H. Ibarra, L. Zhang (Eds.), Computing and Combinatorics. Proceedings, 2002. XIII, 606 pages. 2002.

Vol. 2388: S.-W. Lee, A. Verri (Eds.), Pattern Recognition with Support Vector Machines. Proceedings, 2002. XI, 420 pages. 2002.

Vol. 2389: E. Ranchhod, N.J. Mamede (Eds.), Advances in Natural Language Processing. Proceedings, 2002. XII, 275 pages. 2002. (Subseries LNAI).

Vol. 2391: L.-H. Eriksson, P.A. Lindsay (Eds.), FME 2002: Formal Methods – Getting IT Right. Proceedings, 2002. XI, 625 pages. 2002.

Vol. 2392: A. Voronkov (Ed.), Automated Deduction – CADE-18. Proceedings, 2002. XII, 534 pages. 2002. (Subseries LNAI).

Vol. 2393: U. Priss, D. Corbett, G. Angelova (Eds.), Conceptual Structures: Integration and Interfaces. Proceedings, 2002. XI, 397 pages. 2002. (Subseries LNAI).

Vol. 2395: G. Barthe, P. Dybjer, L. Pinto, J. Saraiva (Eds.), Applied Semantics. IX, 537 pages. 2002.

Vol. 2396: T. Caelli, A. Amin, R.P.W. Duin, M. Kamel, D. de Ridder (Eds.), Structural, Syntactic, and Statistical Pattern Recognition. Proceedings, 2002. XVI, 863 pages. 2002.

Vol. 2398: K. Miesenberger, J. Klaus, W. Zagler (Eds.), Computers Helping People with Special Needs. Proceedings, 2002. XXII, 794 pages. 2002.

Vol. 2399: H. Hermanns, R. Segala (Eds.), Process Algebra and Probabilistic Methods. Proceedings, 2002. X, 215 pages. 2002.

Vol. 2401: P.J. Stuckey (Ed.), Logic Programming. Proceedings, 2002. XI, 486 pages. 2002.

Vol. 2402: W. Chang (Ed.), Advanced Internet Services and Applications. Proceedings, 2002. XI, 307 pages. 2002.

Vol. 2403: Mark d'Inverno, M. Luck, M. Fisher, C. Preist (Eds.), Foundations and Applications of Multi-Agent Systems. Proceedings, 1996-2000. X, 261 pages. 2002. (Subseries LNAI).

Vol. 2404: E. Brinksma, K.G. Larsen (Eds.), Computer Aided Verification. Proceedings, 2002. XIII, 626 pages. 2002.

Vol. 2405: B. Eaglestone, S. North, A. Poulovassilis (Eds.), Advances in Databases. Proceedings, 2002. XII, 199 pages. 2002.

Vol. 2406: C. Peters, M. Braschler, J. Gonzalo, M. Kluck (Eds.), Evaluation of Cross-Language Information Retrieval Systems. Proceedings, 2001. X, 601 pages. 2002.

Vol. 2407: A.C. Kakas, F. Sadri (Eds.), Computational Logic: Logic Programming and Beyond. Part I. XII, 678 pages. 2002. (Subseries LNAI).

Vol. 2408: A.C. Kakas, F. Sadri (Eds.), Computational Logic: Logic Programming and Beyond. Part II. XII, 628 pages. 2002. (Subseries LNAI).

Vol. 2409: D.M. Mount, C. Stein (Eds.), Algorithm Engineering and Experiments. Proceedings, 2002. VIII, 207 pages. 2002.

Vol. 2410: V.A. Carreño, C.A. Muñoz, S. Tahar (Eds.), Theorem Proving in Higher Order Logics. Proceedings, 2002. X, 349 pages. 2002.

Vol. 2412: H. Yin, N. Allinson, R. Freeman, J. Keane, S. Hubbard (Eds.), Intelligent Data Engineering and Automated Learning – IDEAL 2002. Proceedings, 2002. XV, 597 pages. 2002.

Vol. 2413: K. Kuwabara, J. Lee (Eds.), Intelligent Agents and Multi-Agent Systems. Proceedings, 2002. X, 221 pages. 2002. (Subseries LNAI).

Vol. 2414: F. Mattern, M. Naghshineh (Eds.), Pervasive Computing. Proceedings, 2002. XI, 298 pages. 2002.

Vol. 2415: J. Dorronsoro (Ed.), Artificial Neural Networks – ICANN 2002. Proceedings, 2002. XXVIII, 1382 pages. 2002.

Vol. 2417: M. Ishizuka, A. Sattar (Eds.), PRICAI 2002: Trends in Artificial Intelligence. Proceedings, 2002. XX, 623 pages. 2002. (Subseries LNAI).

Vol. 2418: D. Wells, L. Williams (Eds.), Extreme Programming and Agile Methods – XP/Agile Universe 2002. Proceedings, 2002. XII, 292 pages. 2002.

Vol. 2419: X. Meng, J. Su, Y. Wang (Eds.), Advances in Web-Age Information Management. Proceedings, 2002. XV, 446 pages. 2002.

Vol. 2420: K. Diks, W. Rytter (Eds.), Mathematical Foundations of Computer Science 2002. Proceedings, 2002. XII, 652 pages. 2002.

Vol. 2421: L. Brim, P. Jančar, M. Křetinský, A. Kučera (Eds.), CONCUR 2002 – Concurrency Theory. Proceedings, 2002. XII, 611 pages. 2002.

Vol. 2423: D. Lopresti, J. Hu, R. Kashi (Eds.), Document Analysis Systems V. Proceedings, 2002. XIII, 570 pages. 2002.

Vol. 2430: T. Elomaa, H. Mannila, H. Toivonen (Eds.), Machine Learning: ECML 2002. Proceedings, 2002. XIII, 532 pages. 2002. (Subseries LNAI).

Vol. 2431: T. Elomaa, H. Mannila, H. Toivonen (Eds.), Principles of Data Mining and Knowledge Discovery. Proceedings, 2002. XIV, 514 pages. 2002. (Subseries LNAI).

Vol. 2436: J. Fong, R.C.T. Cheung, H.V. Leong, Q. Li (Eds.), Advances in Web-Based Learning. Proceedings, 2002. XIII, 434 pages. 2002.

Vol. 2440: J.M. Haake, J.A. Pino (Eds.), Groupware – CRIWG 2002. Proceedings, 2002. XII, 285 pages. 2002.

Vol. 2442: M. Yung (Ed.), Advances in Cryptology – CRYPTO 2002. Proceedings, 2002. XIV, 627 pages. 2002.

Vol. 2444: A. Buchmann, F. Casati, L. Fiege, M.-C. Hsu, M.-C. Shan (Eds.), Technologies for E-Services. Proceedings, 2002. X, 171 pages. 2002.